KINETIC EQUATIONS

KINETIC EQUATIONS

Edited by

RICHARD L. LIBOFF

NORMAN ROSTOKER

Cornell University

GORDON AND BREACH SCIENCE PUBLISHERS

New York London Paris

Gordon and Breach, Science Publishers, Inc.
440 Park Avenue South
New York, N.Y. 10016

Editorial office for the United Kingdom

Gordon and Breach, Science Publishers Ltd.
12 Bloomsbury Way
London W.C. 1

Editorial office for France

Gordon & Breach
7–9 rue Emile Dubois
Paris 14e

Library of Congress catalog card number 72–122848. ISBN 0 677 14080 0. Printed in east Germany.

Contents

Introduction xi

PART A **AXIOMATICS**

A–1 *Self-consistent kinetic equations*
(H. S. Green) 3

A–2 *Renormalized kinetic equations*
(J. R. Dorfman) 21

A–3 *Master equations for finite systems*
(C. J. Myerscough) 29

A–4 *An application of the generalized Langevin equation to the study of correlations in simple, classical fluids*
(A. Z. Akcasu and J. J. Duderstadt) 45

A–5 *Kinetic equations with radiation effects*
(R. Hakim and A. Mangeney) 69

A–6 *The Boltzmann–Langevin equation* (abstract)
(R. Zwanzig and M. Bixon) 75

PART B **QUANTUM AND RELATIVISTICS**

B–1 *The self-consistent test particle approach to relativistic kinetic theory*
(W. B. Thompson) 79

B–2 *Kinetic equations in spin systems*
(P. Résibois) 91

B–3 *The role of plasma kinetic theory in spectral line shapes*
(W. R. Chappell, J. Cooper, and E. W. Smith) . . . 99

B–4 *Interaction between atom and radiation field* (F. Henin) 111

B–5 *Upper and lower bounds on transport coefficients as applied to the Boltzmann equation of a Fermi liquid* (H. Smith and J. W. Wilkins) 123

B–6 *The Boltzmann equation and gaseous NMR* (R. F. Snider) 129

PART C **PROPERTIES AND SOLUTIONS**

C–1 *Triple collision effects in the transport properties for a gas of hard spheres* (J. V. Sengers) 137

C–2 *Constants of the linearized motion of Vlasov-plasmas* (B. U. Felderhof) 195

C–3 *A review of the unified theory of relaxations in plasmas* (T. Kihara and O. Aono) 201

C–4 *Some comments on the structure of Boltzmann's H function* (S. Harris) 225

C–5 *Quantum Boltzmann equation for a laser* (C. R. Willis) 235

C–6 *Kinetic theory of a weakly, unstable plasma* (abstract) (C. Oberman) 251

C–7 *Kinetic theory of zero-field gain in a gaseous laser* (abstract) (T. F. Morse and J. Cipolla) 253

PART D **WAVES AND FLUCTUATIONS**

D–1 *Kinetic theory of waves in hot, low density plasma* (R. N. Sudan) 257

D–2 *Relaxation spectrum for phonons* (G. Wannier and F. Claro) 273

D–3 *Kinetic equation methods in quantum fluids: Fermi liquids* (A. Ron) 277

D–4 *Wave-wave interactions due to scattering by electrons* (K. Nishikawa) 287

D–5 *Test-particle method in kinetic theory of a plasma* (K. Matsuda) 299

D–6 *The two-particle correlation function in nonequilibrium statistical mechanics* (A. H. Kritz, G. V. Ramanathan, and G. Sandri) . 307

Index 341

Introduction

Interest in kinetic equations has steadily widened since the turn of the century, when L. Boltzmann[1] first presented his equation. It was hoped that from the study of kinetic equations a microscopic formulation of fluid and thermodynamics would ensue. This goal has not yet been completely realized. Nevertheless, many useful applications of kinetic equations have emerged. In the last few decades kinetic equations has become an integral part of the study of fundamental problems in plasma physics, aerodynamics, fluid dynamics, astrophysics, reactor physics and solid state physics.

A kinetic equation[2] is a single equation for the one-particle distribution function $f(\mathbf{x}, \mathbf{v}, t)$. This function is such that

$$f(\mathbf{x}, \mathbf{v}, t)\, d\mathbf{x}\, d\mathbf{v}$$

is the probability of finding a particle in the volume element $d\mathbf{x}$ about $\mathbf{x}$ and the velocity volume element $d\mathbf{v}$ about $\mathbf{v}$, at the time t. All kinetic equations are in form

$$\frac{\partial f}{\partial t} = \Gamma(f; \mathbf{x}, \mathbf{v}, t)$$

where Γ is a functional of its arguments.

The specific form of the kinetic equation depends on the problem at hand. A kinetic equation for a gas whose particles interact under long-range forces (Coulomb, gravitational) has a different structure than one which pertains to a gas whose particles interact under short-range forces (rigid spheres).

The work of Prigogine[3], Lenard[4], Balescu[5], suggests that the structure of a given kinetic equation evolves in time, together with properties of the gas it pertains to. If a plasma is displaced from equilibrium, then in a time before a characteristic plasma oscillation (before the dielectric shielding effect has manifested itself), particles "see" one another through the long-range, bare

Coulomb interaction. In this early-time domain, the Vlasov (or Landau) equation is appropriate. After this time the dielectric structure of the system comes into play and the Balescu–Lenard equation is appropriate to phenomena with wavelengths sufficiently longer than the close-collision distance (average impact parameter corresponding to right-angle deflection). The related collision integral diverges logarithmically for shorter wavelengths. On the other hand, the Landau equation, which does not include a dielectric shielding mechanism, diverges logarithmically for distant interactions. The related collision integrand is valid for impact-parameter sufficiently shorter than the Debye shielding length. Much attention has recently been given to the problem of obtaining a uniformly valid kinetic equation (Hubbard[6], Kihara and Aono[7], Frieman and Book[8]).

The theory of Bogoliubov[9] has sought to describe the role which kinetic equations play in terms of the time development of a gas, according to which three epochs emerge: the *initial stage* (before two particles have collided), the *kinetic stage* (particles have undergone a few collisions) and the *hydrodynamic stage* (a particle has traversed the confining container). In the kinetic stage the gas is well described by only $f(\mathbf{x}, \mathbf{v}, t)$ and kinetic equations are relevant. After this time the fluid dynamic variables are important, while in the initial stage no less than full knowledge of the trajectories of all the particles suffices to describe the system. From this theory (namely that the higher-order multi-particle distributions are functionals of f, in the kinetic stage) an iteration procedure has emerged which is important as a means of handling short-range inhomogenieties in the system. The validity and domain applicability of this expansion procedure has been the object of some recent extensive study (Cohen[10], Cohen and Sengers[11], Cohen and Dorfman[12], and Sandri and Kritz[13]).

Another endeavor in this field concerns the interactions between particles and waves (sound, plasma oscillations, electromagnetic fluctuations, turbulent modes, etc.). Kinetic equations are derived for the number of waves N_K present in a given k-mode (k = wave number) (Coppi, Rosenbluth and Sudan[14], Dupree[15], Kadomtsev[16], Sagdeev and Galeev[17], Klimontovich[18], and Pines and Schrieffer[19]). One speaks here of wave-wave and wave-particle interactions. The formalism is useful for finding the rates of growth and saturation levels for certain instabilities and decay phenomena of the system.

Also of interest in the theory of kinetic equations are the techniques for deriving them. Classically such endeavors usually center about the Liouville equation[1]. Quantum mechanically, they center about the Pauli equation and the artifice of second quantization[20]. In a less formal approach (a-la Boltzmann) one merely constructs an equation for $\partial f/\partial t$ in an ad-hoc manner. In this study of the ground work of kinetic equations, Klimontovich[18] has introduced a generalized one-particle distribution.

$$F(\mathbf{x}, \mathbf{v}, t; \mathbf{x}, \dots, \mathbf{x}_N; \mathbf{v}_1 \dots, \mathbf{v}_N) = \sum_{i=1}^{N} \delta[\mathbf{x} - \mathbf{x}_i(t)]\, \delta[\mathbf{v} - \mathbf{v}_i(t)]$$

In this formula $\mathbf{x}_i$ and $\mathbf{v}_i$ are the position and velocity of the ith particle, respectively. One obtains the one particle distribution f by averaging F over the full N-particle distribution function. Klimontovich attributes a "particle" description to his formalism, as opposed to the "fluid" picture of the Liouville formalism. This becomes apparent when one constructs the two particle distribution $f_2(\mathbf{x}, \mathbf{v}; \mathbf{x}'\mathbf{v}')$ which is accomplished through forming the average: $\overline{F(\mathbf{x}, \mathbf{v})\, F(\mathbf{x}', \mathbf{v}')}$. One obtains a delta-function contribution in addition to the term f_2. This is the particle quality of the F-description which remains after averaging.

In the ad-hoc scheme of obtaining kinetic equations one encounters the study of kinetic modeling. Here one constructs a "collision term" which hopefully is compatible with the conservation laws, implies irreversible behavior, and is relevant to the phenomena being examined. In this vein one encounters the Krook–Bhatnager–Gross equation, Fokker–Planck and Landau equations. Many extended models (Morse[21], Gross and Krook[22]) have been suggested for two component gases.

Also of interest in the study of kinetic equations is the theory of the spectrum of the collision operator (Grad[23], McLennan[24], Case and Zweifel[25], Inönü and Zweifel[26]). Eigenvalues are related to decay and growth rates of eigenstates of the system.

With this background, early in 1969, it was thought appropriate to organize a meeting bringing together the various research endeavors related to the theory and application of kinetic equations. The meeting was held in June of 1969 and lasted two days. The were four sessions: (A) Axiomatics; (B) Quantum and Relativistic Problems; (C) Properties and Solution; and (D) Wave and Fluctuation Phenomena. This proceedings volume follows the same outline. Of the twenty-five papers presented, twenty-two papers and three abstracts are included herein.

We would like to thank the National Science Foundation and Cornell University for supporting the symposium. Host departments at Cornell were: School of Electrical Engineering, Department of Applied Physics, and the Laboratory of Plasma Studies. Thanks are also due Dr. F. A. Long for a very warm welcoming address. Our sincere gratitude is extended to Dr. Joel Snow of the National Science Foundation, for his encouragement and assistance rendered throughout the meeting.

R. L. LIBOFF, *Chairman*
N. ROSTOKER, *Co-Chairman*

References

1. For extensive references to the early works in kinetic theory see: (a) Boltzmann, L., *Lectures on Gas Theory*, translated by S. G. Brush, University of California Press, Berkeley, California, (1964). (b) Liboff, R. L., *Introduction to the Theory of Kinetic Equations*, John Wiley, New York, New York (1969).
2. For general theory of kinetic equations see reference (1b), and: (a) Wu, T. Y. *Kinetic Equation of Gases and Plasmas*, Addison Wesley, Reading, Mass. (1966). (b) Montgomery, D., and Tidman, D., *Plasma Kinetic Theory*, McGraw Hill, New York (1964).
3. Prigogine, I., *Non-Equilibrium Statistical Mechanics*, John Wiley, New York (1962).
4. Lenard, A., *Ann. Phys.* **3**, 390 (1960).
5. Balescu, R., *Statistical Mechanics of Charged Particles*, John Wiley, New York, New York (1963).
6. Hubbard, J., *Proc. Roy. Soc.*, (London) **A261**, 371 (1961).
7. Kihara, T., and Aono, O., *J. Phys. Soc.* Japan **18**, 837, 1043 (1963).
8. Frieman, E. A., and Book, D. C., *Phys. Fluids* **6**, 1700 (1963).
9. The translation of Bogliubov's article appears in, *Studies in Statistical Mechanics*, Vol. 1, Ed. by J. de Boer and G. E. Uhlenbeck, John Wiley, New York (1962). The theory is also discussed in ref. 1b and 2a.
10. Cohen, E. G. D., *Physica*, **28**, 1025, 1045 (1962).
11. Cohen, E. G. D., and Sengers, J. V., *Physica*, **27**, 230 (1961).
12. Cohen, E. G. D., and Dorfman, J. R., *Math. Phys.* **8**, 282 (1967).
14. Coppi, B., Rosenbluth, M., and Sudan, R., *Ann. Phys.* **55**, 207, 248 (1969).
15. Dupree, T. H., *Phys. Fluids*, **6**, 1714 (1963).
16. Kadomtsev, B., *Plasma Turbulence*, Academic Press, New York (1965).
17. Sagdeev, R. Z., and Galeev, A. A., *Non Linear Plasma Theory*, Benjamin, New York (1969).
18. Klimontovich, Y. L., *The Statistical Theory of Non Equilibrium Processes in a Plasma*, MIT Press, Cambridge, Mass. (1967).
19. Pines, D., Schrieffer, J. R., *Phys. Rev.* **125**, 804 (1962).
20. See, for instance: Bogoliubov, N. N., *Lectures on Quantum Statistics*, Gordon and Breach, New York (1967); Fujita, S., *Introduction to Non-Equilibrium Statistical*

Mechanics, W. B. Saunders, Philadelphia, (1966); Wyld, H. W. Jr., and Fried, B. D., *Ann. Phys.* **23**, 374 (1963).
21. Morse, T. F., *Phys. Fluids*, **7**, 2012 (1964).
22. Gross, E. P., and Krook, M., *Phys. Rev.*, **102**, 593, (1956)..
23. Grad, H., *Phys. Fluids*, **6**, 147 (1963).
24. McLennan, J. A., *Phys. Fluids* **8**, 1580 (1965).
25. Case, K. M., and Zweifel, P. F., *Linear Transport Theory*, Addison-Wesley, Reading, Mass. (1967).
26. *Developments in Transport Theory*, ed. E. Inönü and P. F. Zweifel, Academic Press, New York (1967).

PART A

AXIOMATICS

Contributors

H. S. Green
Department of Mathematical Physics
University of Adelaide
Adelaide, Australia

J. R. Dorfman
Institute for Fluid Dynamics and Applied Mathematics
Department of Physics and Astronomy
University of Maryland
College Park, Maryland

C. J. Myerscough
Churchill College
Department of Applied Mathematics and Theoretical Physics
University of Cambridge
England

A. Ziya Akcasu
Department of Nuclear Engineering
The University of Michigan
Ann Arbor, Michigan

J. T. Duderstadt
Department of Nuclear Engineering
The University of Michigan
Ann Arbor, Michigan

R. Hakim
Départment de Physique Mathématique
Montpellier
France

A. Mangeney
Laboratoire de Radio-Astronomie Spatiale
Observatoire de Meudon
Meudon, France

R. Zwanzig
Institute for Fluid Dynamics
University of Maryland
College Park, Maryland

M. Bixon
Institute for Fluid Dynamics
University of Maryland
College Park, Maryland

PAPER A–1

Self-consistent kinetic equations

H. S. GREEN

University of Adelaide, Australia

Abstract

An extension of the work of the author with D. Hoffman, is described. The basis of this work is the observation that certain two body and multibody correlations, which have hitherto been regarded as irreversible effects, are completely determined by the distribution functions of equilibrium. Unless these correlations are built into the first order distribution functions, the equations which determine second order effects (linear in the gradients or forces) will necessarily be inconsistent. A related problem is the divergence of density correlations to the transport coefficients.

The extensions of this work concern the determination of transport coefficients by auto-correlation methods. One application which has been considered in detail is the determination of transport coefficients in plasmas. A reformulation of Boltzmann's equation consistent with the Debye–Hueckel approximation of equilibrium is described.

1 INTRODUCTION TO KINETIC THEORY

The main object of modern kinetic theory is to provide an explanation of the macroscopic properties of a fluid in terms of its ionic, atomic or molecular constituents. In this introductory section, we shall discuss briefly the relations between the macroscopic and microscopic descriptions of a fluid. Some well known results will be stated without proof, which, however, can be found elsewhere[1].

The macroscopic equations which are used to determine the behaviour of a fluid are of two kinds: conservation equations and constitutive equations. The conservation equations are

(1) The equations of conservation of mass of the various irreducible constituents of the fluid. If ϱ_a is the mass density of the a-th constituent of

a fluid, and $\varrho_a \mathbf{u}_a$ is the mass flux density, or inertial momentum density associated with this constituent, the conservation equation is

$$\partial_t \varrho_a + \nabla \cdot (\varrho_a \mathbf{u}_a) = 0, \tag{1}$$

where $\partial_t = \partial/\partial t$ is the partial time derivative, and $\nabla = \partial/\partial \mathbf{x}$. Summation with respect to a yields the usual continuity equation of hydrodynamics:

$$\partial_t \varrho + \nabla \cdot (\varrho \mathbf{u}) = 0, \tag{2}$$

where $\varrho = \Sigma_a \varrho_a$ and $\varrho \mathbf{u} = \Sigma_a \varrho_a \mathbf{u}_a$. From (1) and (2) we obtain also

$$\varrho \, d_t(\varrho_a/\varrho) + \nabla \cdot (\varrho_a \mathbf{w}_a) = 0 \tag{3}$$

where $d_t = \partial_t + \mathbf{u} \cdot \nabla$ is the convective time derivative and $\mathbf{w}_a = \mathbf{u}_a - \mathbf{u}$ is the diffusion velocity of the a-th constituent.

(2) The equation of conservation of momentum, which is the hydrodynamical equation of motion. If $\mathbf{P}$ is the momentum flux density (a symmetric tensor) and $\mathbf{X}$ is the macroscopic force per unit volume acting on the fluid, this may be written in either of the forms

$$\partial_t(\varrho \mathbf{u}) + \nabla \cdot \mathbf{P} = \mathbf{X},$$

$$\varrho d_t \mathbf{u} + \nabla \cdot \mathbf{p} = \mathbf{X}, \tag{4}$$

where $\mathbf{p}$ is the pressure tensor. Generally

$$\mathbf{X} = -\varrho \nabla \psi_0 + \varepsilon \mathbf{E} + \mathbf{j} \times \mathbf{B}/c, \tag{5}$$

where $-\nabla \psi_0$ is the non-electromagnetic force per unit mass, $\varepsilon = \Sigma_a (e_a/m_a) \varrho_a$ is the electric charge density (for a plasma), $\mathbf{j} = \Sigma_a (e_a/m_a) \varrho_a \mathbf{u}_a$ is the electric current density, $\mathbf{E} = -\nabla \varphi - \partial_t \mathbf{A}/c$ and $\mathbf{B} = \nabla \times \mathbf{A}$ are the electric and magnetic fields with scalar and vector potentials φ and $\mathbf{A}$.

3) The equation of conservation of energy. If W is the energy per unit mass of the fluid (exclusive of potential energy due to macroscopic forces), $\mathbf{Q}$ is the energy flux density and Y is the power of the external forces per unit volume, this can be written in the alternative forms

$$\partial_t(\varrho W) + \nabla \cdot \mathbf{Q} = Y, \tag{6}$$

$$\varrho d_t U + (\mathbf{p} \cdot \nabla) \cdot \mathbf{u} + \nabla \cdot \mathbf{q} = Y - \mathbf{X} \cdot \mathbf{u}$$

where $U = W - \frac{1}{2}\mathbf{u}^2$ is the internal energy per unit mass and $\mathbf{q} = \mathbf{Q} - \varrho W \mathbf{u} - \mathbf{p} \cdot \mathbf{u}$ is the thermal flux vector. The power is

$$Y = -\varrho \mathbf{u} \cdot \nabla \psi_0 + \mathbf{j} \cdot \mathbf{E} \tag{7}$$

The constitutive equations are:

(4) The generalized laws of diffusion. If

$$\varkappa_b = \psi_0 - \tfrac{1}{2}\mathbf{u}^2 + \mu_b + (e_b/m_b)(\varphi - \mathbf{u}_b \cdot \mathbf{A}) \tag{8}$$

is the resultant potential, including the chemical potential per unit mass μ_b, these diffusion laws may be written

$$\varrho_a \mathbf{w}_a = -\lambda_a \nabla T/T - \Sigma_b \lambda_{ab} \nabla \varkappa_b \tag{9}$$

where T is the absolute temperature.

(5) The generalization of Newton's laws of viscosity. If $\nabla_s \mathbf{u}$ is the rate of strain tensor, whose components are (for λ, $\mu = 1, 2, 3$)

$$(\nabla_s \mathbf{u})_{\lambda\mu} = \frac{1}{2}\left(\frac{\partial u_\lambda}{\partial x_\mu} + \frac{\partial u_\mu}{\partial x_\lambda} - \frac{2}{3}\nabla \cdot \mathbf{u}\, \delta_{\lambda\mu}\right) \tag{10}$$

they may be written

$$\mathbf{p} - p\boldsymbol{\delta} = -\eta \nabla_s \mathbf{u} - \eta_v \nabla \cdot \mathbf{u}\, \boldsymbol{\delta} \tag{11}$$

where p is the hydrostatic pressure and $\boldsymbol{\delta}$ is the unit tensor with components $\delta_{\lambda\mu}$. Then η and η_v are the coefficients of shear and volume viscosity.

(6) The generalization of Fourier's law of thermal conduction. If

$$\mathbf{q} = \Sigma_a \varrho_a \mathbf{w}_a \chi_a + T\boldsymbol{\sigma}, \tag{12}$$

so that $\boldsymbol{\sigma}$ is the entropy flux density,

$$T\boldsymbol{\sigma} = -\lambda \nabla T/T - \Sigma_a \lambda_a \nabla \chi_a \tag{13}$$

The above laws are valid only if the gradients ∇T, $\nabla \chi_b$, $\nabla_s \mathbf{u}$ and $\nabla \cdot \mathbf{u}$, and the perturbative contribution $-(\partial_t + \mathbf{u}_b \cdot \nabla)\mathbf{A} - (\nabla \mathbf{u}_b) \cdot \mathbf{A}$ to the electromagnetic field are not too large.

Kinetic theory assimilates the results of equilibrium statistical mechanics, which allows us in principle to calculate T, p and the μ_a as functions of U and the ϱ_a. But is also has, as one of its practical aims, the determination of the transport coefficients λ, λ_a, λ_{ab}, η and η_v which appear in the constitutive equations. All these physical quantities can be calculated very easily for any fluid for which the *velocity distribution functions* f_a, and the *pair velocity distribution functions* f_{ab} are known. The function f_a depends in general on the time t, a position vector $\mathbf{x}$, and a velocity vector $\boldsymbol{\xi}$; it may be defined in such a way that $f_a\, d^3x\, d^3\xi$ is the probability that a particle of the a-th constituent (or its mass center, if it has structure) will be found in the element of volume d^3x, enclosing the point $\mathbf{x}$, at time t, and that its velocity will be in the elementary range $d^3\xi$ containing the value $\boldsymbol{\xi}$. Similarly, f_{ab} may be defined so that $f_{ab}\, d^3x\, d^3\xi\, d^3x'\, d^3\xi'$ is the probability of finding a particle of the

a-th type in the elements d^3x and $d^3\xi$ at time t, and a particle of the b-th type in the elements d^3x' and $d^3\xi'$. The pair distribution f_{ab} is obviously a function of the two positions $\mathbf{x}, \mathbf{x}'$ and velocities ξ, ξ' as well as the time, which approaches the value $f_a f_b$, i.e. $f_a(t, \mathbf{x}, \xi) f_b(t, \mathbf{x}', \xi')$ when the distance between the points $\mathbf{x}$ and $\mathbf{x}'$ becomes sufficiently great. The functions f_a and f_{ab} are the first two members of an hierarchy of functions $f_a, f_{ab}, f_{abc}, \ldots$ which together provide a complete description, in statistical terms, of the ionic, atomic or molecular constitution of the fluid. The definitions given assume the validity of classical mechanics, but there is a simple quantal generalization[2].

In thermodynamical equilibrium,

$$f_a = \varrho_a(\beta/2\pi m_a)^{3/2} \exp\left[-\tfrac{1}{2}\beta m_a(\xi - \mathbf{u})^2\right]$$
$$f_{ab} = f_a f_b \exp(-\beta\psi_{ab}) \tag{14}$$

where $\beta = 1/(kT)$ is inversely proportional to the absolute temperature and ψ_{ab} is the mean interaction energy of two particles, of the a-th and b-th types, at the points $\mathbf{x}$ and $\mathbf{x}'$, in the fluid environment. Because of the influence of other particles of the environment, ψ_{ab} depends in general on the thermodynamical variables, e.g., ϱ_a and T, as well as the distance r between ξ and $\mathbf{x}'$. If the particles have structure, ψ_{ab} may also depend on the structural parameters and orientations of the two particles; but we shall not consider such a complication here. The unknown thermodynamical functions of equilibrium may be computed, when ψ_{ab} has been determined, with the help of the formulas

$$p = \Sigma_a n_a[1/\beta - \Sigma_b n_b \textstyle\int (g_{ab} - 1)\, \varphi'_{ab}(r) r\, d^3r/6]$$
$$\varrho U = \Sigma_a n_a[3/(2\beta) + \Sigma_b n_b \textstyle\int (g_{ab} - 1)\, \varphi_{ab}(r)\, d^3r/2]$$
$$\partial\varrho_a/\partial\mu_b = \beta m_b \varrho_a[\delta_{ab} + n_b \textstyle\int (g_{ab} - 1)\, d^3r] \tag{15}$$

where $n_a = \varrho_a/m_a$ is the number density of particles of the a-th type, $g_{ab} = \exp(-\beta\psi_{ab})$ is the radial distribution function, $\varphi_{ab}(r)$ is the interaction energy of the two particles at the separation $\mathbf{r} = \mathbf{x}' - \mathbf{x}$, excluding contributions from the fluid environment, and $\varphi'_{ab}(r)$ is its derivative with respect to r.

Of course we are more interested here in situations where thermodynamical equilibrium does not exist; there, the functions f_a and f_{ab} are more complicated, and depend (to a first approximation, linearly) on the gradients ∇T, $\nabla_s \mathbf{u}$, $\nabla \cdot \mathbf{u}$ and $\nabla\chi_a$ (which vanish in equilibrium). This is evident from the formula

$$\varrho_a \mathbf{w}_a = m_a \textstyle\int f_a \mathbf{v}\, d^3\xi, \tag{16}$$

in which $\mathbf{v} = \boldsymbol{\xi} - \mathbf{u}$, as this must yield the diffusion laws (9), and incidentally the coefficients λ_a and λ_{ab}, when f_a has been correctly determined. The laws of viscosity, and also the coefficients of viscosity, are obtained in a similar way from the formula

$$\mathbf{p} = \Sigma_a[m_a \int f_a \mathbf{vv}\, d^3\xi - \tfrac{1}{2}\Sigma_b \int (n_{ab} - n_a n_b)\, \varphi'_{ab}(r) \cdot (\mathbf{rr}/r)\, d^3r],$$
$$n_{ab} = \iint f_{ab}\, d^3\xi\, d^3\xi', \tag{17}$$

where n_{ab} is the pair density, which reduces to $n_a n_b g_{ab}$ only in equilibrium. Finally, the laws of thermal conduction, and the coefficients of thermal conduction are obtained from the formula

$$\mathbf{q} = \Sigma_a[\tfrac{1}{2}m_a \int f_a \mathbf{v}^2\mathbf{v}\, d^3\zeta + \tfrac{1}{2}\Sigma_b \int n_{ab}(\mathbf{u}_{ab} - \mathbf{u})\, \varphi_{ab}\, d^3r$$
$$-\tfrac{1}{4}\Sigma_b \int n_{ab}(u_{ab} + \mathbf{u}_{ba} - 2\mathbf{u}) \cdot (\mathbf{rr}/r)\, \varphi'_{ab}(r)\, d^3r],$$
$$n_{ab}\mathbf{u}_{ab} = \iint f_{ab}\boldsymbol{\xi}\, d^3\xi\, d^3\xi',$$
$$n_{ab}\mathbf{u}_{ba} = \iint f_{ab}\boldsymbol{\xi}'\, d^3\xi\, d^3\xi', \tag{18}$$

where the velocities $\mathbf{u}_{ab}$ and $\mathbf{u}_{ba}$ differ from $\mathbf{u}$ in equilibrium.

It is remarkable that all the formulas listed above involve f_a and f_{ab}, but not functions such as f_{abc} higher in the hierarchy. This suggests that, at least when the state of the fluid is defined macroscopically, the higher functions are themselves determined by the f_a and f_{ab}. Bogoliubov[3] has gone further, in suggesting that f_{ab} is determined by the f_a; however, we shall argue below that this suggestion cannot be correct when the particles of the fluid are able to fall into bound states.

2 METHODS OF KINETIC THEORY

In this section we shall outline some of the more important techniques which have been developed to determine the velocity distribution functions of a fluid. We shall begin with a direct method which is of considerable theoretical importance, initiated by M. S. Green[4]. The application to the kinetic theory of plasmas in a magnetic field has been described by Green and Leipnik[5]. This method depends on the fact that f_a, f_{ab}, and other members of the hierarchy can all be expressed in terms of a set of phase space distribution functions. A typical function F_N of this set depends in general on the time t, and the positions $\mathbf{x}_i$ and velocities $\boldsymbol{\xi}_i$ ($i = 1, 2, \ldots N$) of all the particles in the fluid in a prescribed configuration. We suppose that N_a of the particles are of the a-th type, so if δ_{ia} has the value 1 if the i-th particle is of the a-th type, and 0 otherwise, $\Sigma_i\, \delta_{ia} = N_a$; also $\Sigma_a N_a = N$.

The momentum of the i-th particle in a plasma, neglecting magnetic effects, is $\mathbf{p}_i = m_i\xi_i$, where $m_i = \Sigma_a m_a\, \delta_{ia}$. If

$$d\Omega_{\mathrm{N}} = \Pi_i\, d^3x_i\, d^3p_i/(\Pi_a N_a!), \tag{19}$$

we define F_{N} by saying that $F_{\mathrm{N}}\, d\Omega_{\mathrm{N}}$ is the probability that at time t the volume elements d^3x_i are occupied by particles of prescribed types, with momenta in the elementary ranges $d^3p_i = m_i^3\, d^3\xi_i$. In thermodynamic equilibrium, it follows from considerations of the kind first made by Willard Gibbs in his theory of the grand ensemble that F_{N} has the form

$$\begin{aligned} F_{\mathrm{N}} &= Z^{-1} \exp\,[-\beta Q_{\mathrm{N}}(\mathbf{x}, \xi)], \\ Z &= \exp\,[\beta \textstyle\int (p + \tfrac{1}{2}\varepsilon\varphi)\, d^3x], \\ Q_{\mathrm{N}}(\mathbf{x}, \xi) &= \Sigma_i(E_i - \mathbf{u}_i \cdot \mathbf{p}_i - m_i\chi_i), \\ E_i &= \tfrac{1}{2}m_i\xi_i^2 + \tfrac{1}{2}\Sigma_j'\varphi_{ij} + m_i\psi_i, \end{aligned} \tag{20}$$

where $\mathbf{u}_i$, the fluid velocity at $\mathbf{x}_i$ has the form $\eta + \omega \times \mathbf{x}_i$ consistent with equilibrium, $\varphi_{ij} = \Sigma_{a,b}\, \varphi_{ab}\, \delta_{ia}\, \delta_{jb}$ is the interaction energy of the i-th and j-th particles, $\chi_i = \Sigma_a \chi_a\, \delta_{ia}$ and $\psi_i = \psi_{0i} + (e_i/m_i)\varphi_i^{\mathrm{ext}}$ is the potential energy due to external non-electromagnetic and electrostatic forces.

If the system is not in thermodynamical equilibrium at time t, we may nevertheless suppose that it was in equilibrium at some initial time t_0, and has been brought into its present non-equilibrium state by the action of time dependent external forces. We may imagine that a particle whose position and velocity at time t_0 were $\mathbf{x}_{i0}$ and ξ_{i0} attains the position $\mathbf{x}_i$ and velocity ξ_i at time t. If the system were initially in the element $d\Omega_{\mathrm{N}0} = \Pi_i\, d^3x_{i0}\, d^3p_{i0}/(\Pi_a N_a!)$ of phase space, it will be in the element $d\Omega_{\mathrm{N}}$ at time t, where, according to Liouville's theorem, $d\Omega_{\mathrm{N}} = d\Omega_{\mathrm{N}0}$; also, if

$$F_{\mathrm{N}0} = Z_0^{-1} \exp\,[-\beta_0 Q_{\mathrm{N}}(\mathbf{x}_0, \xi_0)] \tag{21}$$

is the value of F_{N} at time t_0, it follows from $F_{\mathrm{N}}\, d\Omega_{\mathrm{N}} = F_{\mathrm{N}0}\, d\Omega_{\mathrm{N}0}$ that $F_{\mathrm{N}} = F_{\mathrm{N}0}$. To make use of this result, we must express $Q_{\mathrm{N}}(\mathbf{x}_0, \xi_0)$ in terms of the positions and velocities at time t with the help of

$$\begin{aligned} \Sigma_i E_{i0} &= \Sigma_i(E_i - \int_{t_0}^{t} \partial_t\psi_i\, dt) \\ \Sigma_i \mathbf{p}_{i0} &= \Sigma_i(\mathbf{p}_i + \int_{t_0}^{t} \nabla_i\psi_i\, dt) \\ \Sigma_i \mathbf{x}_{i0} \times \mathbf{p}_{i0} &= \Sigma_i(\mathbf{x}_i \times \mathbf{p}_i + \int_{t_0}^{t} \mathbf{x}_i \times \nabla_i\psi_i\, dt) \end{aligned} \tag{22}$$

Hence

$$F_N = \exp[-\Sigma_i \beta_0 (E_i - \mathbf{u}_{i0} \cdot \mathbf{p}_i - m_i \chi_{i0} - \Delta_i)]/Z_0,$$

$$\Delta_i = \int_{t_0}^{t} (\partial_t \psi_i + \mathbf{u}_{i0} \cdot \nabla_i \psi_i)\, dt. \tag{23}$$

If we write also

$$\beta_0 = \beta_i - \delta\beta_i, \quad \delta\beta_i = \int_{t_0}^{t} (d_t \beta_i + \mathbf{v}_i \cdot \nabla \beta_i)\, dt,$$

$$\mathbf{u}_{i0} = \mathbf{u}_i - \boldsymbol{\delta}\mathbf{u}_i, \quad \boldsymbol{\delta}\mathbf{u}_i = \int_{t_0}^{t} (d_t \mathbf{u}_i + \mathbf{v}_i \cdot \nabla \mathbf{u}_i)\, dt,$$

$$\chi_{i0} = \chi_i - \delta\chi_i, \quad \delta\chi_i = \int_{t_0}^{t} (d_t \chi_i + \mathbf{v}_i \cdot \nabla \chi_i)\, dt, \tag{24}$$

where β_i, $\mathbf{u}_i$ and χ_i are local values of $\beta\mathbf{u}$ and χ_a at the point $\mathbf{x}_i$ at time t, and $\mathbf{v}_i = \boldsymbol{\xi}_i - \mathbf{u}_i$ we have an exact and readily intelligible expression for F_N. Although F_N still depends on the entire previous history of the system, through the time integrals which it contains, there is good reason to expect that the velocity distribution functions derived from it will depend only on values of β, $\mathbf{u}$ and χ_a near to the time t, i.e., that averaging operations will result in a loss of "memory" of the past history of the system. We shall now examine how such averaging operations are to be performed.

If G is any function of the $\mathbf{x}_i$ and $\boldsymbol{\xi}_i$, we may define a mean value $\langle G \rangle$ of G by

$$\langle G \rangle = \Sigma_N \int G F_N \, d\Omega_N, \tag{25}$$

where the summation is over all possible values of each of the N_a, extending in principle from 0 to ∞. In particular, the velocity distribution functions may be defined by

$$f_a = \langle \Sigma_i \delta_{ia}\, \delta(\mathbf{x}_i - \mathbf{x})\, \delta(\boldsymbol{\xi}_i - \boldsymbol{\xi}) \rangle,$$

$$f_{ab} = \langle \Sigma_i \delta_{ia}\, \delta(\mathbf{x}_i - \mathbf{x})\, \delta(\boldsymbol{\xi}_i - \boldsymbol{\xi})\, \Sigma'_j \delta_{jb}\, \delta(\mathbf{x}_j - \mathbf{x}')\, \delta(\boldsymbol{\xi}_j - \boldsymbol{\xi}') \rangle, \tag{26}$$

with similar formulas for higher members of the hierarchy. The substitution of (26), together with (25), (23) and (24), into (16) yields "auto-correlation" formulas, of the type obtained by M. S. Green, for the diffusion coefficients:

$$\lambda_a = \tfrac{1}{3}\beta \langle \Sigma_i m_i \mathbf{v}_i \delta_{ia}\, \delta(\mathbf{x} - \mathbf{x}_i) \cdot \int_{t_0}^{t} \boldsymbol{\tau}\, dt \rangle,$$

$$\lambda_{ab} = \tfrac{1}{3}\beta \langle \Sigma_i m_i \mathbf{v}_i \delta_{ia}\, \delta(\mathbf{x} - \mathbf{x}_i) \cdot \int_{t_0}^{t} \boldsymbol{\tau}_b\, dt \rangle,$$

$$\boldsymbol{\tau} = \Sigma_j [m_j(\tfrac{1}{2}\mathbf{v}_j^2 - \nu_j) + \tfrac{1}{2}\Sigma'_k(\varphi_{jk} - \langle \varphi_{jk} \rangle)]\, \mathbf{v}_j$$

$$\boldsymbol{\tau}_b = \Sigma_j m_j (\delta_{jb} - \varrho_b/\varrho)\, \mathbf{v}_j \tag{27}$$

where $\nu_j = \mu_j + T_j S_j$ is the partial enthalpy per unit mass of a particle at the point $\mathbf{x}_j$. The averages can be evaluated with the phase space distributions of equilibrium. Similar formulas can be written down for the other transport coefficients. It is of course entirely impracticable to evaluate them directly, because of the multiple integrations and summations, and the implied need to integrate the equations of motion of N particles over at least a finite time prior to the time t. For this reason, we proceed to an indirect method, which coalesces with another method which has been used successfully to determine transport coefficients — the method of integral equations.

As the phase space distribution function is a constant of the motion, it satisfies Liouville's equation

$$\frac{\partial F_{\mathbf{N}}}{\partial t} + \Sigma_i \left[\frac{\partial}{\partial \mathbf{x}_i} \cdot (F_{\mathbf{N}} \boldsymbol{\xi}_i) + \frac{\partial}{\partial \boldsymbol{\xi}_i} \cdot (F_{\mathbf{N}} \boldsymbol{\eta}_i) \right] = 0 \tag{28}$$

where

$$\boldsymbol{\eta}_i = -\nabla_i \psi_i - \Sigma_j' \nabla_i \varphi_{ij} / m_i \tag{29}$$

is the acceleration of the i-th particle. From this it follows that the mean value defined in (25) satisfies

$$\frac{\partial \langle G \rangle}{\partial t} = \left\langle \Sigma_i \left(\boldsymbol{\xi}_i \cdot \frac{\partial G}{\partial \mathbf{x}_i} + \eta_i \cdot \frac{\partial G}{\partial \boldsymbol{\xi}_i} \right) \right\rangle. \tag{30}$$

In particular, the velocity distribution functions defined by (26) satisfy

$$D_a f_a = \frac{\partial f_a}{\partial t} + \xi \cdot \frac{\partial f_a}{\partial \mathbf{x}} + \mathbf{F}_a \cdot \frac{\partial f_a}{\partial \boldsymbol{\xi}} = \Sigma_b \iint O_{ab} (f_{ab} - f_a f_b) \, d^3 x' \, d^3 \xi' \tag{31}$$

and

$$\begin{aligned} &(D_a + O_{ab} + D_b + O_{ba}) f_{ab} \\ &= \Sigma_c \iint (O_{ac} + O_{bc}) (f_{abc} - f_{ab} f_c) \, d^3 x'' d^3 \xi'' \end{aligned} \tag{32}$$

where

$$O_{ab} = \frac{1}{m_a} \frac{\partial \varphi_{ab}}{\partial \mathbf{x}} \cdot \frac{\partial}{\partial \boldsymbol{\xi}}, \quad O_{ba} = \frac{1}{m_b} \frac{\partial \varphi_{ab}}{\partial \mathbf{x}'} \cdot \frac{\partial}{\partial \boldsymbol{\xi}'},$$

etc., and $\mathbf{F}_a$ is the total macroscopic force per unit mass acting on a particle of the a-th type, including long-range forces originating in the fluid. It is separation of forces of the latter type which has led to the subtraction terms in the integrals of (31) and (32).

Clearly (31) and (32) are just the first two of a chain of equations to determine $f_a, f_{ab}, f_{abc}, \ldots$, each of which however contains in its integral a distribution function one step higher in the hierarchy. It follows that these equations can only be solved by adopting an approximation which allows one to express the f_{ab} in terms of the f_a, or the f_{abc} in terms of the f_a and f_{ab}, etc. However, care must be exercised in the introduction of such a procedure, that the approximations applied to the different members of the hierarchy are consistent with one another, and also with the macroscopic conservation equations which, as Hilbert first observed in connection with Boltzmann's equation[6], are integrability conditions for (31). For instance, it is well known that Boltzmann's equation is consistent with the perfect gas laws, but not with the imperfect gas laws which result from the substitution of the approximate forms of f_a and f_{ab} in (15). A schema for avoiding this type of inconsistency has been proposed by Green and Hoffman[7], and is further developed below.

3 GENERALIZED INTEGRABILITY CONDITIONS

We shall now discuss the generalization of Hilbert's integrability conditions for fluids in which Boltzmann's assumption of binary encounters cannot be sustained. The solution of the Eqs. (31) and (32) depends on the separation of f_a and f_{ab} into two parts:

$$f_a = f_a^0 + f_a',$$
$$f_{ab} = f_{ab}^0 + f_{ab}' \tag{33}$$

where f_a^0 and f_{ab}^0 have a form designed to yield the macroscopic results of equilibrium, as shown in (15), without any contribution from f_a' and f_{ab}', while the irreversible phenomena are associated with f_a' and f_{ab}'. Thus, if

$$I_a^0 = \Sigma_b \iint O_{ab}(f_{ab}^0 - f_a^0 f_b^0)\, d^3x'\, d^{3'}\xi \tag{34}$$

is the result of substuting the functions f_{ab}^0 and f_a^0 for f_{ab} and f_a in the right side of (31), we require that the conditions

$$\Sigma_a \int (\delta_{ac},\, m_a\xi,\, \tfrac{1}{2}m_a\xi^2)\,(D_a f_a^0 - I_a^0)\, d^3\xi = 0 \tag{35}$$

should reduce to the correct equations of conservation of mass, momentum and energy, at least when diffusion, viscosity and thermal conduction are neglected. On the other hand, whatever approximation is adopted for f_{ab} or f_{abc} in (32) must yield a solution which is consistent with the proposed separation of f_{ab} and f_a shown in (33).

We substitute

$$f_a^0 = \varrho_a(\beta/2\pi m_a)^{3/2} \exp\left[-\tfrac{1}{2}\beta m_a(\xi - \mathbf{u})^2\right], \tag{36}$$

where ϱ_a, $\beta = 1/kT$ and $\mathbf{u}$ are the correct local values at the point $\mathbf{x}$ and the time t, in the non-equilibrium state. This ensures that the equations of conservation of mass and momentum derived from (35) are automatically "correct" (when diffusion and viscosity are neglected), irrespective of the choice of f^0_{ab}. However, to obtain a correct equation of conservation of energy it is necessary to satisfy the pair conservation equation

$$\frac{\partial n_{ab}}{\partial t} + \frac{\partial}{\partial \mathbf{x}} \cdot (n_{ab}\mathbf{u}_{ab}) + \frac{\partial}{\partial \mathbf{x}'} \cdot (n_{ab}\mathbf{u}_{ba}) = 0, \tag{37}$$

where n_{ab}, $\mathbf{u}_{ab}$ and $\mathbf{u}_{ba}$ are the functions defined in (17) and (18). This equation is a rigorous consequence of (32), obtained by integrating with respect to the velocities $\boldsymbol{\xi}$ and $\boldsymbol{\xi}'$. But it is not satisfied if we adopt the plausible approximation

$$f^0_a f^0_b (n_{ab}/n_a n_b),$$

for f_{ab}, even when irreversible effects are neglected; so the identification of this expression with f^0_{ab} would conflict with our requirement that the third equation of (35) should yield a "correct" equation of conservation of energy. Instead, we must define f_{ab} by

$$f^0_{ab} = f^0_{a(b)} f^0_{b(a)} (n^0_{ab}/n_a n_b), \tag{38}$$

where $f^0_{a(b)}$ and $f^0_{b(a)}$ are chosen so that

$$\int f^0_{a(b)}\xi \, d^3\xi = n_a \mathbf{u}^0_{ab}$$
$$\int f^0_{b(a)}\xi' \, d^3\xi' = n_b \mathbf{u}^0_{ba}, \tag{39}$$

and

$$\frac{\partial n^0_{ab}}{\partial t} + \frac{\partial}{\partial \mathbf{x}} \cdot (n^0_{ab}\mathbf{u}^0_{ab}) + \frac{\partial}{\partial \mathbf{x}'} \cdot (n^0_{ab}\mathbf{u}^0_{ba}) = 0 \tag{40}$$

If we take

$$m_a \mathbf{u}_{ab} + m_b \mathbf{u}_{ba} = m_a \mathbf{u} + m_b \mathbf{u}', \tag{41}$$

where $\mathbf{u}'$ is the value of $\mathbf{u}$ at the point $\mathbf{x}'$, (39) reduces to a differential equation which can be solved for the difference $\mathbf{u}^0_{ba} - \mathbf{u}^0_{ab}$, given the equilibrium form of the pair density, n^0_{ab}. The solution is discussed in the following section.

The question now arises whether the T^0_{ab} and T^0_{ba} defined by

$$\int f^0_{a(b)} \tfrac{1}{2} m_a (\xi - \mathbf{u}^0_{ab})^2 \, d^3\xi = 3 n_a k T_{ab}/2,$$
$$\int f^0_{b(a)} \tfrac{1}{2} m_b (\xi' - \mathbf{u}^0_{ba})^2 \, d^3\xi = 3 n_b k T_{ba}/2 \tag{42}$$

may be identified with the local temperatures T and T' at $\mathbf{x}$ and $\mathbf{x}'$, or whether they are analogous to the velocities defined in (39). So far as the conditions

(35) are concerned, there is no reason why they should differ from T and T'. However, there are similar integrability conditions

$$\Sigma_{a,b} \iiint (\delta_{ac}\delta_{bd},\ m_a\xi + m_b\xi',\ \tfrac{1}{2}m_a\xi^2 + \tfrac{1}{2}m_b\xi'^2)$$
$$[(D_a + O_{ab} + D_b + O_{ba}) f^0_{ab} - I^0_{ab}]\, d^3\xi d^3\xi' d^3r = 0,$$
$$I^0_{ab} = \Sigma_c \iint (O_{ac} + O_{bc}) (f^0_{abc} - f^0_{ab}f^0_c)\, d^3x''\, d^3\xi''. \tag{43}$$

for the Eq. (32) satisfied by f_{ab}. These conditions should reduce, in first order, to correct conservation equations of mass, momentum and energy for pairs of particles in the fluid. The first of the conditions is, in fact, automatically satisfied, provided f^0_{ab} is chosen in accordance with (38) and (39). However, the second (relating to conservation of momentum) is satisfied only if the correlation temperatures T^0_{ab} and T^0_{ba} are appropriately chosen, so as to satisfy the equation

$$\frac{\partial}{\partial \mathbf{x}} (n^0_{ab}kT^0_{ab}) + \frac{\partial}{\partial x'} (n^0_{ab}kT^0_{ba}) = kT\left(\frac{\partial n^0_{ab}}{\partial \mathbf{x}} + \frac{\partial n^0_{ab}}{\partial \mathbf{x}'}\right)$$
$$+ \frac{(m_a + m_b)\, n^0_{ab}}{\varrho} \frac{\partial p}{\partial \mathbf{x}}, \tag{44}$$

where p is the hydrostatic pressure defined in (15). If we agree to choose the correlation temperatures so that

$$m_aT^0_{ab} + m_bT^0_{ab} = m_aT + m_bT',$$

the solution of (44) for $T^0_{ab} - T^0_{ba}$ will completely determine them.

The Eqs. (40) and (44) are obviously only representatives of a hierarchy of equations satisfied by a set of correlated velocities and temperatures, which are associated with the functions f^0_{ab}, f^0_{abc}, etc. The correlated velocities and temperatures can be determined, with the help of suitable boundary conditions, from the correlated densities n^0_{ab}, n^0_{abc}, ..., which in turn can be deduced from the statistical mechanics of equilibrium. A direct solution of (40) was obtained by Green and Hoffman in the paper already referred to. In the next section we shall obtain an equivalent solution by another method, which is easily extended for the determination of $T^0_{ab} - T^0_{ba}$, and many-particle velocities and temperatures. This method proceeds by the 'erasure' of gradients of the macroscopic variables ϱ_a, $\mathbf{u}$ and T, built into the functions f^0_{ab}, f^0_{abc}, etc. By such means it can be ensured that an exact solution is obtained of that part of Liouville's equation which is concerned with the relative motion; this solution can be used to determine $\mathbf{u}^0_{ab} - \mathbf{u}^0_{ba}$, $T^0_{ab} - T^0_{ba}$, etc., exactly. That such a procedure is sound in principle can be seen from the theory of the auto-correlation method sketched in Section 1. But it also suggests the possibility of a significant improvement in the determination of the transport coefficients by auto-correlation methods.

4 THE METHOD OF ERASURE

Let us first examine the application of the method of erasure to the determination of correlated velocities and temperatures. The idea is that these can be determined from modified distribution functions of the form

$$f_{ab\ldots}^{(m)} = f_a^{(m)} f_b^{(m)} \ldots \exp\left(-\beta^{(m)} \psi_{ab\ldots}^{(m)}\right), \tag{45}$$

where $f_a^{(m)}, f_b^{(m)}, \ldots$ are singlet velocity distribution functions of equilibrium, corresponding to the macroscopic parameters $\varrho_a^{(m)}$, $\mathbf{u}^{(m)}$ and $T^{(m)}$, $\beta^{(m)} = 1/(kT^{(m)})$, and $\psi_{ab\ldots}^{(m)}$ is the mean interaction energy of the group of particles in an equilibrium environment with the same parameters, defined by

$$n_{ab\ldots}^{(m)} = n_a^{(m)} n_b^{(m)} \ldots \exp\left(-\beta^{(m)} \psi_{ab\ldots}^{(m)}\right).$$

We choose as $\varrho_a^{(m)}$, $\mathbf{u}^{(m)}$ and $T^{(m)}$ the actual values at the point $\mathbf{X}^{(m)}$ and time $t^{(m)}$, where $t^{(m)}$ is some fixed time, and $\mathbf{X}^{(m)}$ is the position of the centre of mass of the particles at time $t^{(m)}$. The choice of $t^{(m)}$ is probably not of vital importance, but here we agree that it shall be chosen so that the moment of inertia of the particles about their centre of mass, in the course of their mean natural motion, has its minimum value at time $t^{(m)}$. For a pair of particles, $t^{(m)}$ will be time of closest approach of the particles, moving in the potential $\psi_{ab}^{(m)}$. We note that, in a plasma,

$$\psi_{ab} = e_a e_b \exp\left(-r/a_D\right)/(4\pi r),$$
$$a_D^{-2} = 4\pi\beta\Sigma_a e_a^2 n_a, \tag{46}$$

in the approximation of Debye and Hueckel, which is very good provided the proportion of charged particles in bound states is small. To take account of quantal effects, which is necessary to avoid certain divergences, the denominator $(4\pi r)$ should be replaced by $(4\pi a_B)$, where a_B is the Bohr radius, when $r < a_B$.

Since $t^{(m)}$ and $\mathbf{X}^{(m)}$ are constants of the motion, the modified distribution function defined in (45) satisfies Liouville's equation for the relative motion,

$$\boldsymbol{\varrho} \cdot \frac{\partial f_{ab}^{(m)}}{\partial \mathbf{r}} = \frac{1}{m_{ab}} \frac{\partial \psi_{ab}}{\partial \mathbf{r}} \cdot \frac{\partial f_{ab}^{(m)}}{\partial \boldsymbol{\varrho}}, \tag{47}$$

where $\mathbf{r} = \mathbf{x} - \mathbf{x}'$, $\varrho = \boldsymbol{\xi} - \boldsymbol{\xi}'$ and $m_{ab} = m_a m_b/(m_a + m_b)$, exactly. It also satisfies the complete Liouville's equation, provided the difference in the macroscopic force between $\mathbf{x}$ and $\mathbf{x}'$ is neglected. The same is not true of the function $f_{ab}^{(c)}$, which we now define as identical with $f_{ab}^{(m)}$, except that the macroscopic parameters $\varrho_a^{(m)}$, $\mathbf{u}^{(m)}$ and $T^{(m)}$ are replaced by $\varrho_a^{(c)}$, $\mathbf{u}^{(c)}$ and $T^{(c)}$, the values at the centre of mass

$$\mathbf{X} = (m_a\mathbf{x} + m_b\mathbf{x}')/(m_a + m_b)$$

at time t. Because the time $t^{(m)} - t$ is short by macroscopic standards, provided the two particles are near enough to interact, we may write

$$f_{ab}^{(m)} = f_{ab}^{(c)} + (t^{(m)} - t)\frac{\partial f_{ab}^{(c)}}{\partial t} + (\mathbf{X}^{(m)} - \mathbf{X}) \cdot \frac{\partial f_{ab}^{(c)}}{\partial \mathbf{X}}. \tag{48}$$

The neglect of higher derivatives with respect to t and $\mathbf{X}$ does not imply any error in determining the transport coefficients. According to standard dynamical theory,

$$t^{(m)} - t = -\operatorname{sgn}(\boldsymbol{\varrho}\cdot\mathbf{r}) \int_{r^{(m)}}^{r} (m)\,(ds/dt)^{-1}\,ds,$$

$$(ds/dt)^2 = 2[\psi_{ab}(r) - \psi_{ab}(s)]/m_{ab} + \varrho^2(1 - r^2/s^2) + (\mathbf{r}\cdot\boldsymbol{\varrho})^2/s^2 \tag{49}$$

where $\operatorname{sgn}(\boldsymbol{\varrho}\cdot\mathbf{r}) = \pm 1$ depending on the sign of $\boldsymbol{\varrho}\cdot\mathbf{r}$, and $r^{(m)}$ is the distance of closest approach of the orbiting particles. Also,

$$\mathbf{X}^{(m)} - \mathbf{X} = (m_a\boldsymbol{\xi} + m_b\boldsymbol{\xi}')\,(t^{(m)} - t)/(m_a + m_b) \tag{50}$$

if we neglect the action of the macroscopic force during the time $t^{(m)} - t$; the correction is easily inserted if required. Obviously,

$$\iint f_{ab}^{(m)}\,d^3\xi\,d^3\xi' = \iint f_{ab}^{(c)}\,d^3\xi\,d^3\xi' = \overset{0}{n}_{ab}. \tag{51}$$

This result is, of course, one of the considerations which influenced our choice of $f_{ab}^{(m)}$ and $f_{ab}^{(c)}$. It implies that we may determine the correlated velocities and temperatures directly from the distribution function $f_{ab}^{(m)}$, expressed as in (48). Thus

$$\overset{0}{n}_{ab}(\overset{0}{\mathbf{u}}_{ab} - \overset{0}{\mathbf{u}}_{ba}) = \iint f_{ab}^{(m)}\boldsymbol{\varrho}\,d^3\xi\,d^3\xi'. \tag{52}$$

The integrations are easily carried out, first over the angle between $\boldsymbol{\varrho}$ and $\mathbf{r}$, then over ϱ, and finally over the velocity of the centre of mass. Thus

$$\boldsymbol{\nu}_{ab} = \iint f_{ab}^{(c)}\,(t^{(m)} - t)\,\boldsymbol{\varrho}\,d^3\xi\,d^3\xi' = -(\mathbf{r}/r^3)\int_0^r \overset{0}{n}_{ab} r^2\,dr$$

and

$$\overset{0}{n}_{ab}(\overset{0}{\mathbf{u}}_{ab} - \overset{0}{\mathbf{u}}_{ba}) = \frac{\partial}{\partial t}\boldsymbol{\nu}_{ab} + \frac{\partial}{\partial \mathbf{x}}\cdot(\mathbf{u}\boldsymbol{\nu}_{ab}) \tag{53}$$

The difference between this result and that obtained previously by Green and Hoffman is easily seen to be the result of a slightly different definition of n^0_{ab}, and the conversion is made very simply if desired. The correlated temperatures can be obtained by a very similar calculation:

$$n^0_{ab}(T^0_{ab} - T^0_{ba}) = \frac{4m_a m_b}{3(m_a + m_b)^2} \frac{\partial}{\partial \mathbf{x}} \cdot (T^{(c)} \boldsymbol{\nu}_{ab}) \tag{54}$$

Obviously correlated velocities and temperatures for groups of more than two particles can be determined formally by a similar method. Although the solution of the three-body problem, or many-body problem, cannot be carried out exactly (except for the hard sphere potential) it is clear that the results should not depend on the details of the multiple encounter so that an explicit solution is not required. It should be noticed that the function $f^{(m)}_{ab\ldots}$ is introduced only to determine $\mathbf{u}^0_{ab\ldots}$ and $T^0_{ab\ldots}$, and that when these have been found they can be used in the construction of the quite different function $f^0_{ab\ldots}$ defined by

$$f^0_{ab\ldots} = f^0_{a(b..)} f^0_{b(a..)} n^0_{ab\ldots}/(n_a n_b \ldots), \tag{55}$$

where $f^0_{a(b..)}$ is a Maxwellian distribution function modified by the substitution of $\mathbf{u}^0_{ab\ldots}$ and $T^0_{ab\ldots}$ for $\mathbf{u}$ and T.

We now discuss the application of correlated velocities and temperatures to the determination of transport coefficients by auto-correlation methods. Firstly it should be noticed that the method of erasure has much in common with the auto-correlation method and that the determination of correlated velocities and temperatures is an essential feature of the evaluation of transport coefficients in the form illustrated above. However, what is needed eventually is the evaluation of correlations in the resultant momenta and energies of groups of particles, which are *not* completely determined by the statistical mechanics of equilibrium. Thus we should begin by expressing the phase space distribution function F_{N} in the form

$$F_{\mathrm{N}} = F^0_{\mathrm{N}} + F'_{\mathrm{N}}, \tag{56}$$

where F^0_{N}, like $f^0_{ab..}$ has been constructed with the help of correctly correlated velocity differences and temperature differences. In order to determine the transport coefficients it is still necessary to find F'_{N}, which represents the irreversible effects. For this purpose, we may make use of the equation

$$F'_{\mathrm{N}} = -\int_{t_0}^{t} (DF^0_{\mathrm{N}}/Dt)\, dt, \tag{57}$$

where (D/Dt) is Liouville's operator, which is applied to F_{N} in (28). The right side of (57) can be found by a calculation similar to that appearing in an earlier paper by the author[8], but has the advantage that the divergent terms resulting there (loc. cit., Section 5) have been eliminated. This is a further indication that the use of correlated velocities and temperatures are required for the development of a consistent theory of irreversible processes.

Secondly, it should be noticed that, as shown in references already cited[1,8], the evaluation of the auto-correlation formulas almost inevitably reduces to the solution of equations of the type of (31) and (32). Therefore, the practical consequences of the use of (57) are already implicit in (33). The auto-correlation method has important theoretical uses, not the least of which is that it shows what boundary conditions to apply to equations of the hierarchy whose first members are (31) and (32). It is clear from (27), for instance, that, in "precollision" configurations, correlations between the velocities of separate groups of particles should disappear. This is, of course, what has usually been assumed in transforming (31) to variants of Boltzmann's equation[9]; the assumption, however, does need justification.

5 APPLICATION TO PLASMAS

To conclude, we shall discuss briefly the application of these methods to the determination of the transport coefficients of plasmas. A plasma is an example of a system in which the assumption of binary encounters, implicit in the usual derivation of Boltzmann's equation, is not valid at low densities. Indeed, it is well known that unless the shielding of the Coulomb forces is taken into account, the transport coefficients derived from Boltzmann's equation are divergent. It is not correct however, to apply the method of Chapman and Enskog, simply using the Debye potential defined in (46) instead of the Coulomb potential. Such a procedure fails to take account correctly of the many-body forces which contribute to the Debye potential, and also leaves out of account the contribution of the closed trajectories which are possible with particles of opposite charge.

The simplest approximation consistent with the Debye potential is to replace (32) by

$$(D_a + O'_{ab} + D_b + O'_{ba}) f_{ab} = 0,$$

$$O'_{ab} = \frac{1}{m_a} \frac{\partial \psi_{ab}}{\partial \mathbf{x}} \cdot \frac{\partial}{\partial \xi} \tag{58}$$

A still better approximation (called the disjunctive approximation) was described by Green and Leipnik[5], and a general account of methods of approximation by Green and Hoffman[7]; but here we shall consider (58) because it admits of an exact solution and therefore allows a better illustration of general principles.

The required solution of (58) is one which satisfies the boundary condition $f_{ab} = f_a f_b$ in pre-collision scattering states, i.e. for $\mathbf{r} \cdot \boldsymbol{\varrho} < 0$ and values of r a few times a_D, the Debye shielding distance. This boundary condition, applied to the solution of (58), gives a value of f_{ab}, depending on f_a, which can be substituted into (32), provided $\frac{1}{2}m_a\xi^2 + \frac{1}{2}m_b\xi'^2 + \psi_{ab}(r) > 0$. It is *not* possible, however, to transform the right side of (32) into a form resembling Boltzmann's collision integral, by the usual methods, because O'_{ab}, as defined in (58), is different from O_{ab}. Nor is it correct to leave out contributions from the closed trajectories, for which $\frac{1}{2}m_a\xi^2 + \frac{1}{2}m_b\xi'^2 + \psi_{ab}(r) < 0$. It is true that, in the approximation we have adopted, the closed trajectories do not contribute to irreversible processes. However, they do contribute to f^0_{ab}, and if their contribution were dropped the resulting equation would be inconsistent with the macroscopic laws. To summarize, therefore, we use

$$f_{ab} = f_a(t^{(0)}, \mathbf{x}^{(0)}, \xi^{(0)}) f_b(t^{(0)}, \mathbf{x}^{(0)\prime}, \xi^{(0)\prime}),$$
$$\tfrac{1}{2}m_a\xi^2 + \tfrac{1}{2}m_b\xi'^2 + \psi_{ab} \geqq 0 \tag{59}$$

where $\mathbf{x}^{(0)}$, $\mathbf{x}^{(0)\prime}$, $\xi^{(0)}$ and $\xi^{(0)\prime}$ are positions on the same trajectory, at the time $t^{(0)}$, earlier than t, when the particles were beginning to interact. But

$$f_{ab} = f^0_{ab},$$
$$\tfrac{1}{2}m_a\xi^2 + \tfrac{1}{2}m_b\xi'^2 + \psi_{ab} < 0. \tag{60}$$

The resulting integral equation is a little more difficult to solve than Boltzmann's equation, since values of f_{ab} are needed throughout the interaction region and not simply on the precollision and post-collision hemispheres. However, the methods of Chapman and Enskog are still applicable and there is no real difficulty of principle. The substitutions (59) and (60) result in a self-consistent equation from which the f_a, and hence the transport coefficients can be determined.

References

1. See, e.g., Chapter 4 of *Research Frontiers in Fluid Dynamics*, Ed. R. J. Seeger and G. Temple, Interscience, New York (1965).
2. E. P. Wigner, *Phys. Rev.* **40**, 749 (1932); H. S. Green, *J. Chem. Phys.* **19**, 955 (1951); **20**, 1274 (1952); J. H. Irving and R. W. Zwanzig, *J. Chem. Phys.* **19**, 1173 (1951).
3. N. N. Bogoliubov, Chapter 1 of *Studies in Statistical Mechanics*, Vol. I (Ed. J. de Boer and G. E. Uhlenbeck), Interscience, New York (1962).
4. M. S. Green, *J. Chem. Phys.* **22**, 398 (1954).
5. H. S. Green and R. B. Leipnik, *Int. J. Engng. Sci.* **3**, 491 (1965).

6. D. Hilbert, *Math. Ann.* **72**, 562 (1912); also, S. Chapman and T. G. Cowling, *Mathematical Theory of Non-Uniform Gases*, p. 111 Cambridge Univ. Press, Cambridge (1939).
7. H. S. Green and D. Hoffman, *J. Chem. Phys.* **49**, 2600 (1968).
8. H. S. Green, *J. Math. Phys.* **2**, 344 (1961).
9. H. Grad, *Encyclopedia of Physics*, Vol. **12** Springer, Berlin (1958); or pp. 216–225 of H. S. Green, *Kinetic Theory of Fluids*, North-Holland, Amsterdam (1952); reprint by Dover, New York (1969).

PAPER A–2

*Renormalized kinetic equations**

J. R. DORFMAN†

Institute for Fluid Dynamics and Applied Mathematics

and

Department of Physics and Astronomy University of Maryland

College Park, Maryland

The basic theory for the density expansion of the pair distribution function, $F_2(1, 2, t)$, for a dense gas, not in equilibrium, was given by Green and Piccirelli[1] and by Dorfman and Cohen[2]. It was shown by these authors that a formally exact density expansion of the pair distribution could be derived from the Liouville equation and expressed in the form‡

$$F_2(1, 2, t) = F_2^{(0)}(1, 2, t) + nF_2^{(1)}(1, 2, t) + n^2F_2^{(2)}(1, 2, t) + \cdots \quad (1a)$$

$$F_2^{(0)}(1, 2, t) = \tilde{\mathscr{S}}_t(1, 2)\, F_1(1, t)\, F_1(2, t) \quad (1b)$$

$$F_2^{(1)}(1, 2, t) = \int d\mathbf{r}_3 \int d\mathbf{p}_3[\tilde{\mathscr{S}}_t(1, 2, 3) - \tilde{\mathscr{S}}_t(1, 2)\, \tilde{\mathscr{S}}_t(1, 3) - \tilde{\mathscr{S}}_t(1,2) \tilde{\mathscr{S}}_t(2, 3) + \tilde{\mathscr{S}}_t(1, 2)]\, F_1(1, t)\, F_1(2, t)\, F_1(3, t) \quad (1c)$$

where

$$\tilde{\mathscr{S}}_t(1, 2, \ldots, j) = e^{-tH_j(1, 2, \ldots, j)}\, a_j(1, 2, \ldots, j, o) \prod_{i=1}^{j} e^{tH_1(i)} \quad (2a)$$

where

$$H_j(1, 2, \ldots, j) = \sum_{i=1}^{j} \frac{\mathbf{p}_i}{m} \cdot \frac{\partial}{\partial \mathbf{r}_i} - \sum_{k<l}^{j} \frac{\partial \Phi_{kl}}{\partial \mathbf{r}_k} \cdot \left(\frac{\partial}{\partial \mathbf{p}_k} - \frac{\partial}{\partial \mathbf{p}_l} \right) \quad (2b)$$

---

* This work was supported by the U.S. Army Research Office, Grant Number AROD-31-124-G999.

† Address during 1969–70, The Rockefeller University, New York, N. Y., 10021.

‡ The notation used here is largely that of Reference 2, to which the reader is directed for a more complete explanation.

and $a_j(1, 2, \ldots, j, 0)$ characterizes the initial correlations between the particles, such that $a_j \equiv 1$ represents no initial correlation between the j particles.

If it were possible to prove, for times long compared to the duration of a collision, that

(a) the initial state is forgotten and all the a_j's may be set equal to unity in Eq. (1) and

(b) the time t that appears in the operator $\tilde{\mathscr{S}}_t$ may be replaced by $t = \infty$, but the time t that appears in the single particle distribution function $F_1(i, t)$ may be retained, in Eq. (1),

then the formal expression could be brought into coincidence with the equation for $F_2(1, 2, t)$ given by Bogoliubov on the basis of the so-called "functional assumption,"[3] and which may be expressed in the form[4]

$$\begin{aligned} F_2^{(B)}(1, 2, t) &= \mathscr{S}_\infty(1, 2)\, F_1(1, t)\, F_1(2, t) \\ &+ n \int d\mathbf{r}_3 \int d\mathbf{p}_3 [\mathscr{S}_\infty(1, 2, 3) - \mathscr{S}_\infty(1, 2)\, \mathscr{S}_\infty(1, 3) - \mathscr{S}_\infty(1, 2)\, \mathscr{S}_\infty(2, 3) \\ &+ \mathscr{S}_\infty(1, 2)]\, F_1(1, t)\, F_1(2, t)\, F_1(3, t) + \cdots \end{aligned} \tag{3}$$

where

$$\mathscr{S}_\infty(1, 2), \ldots, j) = \lim_{t\to\infty} \mathscr{S}_t(1, 2, \ldots, j) \tag{3a}$$

and

$$\mathscr{S}_\tau(1, 2, \ldots, j) = e^{-tH_j(1, 2, \ldots, j)} \prod_{i=1}^{j} e^{tH_1(i)} \tag{3b}$$

However, it was noticed, by Weinstock, and independently by Cohen and Dorfman[2,6] and by Goldman and Frieman[7], that all but the first few terms in Eq. (3) do not exist and hence that limiting processes outlined in (b) above is not possible. This divergence in the Bogoliubov equations is connected to sequences of j binary collisions among j particles and is due to the failure of the phase space volume for such sequences to remain bounded as the time in which such sequences occur grows infinitely large. A quantitative discussion of the divergence was given by Sengers[8] and is further discussed in his lecture at this Symposium.

It is interesting to consider the type of behavior exhibited by several different types of systems. For a gas of hard rods, constrained to move on a line, the contribution to the pair function from three particle dynamics, given by expression (1c) above, grows proportionally to t for large t, and is due to three successive binary collisions of the rods. The theory of this system was worked out in considerable detail by Lebowitz and Percus[9]. Another class of models which have been extensively investigated in this connection are the Lorentz models in which a light particle moves in a system of heavy random scatterers. Here one considers, instead of the pair function, the diffusion coefficient or mean square displacement of the light particle, for which formally exact expansions in powers of the density of the scatterers may be derived and which also exhibit the divergent behavior. In the two dimensional "wind-tree" version of the model, studied by Hauge and Cohen[10], in which the particle can move only in the x or y direction and is scattered by squares whose diagonals are parallel to the x and y axes, the first divergent term in the expansion of the diffusion coefficient also grows proportionally to the time t. If, instead, the particle can move in any direction in two or three dimensions and is scattered by hard discs or spheres instead of squares, the first divergent term grows like log t instead of t. The difference in the large t behavior of the two models is due to purely geometrical factors in the scattering process. Those models in which there is only a discrete number of directions available to the particle velocity exhibit a divergence proportional to t, otherwise the first divergent term is proportional to log t. A similar situation holds in the density expansion of the pair function of a real gas with short range repulsive forces. The first appearance of the divergence is in the three body term in two-dimensions and in the four body term in three dimensions[2]. In both cases the long time behavior in the pair function is proportional to log t[2,7]. One open problem in the area is to establish the types of potentials for which the above discussion is valid. Most of the work has been done using hard potentials, and the discussion of the divergence by Cohen and Dorfman is correct for any potential with finite total cross section. However, it seems that the main points above are correct for any repulsive potential $\phi(r)$ which is hard in the sense of Grad[13], that is, $\phi(r) \sim r^{-n}$, with $n > 4$, for large r. Otherwise, the system should exhibit the characteristic plasma divergences[14]. Moreover, Weinstock[15], Murase[16] and Langer and Neal[17] have asserted that for quantum mechanical systems the situation is similar to the classical case. However, a full discussion of the quantum mechanical problem as it applies, say, to neutron scattering has not yet been given.

The missing ingredient in the density expansion, as typified by Eq. (1), is the mean free path, and its absence in each of the terms is responsible for the divergence. The inclusion of the requirement that a particle can only travel a certain distance before it suffers a collision is enough to make all the divergent integrals finite. However, in the integrals that appear in Eq. (1) there is no such mechanism. Only the dynamics of a few isolated particles contribute to each term, while the mean free path is a collective property of the gas as a whole.

There are two main ways to incorporate the mean free path into the theory. One method, first used in this connection by Kawasaki and Oppenheim[18], employs the binary expansion to sum through the density series, combining terms from every order of the density in such a way that originally divergent integrals now contain a damping factor which, presumably, renders them finite. This method has been discussed in great detail elsewhere[19,20] and will not be further discussed here, except to say that the work of Hauge and Cohen, van Leeuwen and Weyland, and Hoegy on Lorentz models rely heavily on the binary collision expansion.

The other method, used first by Goldman and Frieman[7], is based upon a more detailed use of the B.B.G.K.Y. hierarchy equations than the binary collision expansion method, although the results are the same. However, the method of Goldman and Frieman is not as well known and *I* will discuss it here in a little more detail. The basic idea is not to determine the pair function by means of the density expansion, Eq. (1), but instead to solve the second heirarchy equation to obtain F_2, i.e.

$$\frac{\partial F_2(1, 2, t)}{\partial t} + H_2(1, 2)\, F_2(1, 2, t) = n \int d\mathbf{r}_3 \int d\mathbf{p}_3 (\Theta_{13} + \Theta_{23})\, F_3(1, 2, 3, t) \tag{4a}$$

where

$$\Theta_{ij} = \frac{\partial \Phi_{ij}}{\partial \mathbf{r}_i} \cdot \frac{\partial}{\partial \mathbf{p}_i} - \frac{\partial}{\partial \mathbf{p}_j} \tag{4b}$$

and $F_3(1, 2, 3, t)$ is the three particle distribution function. The methods that have been used to obtain a density expansion for $F_2(1, 2, t)$ in terms of $F_1(t)$, can similarly be applied to obtain a density expansion for $F_3(1, 2, 3, t)$ in terms of $F_1(t)$ and $F_2(t)$. If this is done one can obtain an equation for $F_2(t)$ in terms of $F_2(t)$ and $F_1(t)$. Frieman and Goldman actually consider the equation for $G_2(1, 2, t)$ which is

$$G_2(1, 2, t) = F_2(1, 2, t) - F_1(1, t)\, F_1(2, t) \tag{5a}$$

and satisfies

$$\frac{\partial G_2(1, 2, t)}{\partial t} + H_2(1, 2)\, G_2(1, 2, t) - \Theta_{12} F_1(1, t)\, F_1(2, t)$$
$$= n \int d\mathbf{r}_3 \int d\mathbf{p}_3 [\Theta_{13}(F_3(1, 2, 3, t) - F_2(1, 3, t)\, F_1(2, t))$$
$$+ \Theta_{23}(F_3(1, 2, 3, t) - F_2(2, 3, t)\, F_1(1, t))]. \tag{5b}$$

Then, using cluster expansion methods, it is possible to show that, if the $(a(1, 2, \ldots, j, 0) - 0)$ vanish wherever two particles in the set $(1, 2, \ldots, j)$ are far apart*,

$$\Theta_{13}[F_3(1, 2, 3, t) - F_2(1, 3, t)\,F_1(2, t)] = \Theta_{13}\mathscr{S}_\infty(13)\,[G_2(1, 2, t)\,F_1(3, t) + G_2(2, 3, t)\,F_1(1, t)] + \text{terms of order } n \tag{6}$$

leading to

$$\frac{\partial G_2(1, 2, t)}{\partial t} + H_2(1, 2)\,G_2(1, 2, t) - \Theta_{12}F_1(1, t)\,F_1(2, t) = n \int dx_3 \lambda_3(1, 2)\,G_2(1, 2, t) + 0(n^2) \tag{7a}$$

where

$$\lambda_3(1, 2) = [\Theta_{13}\mathscr{S}_\infty(13)\,(1 + P_{13}) + \Theta_{23}\mathscr{S}_\infty(23)\,(1 + P_{23})]\,F_1(3, t) \tag{7b}$$

and P_{ij} is the permutation operator which interchanges particles i and j. Equation (7a) is an equation for $G_2(1, 2, t)$ which can be solved, at least formally. The quantity

$$\int dx_3 \lambda_3(1, 2)$$

plays the role of a collisional damping operator, and it appears in both the binary collision expansion method and the hierarchy equation method[7,19].

As a result of either approach, it is possible to incorporate the mean free path into the theory in the form of a collisional damping operator, which is similar in structure but not identical to the Boltzmann collision operator. In the Lorentz models, calculations can be performed using the renormalized equations, and the results obtained for the diffusion coefficients are free of divergences. In these cases, if the first singular term behaves like $\log t$ as $t \to \infty$, the result of the corrected theory is proportional to $\log (t_{m \cdot F \cdot p})$ where $t_{m \cdot F \cdot p}$ is the time it takes for a particle to travel a mean free path. This time is inversely proportional to the density, and as a result, there is a logarithm of the density in the diffusion coefficient. If, however, the first singular term in proportional to $t \to \infty$, the renormalized theory provides a term proportional to $(t_{m \cdot F \cdot p})$. In other words, for the Lorentz models, the renormalized theory has a natural cut off in the originally divergent time integrations.

---

* The Θ_{13} is retained in equation (6) since this equation is only valid for configuration of three particles (1, 2, 3), in which particle 1 and 3 are close together. This is ensured by Θ_{13} which vanishes wherever 1 and 3 are not interacting.

The situation for a gas of equal particles is not yet entirely clear. There does not yet exist a proof that the renormalized theory leads to finite results. In fact, Pomeau[21] claims that for a two dimensional gas, the renormalized expressions still diverge. Clearly the question of whether or not the present renormalized theory is convergent is the most outstanding problem in this area at the moment.

If one assumes that the effect of the renormalization is the same for a gas as it is for the Lorentz models, then one can argue that the transport coefficients, such as the viscosity, η, must behave with density, in three dimensions, as

$$\eta = \eta_0 + n\eta_1 + (\eta_L n^2 \log n + n^2 \eta_2) + \cdots \tag{8}$$

where

η_0 is the result of the Boltzmann equation
η_1 is the result of the Choh–Uhlenbeck equation and is discussed in the Symposium by Sengers
η_L is the coefficient of the non-analytic term $n^2 \log n$ and is presently unknown, as is η_2.

Another result of the renormalization procedure, which may be easily seen from the solution of Eq. (7a), is that the pair function at time t, $F_2(1, 2, t)$, no longer depends on the value of single particle distribution functions at the same time t, as in Eq. (1), but instead depends upon the single particle distribution functions at earlier times, in a way which reflects the history of the system. This was first noticed by Goldman and Frieman[7], and is discussed at some length by Cohen[22].

Two further points which are of some interest in these connections. One is the very peculiar behavior of the diffusion process in the wind-tree model, found by Hauge and Cohen[10]. In the model, under certain circumstances, the coefficient of self diffusion vanishes, and the mean square displacement of the particle grows with time more slowly than t. Whether the property is a pathological feature of the model or is more general is at present an open question. Cohen and Hauge conjecture that the vanishing of the diffusion coefficient in the model is connected to the vanishing of the diffusion coefficient in a binary mixture, at the critical point. The wind tree model has also recently been treated by Gallovotti[23] who has given a very careful discussion of many points using more rigorous methods than have been used heretofore.

Finally, there is a very close connection between the renormalized kinetic equation for a gas and the Balescu–Lenard–Guernsey equation for a plasma.

This was first pointed out by Goldman[24] and has been mentioned by Pomeau[21]. Basically, the connection consists in the fact that the divergent terms for a gas with a short range potential are still divergent if the potential is long range, as is the Coulomb potential. Therefore the resummations that must be performed for the gas resemble the divergent plasma terms that must be resummed to obtain the B.G.L. equation.* In fact, the collision damping operator that appears in the B.G.L. equation[25,26], $\lambda_3^{(p)}(1, 2)$, given by

$$\lambda_3^{(p)}(1, 2) = [\Theta_{13}(1 + P_{13}) + \Theta_{23}(1 + P_{23})] F_1(3, t)$$

bears a striking resemblance to the damping operator for the gas, eq. (7b), and is obtained from expression (7b) by setting $\mathscr{S}_\infty(1, 3)$ and $\mathscr{S}_\infty(2, 3)$ equal to unity, which is the appropriate limit for a weak potential. This connection will be discussed further in a future publication[27].

References

1. M. S. Green and R. A. Piccirelli, *Phys. Rev.* **132**, 1388 (1963).
2. J. R. Dorfman and E. G. D. Cohen, *J. Math. Phys.* **8**, 282 (1967).
3. N. N. Bogoliubov, in *Studies in Statistical Mechanics*, Vol. **I**, North Holland Pub. Co., Amsterdam, 1961.
4. E. G. D. Cohen, *Physica* **28**, 1045 (1962).
5. J. Weinstock, *Phys. Rev.* **140**, A460 (1965).
6. J. R. Dorfman and E. G. D. Cohen, *Phys. Letters* **16**, 124 (1965).
7. E. A. Frieman and R. Goldman, *J. Math. Phys.* **7**, 2153 (1966); **8**, 1410 (1967), R. Goldman, in *Statistical Mechanics*. T. Bak, ed. W. A. Benjamin, Inc., New York, 1967.
8. J. V. Sengers, *Phys. Rev. Letters* **15**, 515 (1965), *Phys. Fluids* **9**, 1685 (1966).
9. J. L. Lebowitz and J. K. Percus, *Phys. Rev.* **155**, 122 (1967), J. L. Lebowitz, J. K. Percus, and J. Sykes, *Phys. Rev.* **171**, 224 (1968).
10. E. Hauge and E. G. D. Cohen, *Phys. Letters* **25A**, 78 (1967) and *J. Math. Phys.* **10**, 397 (1969).
11. J. M. J. van Leeuwen and A. Weyland, *Physica* **36**, 457 (1967), **38**, 35, (1968).
12. W. Hoegy, Ph. D. thesis, University of Michigan, 1963; and *Phys. Rev.* **185**, 210 (1969).
13. H. Grad, in *Rarefied Gas Dynamics*, 1963, Vol. I. Academic Press, New York (1963).
14. See, for example, D. Montgomery and D. A. Tidman, *Plasma Kinetic Theory*, McGraw-Hill Book Company, Inc., New York (1964).
15. J. Weinstock, *Phys. Rev. Letters* **17**, 130 (1966).
16. C. Murase, *J. Phys. Soc.* Japan **21**, 1737 (1966).
17. J. S. Langer and T. Neal, *Phys. Rev. Letters* **16**, 984 (1966).
18. K. Kawasaki and J. Oppenheim, *Phys. Rev.* **139**, A1763 (1965); and in *Statistical Mechanics*. T. Bak, ed., W. A. Benjamin, Inc. New York, 1967.

* The fact that the plasma divergence already appears in the Boltzmann equation does not seriously alter the connection.

19. J. R. Dorfman in *Lectures in Theoretical Physics*, Vol. **9C**, Gordon and Breach Science Publishers, New York, 1967.
20. M. H. Ernst, J. R. Dorfman, W. Hoegy, and J. M. J. van Leeuwen, *Physica* **45**, 127 (1969).
21. Y. Pomeau, *Phys. Letters*, **27A**, 601 (1968).
22. E. G. D. Cohen, in *Fundamental Problems in Statistical Mechanics, II*, North Holland Pub. Co., Amsterdam, 1968.
23. G. Gallavotti, *Phys. Rev.* **185**, 308 (1969).
24. R. Goldman, *Phys. Rev. Letters* **17**, 910 (1966).
25. T. Dupree, *Phys. Fluids* **4**, 696 (1961).
26. R. Guernsey, in *Lectures in Theoretical Physics*, Vol. **9C**, Gordon and Breach Science Publishers, New York, 1967, and Can. J. Phys. **45**, 179 (1967).
27. E. G. D. Cohen and J. R. Dorfman, (to be published).

PAPER A–3

Master equations for finite systems

C. J. MYERSCOUGH

Churchill College and Department of Applied Mathematics and Theoretical Physics, University of Cambridge, England

SUMMARY AND INTRODUCTION

The essential part of any derivation from the Liouville equation of a master equation for the evolution of a classical system is the introduction of phase mixing. The behaviour of the probability density itself becomes more and more contorted as time proceeds, but these oscillations have less and less physical significance, and must be eliminated by some kind of 'smoothing' process. Most previous work has introduced this by allowing the volume of the system to become infinite. (The number of particles in the system must also become infinite, to keep the average density finite.) Sums over Fourier wave vectors then become integrals, and almost periodic expressions become decaying expressions (the Poincaré recurrence time becomes infinite).

In this work we shall suggest an alternative method. We consider the evolution not of the probability density itself but of the result of 'smoothing' it over a small range of velocities. One such expression satisfies a closed and exact equation for its evolution differing from the Liouville equation by the addition of one term. This equation has phase mixing built in; we derive it in section 1 and discuss its solution in the case of no interactions in section 2.

It might therefore be possible using this modified Liouville equation to derive master equations for the evolution of various components of the smoothed density without requiring that the volume of the system become infinite. Such equations could establish a closer connection with ergodic theory, for many results of this theory are concerned with finite systems: in particular, the important theorem of Fermi (see e.g. ter Haar 1955) that a kanonisches normalsystem—a system with periodic boundary conditions—is quasi-ergodic, and admits of no integrals of motion other than the total energy. In sections 3 and 4 we study the evolution of such systems.

In section 3 we confine ourselves to the weak interaction case, using the 'quasi-linear' techniques due to Zwanzig 1960, 1961, Montroll 1961, and Sandri 1963. We find that we can indeed obtain a master equation, closely resembling that first derived by Brout and Prigogine 1956, and describing to first order in the interaction strength the tendency of the system towards the microcanonical average of the initial conditions.

In section 4 we attempt the full perturbation—theoretic solution of our modified Liouville equation, as in the work of the Brussels school (Prigogine 1962, Balescu 1962, Résibois 1966). The weak interaction results of the previous section are recovered, but when we consider higher terms we find that we cannot immediately derive, for example, an analogue of the Prigogine–Résibois master equation. As yet we can only suggest a possible way out of these difficulties; this we do in our concluding section 5.

1 PRELIMINARIES

We consider a classical dynamical system of N particles interacting through a potential field; the state of the system is described completely by the position and velocity vectors of all its particles

$$\mathbf{X} = \{\mathbf{x}_1, \ldots, \mathbf{x}_N\}; \quad \mathbf{V} = \{\mathbf{v}_1, \ldots, \mathbf{v}_N\} \tag{1.1}$$

Throughout this work we shall use a capital letter for a $3N$-vector, with the corresponding small letter for its N 3-vector components, as above. We write

$$\left.\begin{aligned} \mathbf{A}\cdot\mathbf{B} &= \mathbf{a}_1\cdot\mathbf{b}_1 + \cdots + \mathbf{a}_N\cdot\mathbf{b}_N \\ \frac{\partial\varphi}{\partial\mathbf{X}} &= \left\{\frac{\partial\varphi}{\partial\mathbf{x}_1}, \ldots, \frac{\partial\varphi}{\partial\mathbf{x}_N}\right\} \end{aligned}\right\} \tag{1.2}$$

The real and positive valued function $\varrho(\mathbf{X}, \mathbf{V}, t)$ is called the probability density of the system at time t if the probability of finding the system in a given subset Ω of the set Γ of possible states $(\mathbf{X}, \mathbf{V})$ is

$$\int_\Omega \varrho(\mathbf{X}, \mathbf{V}, t)\, dX\, dV \tag{1.3}$$

The only possible physical interpretation of ϱ is through such expressions. Whilst $\varrho(\mathbf{x}, \mathbf{v}, t)$ will not in general tend to a limit as $t \to \infty$ or as $t \to -\infty$, (1.3) should do so, for any set Ω of non-zero measure. This point is due to

Gibbs; see van Kampen 1962. If we consider the well-known example of a single particle (or N non-interacting particles) confined to a cuboidal box with perfectly reflecting walls—or, alternatively, subject to periodic boundary conditions—we may demonstrate this explicitly. We also see that it is the smoothing over velocity rather than the smoothing over space components which is important in (1, 3); all expressions of the form

$$\bar{\varrho}(\mathbf{X}, \mathbf{V}, t) = \int d(\Delta V)\, \alpha(\Delta \mathbf{V})\, \varrho(\mathbf{X}, \mathbf{V} - \Delta\mathbf{V}, t) \tag{1.4}$$

tend to limits as $t \to \pm\infty$, for all positive integrable kernels α which are non-zero over a set of non-zero measure. It therefore seems sensible physically to interpret the tendency of a system to equilibrium in the sense of weak convergence of ϱ—i.e. as the tendency of expressions such as $\bar{\varrho}$ above to limits as $t \to \pm\infty$.

If now we set

$$\alpha(\Delta\mathbf{V}) = \frac{1}{(2\pi\sigma^2)^{3N/2}} \exp\left\{\frac{-(\Delta\mathbf{V})^2}{2\sigma^2}\right\} \tag{1.5}$$

it turns out that as ϱ evolves according to Liouville's equation

$$\frac{\partial\varrho}{\partial t} + \mathbf{V}\cdot\frac{\partial\varrho}{\partial\mathbf{X}} + \mathbf{F}\cdot\frac{\partial\varrho}{\partial\mathbf{V}} = 0 \tag{1.6}$$

$\bar{\varrho}$ satisfies the closed equation

$$\frac{\partial\bar{\varrho}}{\partial t} + \mathbf{V}\cdot\frac{\partial\bar{\varrho}}{\partial\mathbf{X}} + \sigma^2\frac{\partial}{\partial\mathbf{V}}\cdot\frac{\partial\bar{\varrho}}{\partial\mathbf{X}} + \mathbf{F}\cdot\frac{\partial\bar{\varrho}}{\partial\mathbf{V}} = 0 \tag{1.7}$$

We assume two-body central interactions, and no external field.

$$\left.\begin{aligned} \mathbf{F}(\mathbf{X}) &= \frac{-\partial H(\mathbf{X})}{\partial\mathbf{X}} \\ H(\mathbf{X}) &= \sum_{p=2}^{N}\sum_{q=1}^{p-1} V(|\mathbf{x}_p - \mathbf{x}_q|) \end{aligned}\right\} \tag{1.8}$$

The existence of the closed equation (1.7) for $\bar{\varrho}$ implies that the smoothing process (1.4) does not reject completely any of the information contained in ϱ; it merely reduces in importance the less physically relevant parts. We shall investigate the behaviour of $\bar{\varrho}$ for systems satisfying periodic boundary conditions in a box side L, so that

$$\bar{\varrho}(\mathbf{X}, \mathbf{V}, t) = \sum_{\mathbf{K}} \bar{\varrho}_{\mathbf{K}}(\mathbf{V}, t)\, e^{i\mathbf{K}\cdot\mathbf{X}} \tag{1.9}$$

where the vector **K** is a member of a $3N$-dimensional lattice of side $2\pi/L$. Also

$$\left.\begin{aligned} \mathbf{F}_{\mathbf{K}} &= -i\mathbf{K}H_{\mathbf{K}} \\ H_{\mathbf{K}} &= \sum_{p=2}^{N}\sum_{q=1}^{p-1} V_{|\mathbf{k}_p-\mathbf{k}_q|} \prod_{\substack{r=1 \\ r\neq p,q}}^{N} \delta_{\mathbf{K}_r} \end{aligned}\right\} \tag{1.10}$$

where $\delta_{\mathbf{k}} = 1(\mathbf{k} = 0)$, 0 otherwise. In other words, $\mathbf{F}_{\mathbf{K}} \neq 0$ only if

$$\mathbf{K} = (0, \ldots, 0, -\mathbf{k}, 0, \ldots, 0, \mathbf{k}, 0, \ldots, 0) \tag{1.11}$$

for some **k**.

2 NON-INTERACTING PARTICLES

When $\mathbf{F} = \mathbf{0}$ the solution of (1.6) reflects the uniform motion of each particle.

$$\left.\begin{aligned} \varrho(\mathbf{X}, \mathbf{V}, t) &= \varrho(\mathbf{X} - \mathbf{V}\,t, \mathbf{V}, 0) \\ \varrho_{\mathbf{K}}(\mathbf{V}, t) &= \varrho_{\mathbf{K}}(\mathbf{V}, 0)\, e^{-i\mathbf{K}\cdot\mathbf{V}t} \end{aligned}\right\} \tag{2.1}$$

Either by applying (1.4) and (1.5) to (2.1) or by solving (1.7) with $\mathbf{F} = \mathbf{0}$, we find

$$\bar{\varrho}_{\mathbf{K}}(\mathbf{V}, t) = \bar{\varrho}_{\mathbf{K}}(\mathbf{V} - i\mathbf{K}\sigma^2 t, 0) \exp\left\{-i\mathbf{K}\cdot\mathbf{V}t - \frac{\sigma^2 K^2 t^2}{2}\right\} \tag{2.2}$$

In deriving (2.2) we assume that ϱ (and hence $\bar{\varrho}$) is an analytic function of the complex variables **V**—no physics is lost by doing this.

Throughout this work we assume that our smoothing radius σ, though non-zero, is small compared with the scale μ upon which all Fourier components of $\bar{\varrho}$ at $t = 0$, or the spatially homogeneous component $\bar{\varrho}_0$ at any time, very with **V**. In this case all non-zero Fourier components of $\bar{\varrho}$ will tend to zero within a time of order L/σ—the "phase mixing time". (In fact, they will always do so provided the initial conditions do not correspond to exact knowledge of the state of the system.) $\bar{\varrho}$ tends to the equilibrium solution of (1.7) that preserves all constants of the motion.

3 WEAKLY INTERACTING SYSTEMS: QUASI-LINEAR THEORY

We discuss the solution of (1.7) when H is of order some small parameter ε. We assume that inhomogeneous components of $\bar{\varrho}$ are of the same order of magnitude as in equilibrium—that is, ε.

$$\mathbf{K} \neq \mathbf{0}, \quad \frac{\bar{\varrho}_{\mathbf{K}}(\mathbf{V}, t)}{\bar{\varrho}_0(\mathbf{V}, t)} = \mathrm{O}(\varepsilon) \tag{3.1}$$

From (1.7), exactly

$$\frac{\partial \bar{\varrho}_0(\mathbf{V}, t)}{\partial t} + \sum_{\mathbf{K}} \mathbf{F}_{-\mathbf{K}} \cdot \frac{\partial \varrho_{\mathbf{K}}(\mathbf{V}, t)}{\partial \mathbf{V}} = 0 \tag{3.2}$$

Hence ϱ_0 changes very slowly, at a rate of order ε^2. To first order in ε

$$\frac{\partial \bar{\varrho}_{\mathbf{K}}(\mathbf{V}, t)}{\partial t} + i\mathbf{K} \cdot \mathbf{V} \bar{\varrho}_{\mathbf{K}} + i\sigma^2 \mathbf{K} \cdot \frac{\partial \bar{\varrho}_{\mathbf{K}}}{\partial \mathbf{V}} = -\mathbf{F}_{\mathbf{K}} \cdot \frac{\partial \bar{\varrho}_0(\mathbf{V}, t)}{\partial \mathbf{V}} \tag{3.3}$$

The solution of (3.3) is

$$\bar{\varrho}_{\mathbf{K}}(\mathbf{V}, t) = \bar{\varrho}_{\mathbf{K}}(\mathbf{V} - i\mathbf{K}\sigma^2 t, 0) \exp\left\{-i\mathbf{K} \cdot \mathbf{V}t - \frac{\sigma^2 k^2 t^2}{2}\right\}$$
$$- \int_0^t d\tau \exp\left\{-i\mathbf{K} \cdot \mathbf{V}\tau - \frac{\sigma^2 k^2 \tau^2}{2}\right\} \mathbf{F}_{\mathbf{K}} \cdot \frac{\partial \bar{\varrho}_0}{\partial \mathbf{V}}(\mathbf{V} - i\mathbf{K}\sigma^2\tau, t - \tau) \tag{3.4}$$

Now $|\mathbf{K}| \sim 1/l$, where l is the range of an interaction (assumed finite). Once a time much larger than l/σ has elapsed

$$\bar{\varrho}_{\mathbf{K}}(\mathbf{V}, t) = - \int_0^\infty d\tau \exp\left\{-i\mathbf{K} \cdot \mathbf{V}\tau - \frac{\sigma^2 K^2 \tau^2}{2}\right\} \mathbf{F}_{\mathbf{K}} \cdot \frac{\partial}{\partial \mathbf{V}} \bar{\varrho}_0(\mathbf{V} - i\mathbf{K}\sigma^2\tau, t - \tau) \tag{3.5}$$

By taking $+\infty$ rather than $-\infty$ in (3.5) we have given a definite direction to time; this is the origin of the irreversibility of our final equation. If the interactions are weak enough we may assume that $\bar{\varrho}_0$ varies little over time l/σ. By writing

$$-i\mathbf{K} \cdot \mathbf{V}\tau - \frac{\sigma^2 K^2 \tau^2}{2} = -\frac{(\mathbf{K} \cdot \mathbf{V})^2}{2\sigma^2 K^2} - \frac{\sigma^2 K^2}{2}\left(\tau + \frac{i\mathbf{K} \cdot \mathbf{V}}{\sigma^2 K^2}\right)^2$$

and performing a contour integration, we obtain

$$\bar{\varrho}_{\mathbf{K}}(\mathbf{V}, t) = -\mathbf{F}_{\mathbf{K}} \cdot (\mathbf{I}_{\mathbf{K}} + \mathbf{J}_{\mathbf{K}}) \tag{3.6}$$

where

$$\mathbf{I}_{\mathbf{K}} = -i \exp\left\{-\frac{(\mathbf{K} \cdot \mathbf{V})^2}{2\sigma^2 K^2}\right\}$$
$$\times \int_0^{\frac{\mathbf{K} \cdot \mathbf{V}}{\sigma^2 K^2}} \mathbf{g}(\mathbf{V} - \sigma^2 \tau'' \mathbf{K}, t) \exp\left\{\frac{\sigma^2 K^2}{2}\left(\tau'' - \frac{(\mathbf{K} \cdot \mathbf{V})}{\sigma^2 K^2}\right)^2\right\} d\tau'' \tag{3.7}$$

$$\mathbf{J_K} = \exp\left\{-\frac{(\mathbf{K}\cdot\mathbf{V})^2}{2\sigma^2 K^2}\right\}$$

$$\times\int_0^\infty d\tau' \exp\left(-\frac{\sigma^2 K^2 \tau'^2}{2}\right)\mathbf{g}\left(\mathbf{V} - \frac{(\mathbf{V}\cdot\mathbf{K})\mathbf{K}}{K^2} - i\mathbf{K}\sigma^2\tau', t\right) \tag{3.8}$$

and

$$g(\mathbf{V}, t) = \frac{\partial\bar{\varrho}_0(\mathbf{V}, t)}{\partial\mathbf{V}}$$

We easily verify that

$$\left.\begin{aligned} i\mathbf{K}\cdot\mathbf{V}(-\mathbf{F_K}\cdot\mathbf{I_K}) + i\mathbf{K}\sigma^2\cdot\frac{\partial}{\partial\mathbf{V}}(-\mathbf{F_K}\cdot\mathbf{I_K}) + \mathbf{F_K}\cdot\frac{\partial\bar{\varrho}_0}{\partial\mathbf{V}} &= 0 \\ i\mathbf{K}\cdot\mathbf{V}(-\mathbf{F_K}\cdot\mathbf{J_K}) + i\mathbf{K}\sigma^2\cdot\frac{\partial}{\partial\mathbf{V}}(-\mathbf{F_K}\cdot\mathbf{J_K}) \qquad\qquad &= 0 \end{aligned}\right\} \tag{3.9}$$

as should be so according to our "quasi-equilibrium" hypothesis.

We have

$$\mathbf{I_{-K}} = -\mathbf{I_K}; \quad \mathbf{J_{-K}} = \mathbf{J_K} + \mathrm{O}\left(\frac{\sigma^2}{\mu^2}\right) \tag{3.10}$$

In fact, provided $\mu \gg \sigma$,

$$\mathbf{J_K} = \exp\left\{-\frac{(\mathbf{K}\cdot\mathbf{V})^2}{2\sigma^2 K^2}\right\}\frac{1}{\sigma|\mathbf{K}|}\sqrt{\frac{\pi}{2}}\,\mathbf{g}\left(\mathbf{V} - \frac{(\mathbf{V}\cdot\mathbf{K})\mathbf{K}}{K^2}, t\right) \tag{3.11}$$

and when $\dfrac{|\mathbf{K}\cdot\mathbf{V}|}{K} \gg \sigma$ we have an asymptotic expansion beginning

$$I_{\mathbf{K}} = \frac{\mathbf{g}(\mathbf{V}, t)}{i\mathbf{V}\cdot\mathbf{K}} + \mathrm{O}\left(\frac{\sigma|\mathbf{K}|}{|\mathbf{K}\cdot\mathbf{V}|}\right) \tag{3.12}$$

From (3.10) the $I_{\mathbf{K}}$ terms cancel when we substitute (3.6) into (3.2). Using (3.11)

$$0 = \frac{\partial\bar{\varrho}_0(\mathbf{V}, t)}{\partial t}$$

$$-\sum_{\mathbf{K}}\frac{H_{\mathbf{K}}H_{-\mathbf{K}}}{\sigma|\mathbf{K}|}\sqrt{\frac{\pi}{2}}\left(\mathbf{K}\cdot\frac{\partial}{\partial\mathbf{V}}\right)e^{-\frac{(\mathbf{K}\cdot\mathbf{V})^2}{2\sigma^2K^2}}\,\mathbf{K}\cdot\mathbf{g}\left(\mathbf{V} - \frac{(\mathbf{V}\cdot\mathbf{K})\,\mathbf{K}, t}{K^2}\right) \tag{3.13}$$

Before discussing the properties of (3.13) we point out in what ways it differs from the equation conventionally derived (Brout & Prigogine 1956, Zwanzig 1960).

If $\sigma = 0$ it is not permissible to go from (3.4) to (3.5): the initial condition terms do not die out and the effects of the indefinite past are propagated through to $\varrho_{\mathbf{K}}(\mathbf{V}, t)$. All we can do is substitute (3.4) in (3.2) and make the size of the system tend to infinity, so that the sum in (3.2) may be replaced by an integration over **K**. The rapidly oscillating exponentials are damped out by the Riemann–Lebesgue lemma, and we are left with the Brout-Prigogine equation:

$$0 = \frac{\partial \varrho_0(\mathbf{V}, t)}{\partial t} - \int dK\, H_{\mathbf{K}} H_{-\mathbf{K}} \left(\mathbf{K} \cdot \frac{\partial}{\partial \mathbf{V}}\right) \pi\, \delta(\mathbf{K} \cdot \mathbf{V}) \left(\mathbf{K} \cdot \frac{\partial}{\partial \mathbf{V}}\right) \varrho_0(\mathbf{V}, t) \quad (3.14)$$

(here we define $\int dK$ to include any constants arising from the limiting process). In effect one can write, in the sense of weak convergence

$$\mathbf{J}_{\mathbf{K}} = \pi \mathbf{g}(\mathbf{V})\, \delta(\mathbf{K} \cdot \mathbf{V}) \quad (3.15)$$

$$\mathbf{I}_{\mathbf{K}} = \frac{\mathbf{g}(\mathbf{V})}{i\mathbf{K} \cdot \mathbf{V}} \quad (3.16)$$

Clearly these results are the limiting forms of (3.11) and (3.12) as $\sigma \to 0$.

It appears therefore that by building our phase mixing into our definition of $\bar{\varrho}$ we may avoid having introduce it by smoothing over a continuously varying **K**. Instead we smooth over **V**, and we have to assume that this variable may vary continuously—i.e. that the energy may vary continuously. Thus our method may not easily carry over to quantum systems (see Golden & Longuet–Higgins 1960).

The effect of neglecting non-Markovian terms in (3.5) may be discussed: it may easily be shown that for weak enough interactions with $\sigma > 0$ this neglect is justified (this is not true for a finite system with $\sigma = 0$; see Montroll 1961.)

We now discuss some properties of (3.13). Write it as

$$\frac{\partial \bar{\varrho}_0}{\partial t} = C \bar{\varrho}_0 \quad (3.17)$$

a) Integrating by parts

$$\int dV \bar{\varrho}_1(\mathbf{V})\, C \bar{\varrho}_2(\mathbf{V}) = -\int dV \sum_{\mathbf{K}} \frac{H_{\mathbf{K}} H_{-\mathbf{K}}}{\sigma |\mathbf{K}|} \sqrt{\frac{\pi}{2}}$$
$$\times \exp\left\{-\frac{(\mathbf{K} \cdot \mathbf{V})^2}{2\sigma^2 K^2}\right\} (\mathbf{K} \cdot \mathbf{g}_1(\mathbf{V})) \left(\mathbf{K} \cdot \mathbf{g}_2\left(\mathbf{V} - \frac{(\mathbf{V} \cdot \mathbf{K})\mathbf{K}}{K^2}\right)\right) \quad (3.18)$$

Now the only contributions to this sum will come from $\mathbf{K}$ such that

$$\frac{\mathbf{K}\cdot\mathbf{V}}{\sigma|\mathbf{K}|} \sim 1 \tag{3.19}$$

So we expand $\mathbf{g}_2\left(\mathbf{V} - \frac{(\mathbf{V}\cdot\mathbf{K})\,\mathbf{K}}{K^2}\right)$ in a Taylor series; each successive term is less than that previous in the ratio σ/μ. In fact, we may write

$$-\frac{(\mathbf{K}\cdot\mathbf{V})\,\mathbf{K}}{K^2}\exp - \frac{(\mathbf{K}\cdot\mathbf{V})^2}{2\sigma^2K^2} = \sigma^2\frac{\partial}{\partial\mathbf{V}}\exp - \frac{(\mathbf{K}\cdot\mathbf{V})^2}{2\sigma^2K^2}$$

and integrate the first order term by parts to show that (3.18) is symmetric in the indices 1, 2 to order σ^2/μ^2.

$$\int dV\bar{\varrho}_1(\mathbf{V})\,C\bar{\varrho}_2(\mathbf{V}) = \left[1 + O\left(\frac{\sigma^2}{\mu^2}\right)\right]\int dV\bar{\varrho}_2(\mathbf{V})\,C\bar{\varrho}_1(\mathbf{V}) \tag{3.20}$$

C is self-adjoint to this order.

b) Immediately we have

$$\frac{d}{dt}\int \bar{\varrho}_0\,dV = 0 \tag{3.21}$$

Also by a proof similar to that of (3.20)

$$\frac{d}{dt}\int \frac{V^2}{2}\bar{\varrho}_0\,dV = O\left(\frac{\sigma^2}{\mu^2}\varepsilon^2\right) \tag{3.22}$$

c) If

$$\bar{\varrho}_0 = \bar{\varrho}_0\left(\frac{V^2}{2}, t\right)$$

we have exactly

$$\mathbf{K}\cdot\mathbf{g}\left(\mathbf{V} - \frac{(\mathbf{V}\cdot\mathbf{K})\,\mathbf{K}}{K^2}, t\right) = 0 \tag{3.23}$$

so that any $\bar{\varrho}_0\left(\frac{V^2}{2}\right)$ is a time-independent solution of (3.13).

d) Expanding $\mathbf{g}$ in a Taylor series as in (a), a little algebraic manipulation shows that to second order in σ/μ the maxima of $\bar{\varrho}_0(\mathbf{V})$ tend to be reduced and the minima increased, except for equilibrium states.

e) Finally, exactly as for (3.14) we may prove that if

$$\bar{S} = -\int \bar{\varrho}_0 \log \bar{\varrho}_0 \, dV \quad \text{then} \quad \frac{d\bar{S}}{dt} = Q + O\left(\frac{\varepsilon^2\sigma^2}{\mu^2}\right) \tag{3.24}$$

where $Q > 0$. Owing to the "error term" in (3.24) the argument that since $\bar{S}$ is bounded above it must tend to a limit, and that all $\bar{\varrho}_0(\mathbf{V})$ such that $\frac{d\bar{S}}{dt} = 0$ are isotropic in $\mathbf{V}$, is rather intuitive; it seems fairly convincing, however. (These error terms seem to be inherent in the use of $\bar{\varrho}$ rather than ϱ: if we work, more tediously, with the exact expression (3.8) rather than the approximation (3.11), we find that they are still present.)

Thus in a relaxation time of order $\varepsilon^2 \varrho_0$ evolves to an equilibrium value $\bar{\varrho}_{0s}\left(\frac{V^2}{2}\right)$. In fact

$$\bar{\varrho}_{0s}\left(\frac{V^2}{2}\right) = \operatorname*{Lt}_{\delta\to 0} \frac{\int_{I_{0\delta}} \bar{\varrho}_0(\mathbf{W}, 0)\, dW}{\int_{I_{0\delta}} dW} \tag{3.25}$$

where $I_{0\delta}$ is the region $V^2 < W^2 < V^2 + \delta$. (3.22) shows that energy is conserved to order σ^2/μ^2 over a time interval of this length.

The asymptotic expression (3.5) for $\varrho_\mathbf{K}$ exists throughout this evolution and

$$\bar{\varrho}_0 = \bar{\varrho}_{0s}\left(\frac{V^2}{2}\right) \Rightarrow \bar{\varrho}_\mathbf{K} = H_\mathbf{K}\bar{\varrho}'_{0s}\left(\frac{V^2}{2}\right) + O\left(\frac{\varepsilon\sigma^2}{\mu^2}\right) \tag{3.26}$$

So to first order in ε

$$\bar{\varrho}(\mathbf{X}, \mathbf{V}, t) \Rightarrow \bar{\varrho}_{0s}\left(\frac{V^2}{2} + H(\mathbf{X})\right) \quad \text{as} \quad t \to \pm\infty \tag{3.27}$$

and again to first order this is the phase average of the initial conditions

$$\frac{\int_{I_\delta} \bar{\varrho}(\mathbf{Y}, \mathbf{W}, 0)\, dY\, dW}{\int_{I_\delta} dY\, dW} \tag{3.28}$$

where I_δ is the region $\frac{V^2}{2} + H(\mathbf{X}) \leqq \frac{W^2}{2} + H(\mathbf{Y}) \leqq \frac{V^2}{2} + H(\mathbf{X}) + \delta$.

If we integrate (3.13) over all particle velocities except two, and use (1.10), we obtain an equation relating the one and two-particle reduced distribution functions (cf. Prigogine 1962 section 4.3)

$$\frac{\partial \varphi_1(\mathbf{V}, t)}{\partial t} = (N-1)\int dv' \sum_{\mathbf{k}} |V_{\mathbf{k}}|^2 \left(\mathbf{k}\cdot\frac{\partial}{\partial \mathbf{v}}\right)\frac{1}{\sigma|\mathbf{k}|}\sqrt{\frac{\pi}{2}}$$

$$\exp\left(-\frac{[\mathbf{k}\cdot(\mathbf{v}-\mathbf{v}')]^2}{2\sigma^2 k^2}\right)\mathbf{k}\cdot\left(\frac{\partial}{\partial \mathbf{v}}-\frac{\partial}{\partial \mathbf{v}'}\right)\varphi_2(\mathbf{v}, \mathbf{v}', t) + O\left(\frac{\varepsilon^2\sigma^2}{\mu^2}\right) \quad (3.29)$$

Provided $\varphi_2(\mathbf{v}, \mathbf{v}')$ varies slowly on scale σ we may replace

$$\frac{1}{\sigma|\mathbf{k}|}\sqrt{\frac{\pi}{2}}\exp\left\{-\frac{[\mathbf{k}\cdot(\mathbf{v}-\mathbf{v}')]^2}{2\sigma^2 k^2}\right\} \quad \text{by} \quad \pi\delta[\mathbf{k}\cdot(\mathbf{v}-\mathbf{v}')]$$

and we recover the Landau ("Boltzmann-like") equation if we also assume

$$\varphi_2(\mathbf{v}, \mathbf{v}', t) = \varphi_1(\mathbf{v}, t)\,\varphi_1(\mathbf{v}', t) \quad (3.30)$$

However, to make this assumption in a self-consistent manner we must take N large. The Landau equation describes steady evolution to a Maxwellian distribution, which is essentially a property of large systems; for these the average (3.25) reduces to Maxwellian form (Nijboer 1962).

4 PERTURBATION THEORY

To discuss the problem without assuming weak interactions or initial correlations, we use perturbation theory

$$\bar{\varrho}_{\mathbf{K}_0}(\mathbf{V}, t) = \sum_{n=0}^{\infty} \bar{\varrho}_{n\mathbf{K}_0}(\mathbf{V}, t) \quad (4.1)$$

$$\left.\begin{aligned}&\frac{\partial \bar{\varrho}_{0,\mathbf{K}_0}(\mathbf{V}, t)}{\partial t} + i\mathbf{K}_0\cdot\mathbf{V}\bar{\varrho}_{0,\mathbf{K}_0}(\mathbf{V}, t) + i\sigma^2\mathbf{K}_0\cdot\frac{\partial}{\partial \mathbf{v}}\bar{\varrho}_{0,\mathbf{K}_0}(\mathbf{V}, t) = 0\\&\bar{\varrho}_{0,\mathbf{K}_0}(\mathbf{V}, 0) = f_{\mathbf{K}_0}(\mathbf{V}), \text{ the initial distribution.}\end{aligned}\right\} \quad (4.2)$$

$$\left.\begin{aligned}n \geqq 1, \quad &\frac{\partial \bar{\varrho}_{n,\mathbf{K}_0}}{\partial t} + i\mathbf{K}_0\cdot\mathbf{V}\bar{\varrho}_{n,\mathbf{K}_0} + i\sigma^2\mathbf{K}_0\cdot\frac{\partial}{\partial \mathbf{V}}\mathbf{V}_{n,\mathbf{K}_0}\\&= -\sum_{\mathbf{K}_1}\mathbf{F}_{-\mathbf{K}_1}\cdot\frac{\partial}{\partial \mathbf{V}}\bar{\varrho}_{1-n,\mathbf{K}_0+\mathbf{K}_1}\\&\bar{\varrho}_{n,\mathbf{K}_0}(\mathbf{V}, 0) = 0\end{aligned}\right\} \quad (4.3)$$

The solution of (4.2) is as (2.2), whilst

$$\bar{\varrho}_{n\mathbf{K}_0}(\mathbf{V}, t) = (-1)^n \int_0^t d\tau_0 \int_0^t d\tau_1 \dots \int_0^t d\tau_n \sum_{\mathbf{K}_1, \dots, \mathbf{K}_n}$$

$$\tau_0 + \tau_1 + \dots + \tau_n + t$$

$$e^{-i\mathbf{K}_0 \cdot \mathbf{V}\tau_0} \mathbf{F}_{-\mathbf{K}_1} \cdot \frac{\partial}{\partial \mathbf{V}} e^{-i(\mathbf{K}_0+\mathbf{K}_1)\cdot \mathbf{V}\tau_1} \mathbf{F}_{-\mathbf{K}_2} \cdot \frac{\partial}{\partial \mathbf{V}} e^{-i(\mathbf{K}_0+\mathbf{K}_1+\mathbf{K}_2)\cdot \mathbf{V}\tau_2}$$

$$\dots \mathbf{F}_{-\mathbf{K}_n} \cdot \frac{\partial}{\partial \mathbf{V}} e^{-i(\mathbf{K}_0+\dots+\mathbf{K}_n)\cdot \mathbf{V}\tau_n} \bar{\varrho}_{0, \mathbf{K}_0+\dots+\mathbf{K}_n}(\mathbf{V} - i\sigma^2 \mathbf{A})\, e^{-\frac{\sigma^2 \mathbf{A}^2}{2}} \tag{4.4}$$

where

$$\mathbf{A} = \mathbf{K}_0 \tau_0 + (\mathbf{K}_0 + \mathbf{K}_1)\,\tau_1 + \dots + (\mathbf{K}_0 + \dots + \mathbf{K}_n)\,\tau_n \tag{4.5}$$

We can derive (4.4) from the usual expression for ϱ_n employed by the Brussels school, by integration as with (2.2).

We can classify the terms of (4.4) diagrammatically using (1.10) exactly as for $\sigma = 0$ (see Prigogine 1962, Ch. 7), but there is one extremely important difference in the behaviour of these diagrams. When $\sigma = 0$ (4.4) is a convolution, and its temporal Laplace transform is a product of relatively simple operators acting on the initial distribution. These operators, corresponding to the vertices of the diagram, may be discussed separately; the behaviour of the whole diagram is known at once given the behaviour of its vertices. All this is no longer true when $\sigma > 0$.

The spatially homogeneous component

If $\mathbf{K}_0 + \mathbf{K}_1 + \dots + \mathbf{K}_r = 0$ for any r, τ_r does not occur in (4.4) and may be integrated out. Either doing this or considering the Laplace transform, we obtain a recurrence relation that leads to

$$\frac{\partial \bar{\varrho}_0}{\partial t}(\mathbf{V}, t) = F(\mathbf{V}, t) + \int_0^t G(\tau)\, \bar{\varrho}_0(\mathbf{V}, t - \tau)\, d\tau \tag{4.6}$$

where

$$G(t) = \sum_{s=1}^{\infty} G_s(t) \tag{4.7}$$

and $G_s(t)$ is an operator defined by

$$G_s(t)f(\mathbf{V}) = (-1)^s \int_0^t d\tau_1 \cdots \int_0^t d\tau_{s-1} \sum_{\substack{\mathbf{K}_1,\dots,\mathbf{K}_s \\ \mathbf{K}_1+\cdots+\mathbf{K}_s=0 \\ \mathbf{K}_1+\cdots+\mathbf{K}_r\neq 0\ (1\leq r\leq s-1)}}$$
$$\tau_1+\cdots+\tau_{s-1}$$

$$\mathbf{F}_{-\mathbf{K}_1}\cdot\frac{\partial}{\partial\mathbf{V}}e^{-i\mathbf{K}_1\cdot\mathbf{V}\tau_1}\mathbf{F}_{-\mathbf{K}_2}\cdot\frac{\partial}{\partial\mathbf{V}}e^{-i(\mathbf{K}_1+\mathbf{K}_2)\cdot\mathbf{V}\tau_2}$$

$$\ldots\mathbf{F}_{-\mathbf{K}_{s-1}}\cdot\frac{\partial}{\partial\mathbf{V}}e^{-i(\mathbf{K}_1+\cdots+\mathbf{K}_{s-1})\cdot\mathbf{V}_{s-1}}\mathbf{F}_{-\mathbf{K}_s}\cdot\frac{\partial}{\partial\mathbf{V}}f(\mathbf{V}-i\sigma^2\mathbf{A}_0)\exp\left\{-\frac{\sigma^2A_0^2}{2}\right\} \tag{4.8}$$

where

$$\mathbf{A}_0 = \mathbf{K}_1\tau_1 + (\mathbf{K}_1+\mathbf{K}_2)\tau_2 + \cdots + (\mathbf{K}_1+\cdots+\mathbf{K}_{s-1})\tau_{s-1} \tag{4.9}$$

whilst

$$F(\mathbf{V},t) = \sum_{\mathbf{K}\neq 0} D_{\mathbf{K}}(t)\,\bar{\varrho}_{\mathbf{K}}(\mathbf{V},0) \tag{4.10}$$

$$D_{\mathbf{K}}(t) = \sum_{s=1}^{\infty} D_{s\mathbf{K}}(t) \tag{4.11}$$

$$D_{s\mathbf{K}}(t)f(\mathbf{V}) = \int_0^t d\tau_1 \cdots \int_0^t d\tau_s (-1)^s \sum_{\substack{\mathbf{K}_1,\dots,\mathbf{K}_s \\ \mathbf{K}_1+\cdots+\mathbf{K}_s=\mathbf{K} \\ \mathbf{K}_1+\cdots+\mathbf{K}_r\neq 0\ (1\leq r\leq s-1)}}$$
$$\tau_1+\cdots+\tau_s=t$$

$$\mathbf{F}_{-\mathbf{K}_1}\cdot\frac{\partial}{\partial\mathbf{V}}e^{-i\mathbf{K}_1\cdot\mathbf{V}\tau_1}\mathbf{F}_{-\mathbf{K}_2}\cdot\frac{\partial}{\partial\mathbf{V}}e^{-i(\mathbf{K}_1+\mathbf{K}_2)\cdot\mathbf{V}\tau_2}$$

$$\cdots\mathbf{F}_{-\mathbf{K}_{s-1}}\cdot\frac{\partial}{\partial\mathbf{V}}e^{-i(\mathbf{K}_1+\cdots+\mathbf{K}_{s-1})\cdot\mathbf{V}\tau_{s-1}}\mathbf{F}_{-\mathbf{K}_s}\cdot\frac{\partial}{\partial\mathbf{V}}e^{-i\mathbf{K}\cdot\mathbf{V}\tau_s}f(\mathbf{V}-i\sigma^2\mathbf{A}')\,e^{-\frac{\sigma^2A'^2}{2}} \tag{4.12}$$

where

$$\mathbf{A}' = \mathbf{K}_1\tau_1 + \cdots + (\mathbf{K}_1+\cdots+\mathbf{K}_{s-1})\tau_{s-1} + \mathbf{K}\tau_s \tag{4.13}$$

(4.6) is exact—it is just a manipulation of the series solution. If now

$$F(\mathbf{V},t)\to 0,\quad G(t)f(\mathbf{V})\to 0\quad \text{as}\quad t\to\infty \tag{4.14}$$

then for large positive times we have the asymptotic form

$$\frac{\partial\bar{\varrho}_0(\mathbf{V},t)}{\partial t} = \int_0^\infty G(\tau)\,\bar{\varrho}_0(\mathbf{V},t-\tau)\,d\tau \tag{4.15}$$

exactly analagous to the Prigogine–Résibois master equation (Prigogine and Résibois 1961; Prigogine 1962, Ch. 11).

Unfortunately we have not been able to establish the properties (4.14) for $\sigma \neq 0$ and a finite system, other than in the weak interaction case. If we assume that $\mathbf{F}$ and the initial non-zero Fourier components $\bar{\varrho}_{\mathbf{K}}(\mathbf{V}, 0)$ are of order ε, the highest order terms in (4.6) are of order ε^2: G_2 and D_1. Once $t \gg \frac{l}{\sigma}$ (4.15) reduces to the equation obtained by combining (3.5) and (3.2); from this we obtain (3.13). However

$$G_3(t)f(\mathbf{V}) = -\sum_{\substack{\mathbf{K}_1, \mathbf{K}_2 \\ \mathbf{K}_1+\mathbf{K}_2 \neq 0}} \int_0^t d\tau\, \mathbf{F}_{-\mathbf{K}_1} \cdot \frac{\partial}{\partial \mathbf{V}} e^{-i\mathbf{K}_1 \cdot \mathbf{V}(t-\tau)} \mathbf{F}_{-\mathbf{K}_2} \cdot \frac{\partial}{\partial \mathbf{V}}$$

$$e^{-i(\mathbf{K}_1+\mathbf{K}_2)\cdot \mathbf{V}\tau} \mathbf{F}_{\mathbf{K}_1+\mathbf{K}_2} \cdot \frac{\partial}{\partial \mathbf{V}} f(\mathbf{V} - i\sigma^2(\mathbf{K}_1 t + \mathbf{K}_2 \tau)) \exp\left\{-\frac{\sigma^2}{2}(\mathbf{K}_1 t + \mathbf{K}_2 \tau)^2\right\} \tag{4.16}$$

Since

$$(\mathbf{K}_1 t + \mathbf{K}_2 \tau)^2 = \mathbf{K}_2^2 \left(\tau + \frac{(\mathbf{K}_1 \cdot \mathbf{K}_2)\, t}{K_2^2}\right)^2 + t^2 \left(\mathbf{K}_1^2 - \frac{(\mathbf{K}_1 \cdot \mathbf{K}_2)^2}{K_2^2}\right) \tag{4.17}$$

all terms of (4.16) except those in which $\mathbf{K}_1$, $\mathbf{K}_2$ are parallel

$$\mathbf{K}_2 + \alpha \mathbf{K}_1 = 0 \quad (\alpha \neq 1) \tag{4.18}$$

will tend to zero. However, if (4.18) holds and $\alpha > 1$, G_3 contains terms increasing linearly with t. In general for $s \geqq 3$ G_s contains terms increasing like t^{2s-5}. These correspond to the extreme case of $\mathbf{K}_1, \ldots, \mathbf{K}_{s-1}$ all parallel; even just one linear relationship between them gives a term increasing like t^{s-2}.

We can see why these objectionable terms do not appear in the infinite system case. Permissible values of $\mathbf{K}_r$ are of the form (1.11), where $\mathbf{k}$ is a member of a cubic lattice of side $2\pi/L$, and $|\mathbf{k}| \sim 1/l$. There are about $(L/l)^3$ possible $\mathbf{k}$, of which at most L/l may be parallel to a given vector. The coefficients of the secular terms in G_s will contain factors $(l/L)^2$. The secular terms correspond to "resonances" caused by our periodic soundary conditions. The fact that it is precisely this resonance effect that enables us to prove that the system is quasi-ergodic suggests that these terms are worth investigating further.

Similar remarks apply to $D_{s\mathbf{K}}$.

Inhomogeneous components

We write exactly as in the $\sigma = 0$ case (Prigogine and Henin 1960, Prigogine 1962, Ch. 11)

$$\bar{\varrho}_{\mathbf{K}_0}(\mathbf{V}, t) = \bar{\varrho}'_{\mathbf{K}_0}(\mathbf{V}, t) + \bar{\varrho}''_{\mathbf{K}_0}(\mathbf{V}, t) \tag{4.19}$$

where $\varrho''_{\mathbf{K}_0}(\mathbf{V}, t)$ (the created component) is the sum of all terms in the series such that $\mathbf{K}_0 + \cdots + \mathbf{K}_r = 0$ for some r.

$$\bar{\varrho}''_{\mathbf{K}_0}(\mathbf{V}, t) = \int_0^t C_{\mathbf{K}_0}(\tau)\, \bar{\varrho}_0(\mathbf{V}, t - \tau)\, d\tau \tag{4.20}$$

where $C_{\mathbf{K}_0}(\tau)$ is defined similarly to G and $D_{\mathbf{K}}$. We want

$$C_{\mathbf{K}_0}(t) \to 0 \quad \text{as} \quad t \to \infty \tag{4.21}$$

Once again this is so in the weak interaction case, and we recover the results of section 3, but higher terms go wrong.

$\bar{\varrho}'_{\mathbf{K}_0}(\mathbf{V}, t)$ (the propagated component) is the sum of all remaining terms. A closed equation for its evolution may easily be found, but its form appears to be less similar to that for $\bar{\varrho}_0$ than is so when $\sigma = 0$. Little progress has been made on the problem of proving that $\bar{\varrho}'_{\mathbf{K}_0} \to 0$ as $t \to \infty$.

5 CONCLUSION

It is clear that as yet we are nowhere near a full discussion of the behaviour of $\bar{\varrho}$; even our apparent success with the weak interaction case now appears rather dubious. A possible method of progressing further might be to set up equations for various terms in G in the same way as we have taken into account the gross secularities in (4.4) (occuring when $\mathbf{K}_0 + \cdots + \mathbf{K}_r = \mathbf{0}$ for some r) into account by forming the equation (4.6). (4.14) and (4.21) might still be true. Alternatively such a technique might lead us to a different partial summation of the perturbation series; and equations like (4.6) but with G etc. defined differently, and satisfying (4.14) and (4.21).

The "most secular terms" in G are for each order in its expansion those in which all wave vectors are parallel to some vector, $\mathbf{J}$ say. For each $\mathbf{J}$ the sum of these terms is

$$\frac{|\mathbf{J}|}{2\pi} \int dx\, F(z) \frac{\partial Z(x)}{\partial x} \tag{5.1}$$

where $Z_{\mathbf{J}}$ satisfies certain initial conditions and

$$\frac{\partial Z(x, u, t)}{\partial t} + u\frac{\partial Z}{\partial x} + \sigma^2 \frac{\partial^2 Z}{\partial x \partial u} + F(x)\frac{\partial Z}{\partial u} = F(x)\frac{|\mathbf{J}|}{2\pi}\int \frac{\partial Z(x')}{\partial u} dx' \quad (5.2)$$

x and u being position and velocity components parallel to **J**, and F suitably defined. Despite the nearness of (5.2) to the one-dimensional problem (easily solved by writing down the equations of motion), we have not yet suceeded in proving that the expression (5.1) tends to zero as $|t| \to \infty$. If we can do this, it will be encouraging, but we shall still need to tackle the other secular terms in G. There is much scope for further work on this problem.

Our basic Eq. (1.7) is useful in a number of other problems bsides those discussed in this paper. For example, we can form a modified Vlasov equation from it, and follow through Landau's linearised solution for a plasma. Provided the initial perturbation is small enough, the linearisation need not break down, as it must do eventually in the usual calculation. The standard results on the behaviour of the electric field are recovered in a self-consistent manner.

References

R. Balescu 1962: *Statistical Mechanics of charged Particles* (Interscience).

R. Brout and I. Prigogine 1956: *Physica* **22**, 621.

S. Golden and H. C. Longuet-Higgins 1960: *J. Chem. Phys.* **33**, 1479.

E. A. Montroll 1961: in *Lectures in Theoretical Physics*, Vol. III ed. W. E. Brittin (Academic Press) p. 309–316.

B. R. A. Nijboer 1962: in *Fundamental Problems in Statistical Mechanics* ed.

E. G. D. Cohen (North-Holland) p. 14.

I. Prigogine 1962: *Non-Equilibrium Statistical Mechanics* (Interscience).

I. Prigogine and F. Henin 1960: *J. Math. Phys.* **1**, 349.

I. Prigogine and P. Résibois, *Physica* **27**, 629 (1961).

P. Résibois 1966: in *Many Particle Physics: theory and applications* ed. E. Meeron (Gordon & Breach).

G. Sandri 1963: *Ann. Phys.* **24**, 332, 380; *Phys. Rev. Lets.* **11**, 178.

D. ter Haar 1955: *Revs. Mod. Phys.* **27**, 289.

E. C. Titchmarsh 1937: *The Theory of Fourier Integrals* (Oxford).

N. G. van Kampen 1962: *Fundamental Problems in Statistical Mechanics*, p. 173–202.

R. Zwanzig 1960: *J. Chem. Phys.* **33**, 1338.

R. Zwanzig 1961: *Lectures in Theoretical Physics*, vol. III, p. 130–135.

PAPER A–4

*An application of the generalized Langevin equation to the study of correlations in simple, classical fluids**

A. Z. AKCASU
J. J. DUDERSTADT
The University of Michigan Ann Arbor, Michigan 48105
Department of Nuclear Engineering

I INTRODUCTION

Time correlation functions[1] of dynamical variables play an extremely important role in the description of many-body systems such as liquids or plasmas. It is well known that transport parameters which characterize the irreversible behavior of such systems can be expressed in terms of time correlation functions calculated under equilibrium conditions. Moreover, the cross sections for scattering of radiation (neutron or photons) from an aggregate of particles can be directly related to the time correlation of density fluctuations in the system.

Most procedures useful for the calculation of such quantities can be classified as one of two general types of approaches—as either calculations based upon a "hydrodynamic-like" description of the many-body system, or alternatively, as a scheme proceeding from a kinetic equation describing the singlet distribution function, $f(\mathbf{x}, \mathbf{p}, t)$, for the system.

The hydrodynamic description has long been in use as a phenomenological theory in the study of correlations in arbitrary continuous media. In its most rudimentary form, this approach uses the conventional hydrodynamic equations to describe the linear response of the system, and this linear response is then related to the correlation functions of interest by means of

* This work was supported in part by the National Science Foundation under Grant GK-4853.

the fluctuation-dissipation theorem[2]. Using the formulation described by Landau and Lifschitz[3], Rytov[4] applied the fluctuation-dissipation theorem to the calculation of correlation functions in distributed parameter systems. This approach has been used by Mountain[5] to interpret light scattering from liquids. More recently, Kadanoff and Martin[6] have developed an elegant and exact generalization of the hydrodynamic approach useful for the calculation of time correlations and transport coefficients. This generalized hydrodynamic approach has been applied by Yip, *et al.*[7,8] to the interpretation of molecular dynamics calculations as well as actual neutron scattering data for liquids with considerable success.

The kinetic description of time correlation functions seeks to obtain an approximate kinetic equation describing correlations among fluctuations in the microscopic phase density, $g(\mathbf{x}, \mathbf{p}, t)$. Provided one can solve this kinetic equation, it is then a simple matter to generate the correlation functions of interest by suitable integrations over momentum. This scheme, which has been developed extensively by Nelkin, *et al.*[9], has been justified theoretically for dilute gases[10] and used successfully to interpret Brillouin scattering from gases [11,12]. More recently, attempts have been made to extend this description to dense fluids[13,14]. However for such systems, the choice of an appropriate kinetic equation is far from obvious, and existing calculations have been based primarily upon a Vlasov-like equation for liquids with limited success.

In this paper, we shall study both the hydrodynamic and kinetic approaches by using the generalized Langevin equation of Zwanzig[15] and Mori[16] to obtain exact equations describing the correlation functions of interest. That is, by suitable choices of the set of dynamical variables appearing in the generalized Langevin equation (G.L.E.), we arrive at either an exact hydrodynamic or kinetic description similar to those described above. To demonstrate the power of such approaches based upon the G.L.E. and also to compare the hydrodynamic and kinetic level of description, we will explicitly perform detailed calculations to obtain the transverse current-current correlation function for argon. These calculations will then be compared with the molecular dynamics data of Rahman[17].

II THE GENERALIZED LANGEVIN EQUATION

Recently there has been considerable interest in the application of the projection operator techniques developed by Zwanzig[15] for the study of irreversible processes. As one of several specific illustrations of projection methods, Zwanzig derived an exact equation of motion for the autocorrelation function of a dynamical variable. This approach was subsequently generalized and extended by Mori[16], who utilized projection operators similar to those of Zwanzig to derive an exact "generalized Langevin equation" describing the time evolution of an arbitrary vector $\mathbf{a}(t)$ whose components $a_j(t)$ are dynamical variables of the phase $(\mathbf{x}^1, \ldots, \mathbf{x}^N, \mathbf{p}^1, \ldots, \mathbf{p}^N)$ of a many body system:

$$\frac{d\mathbf{a}}{dt} - i\mathbf{\Omega} \cdot \mathbf{a}(t) + \int_0^t d\tau\, \boldsymbol{\phi}(\tau) \cdot \mathbf{a}(t - \tau) = \mathbf{f}(t). \tag{2.1}$$

Here the state vector, $\mathbf{a}(t)$, is defined such that it has no time invariant part, e.g.

$$\mathbf{a}(t) \equiv \mathbf{A}(t) - \langle \mathbf{A}(t) \rangle, \tag{2.2}$$

where $\langle \cdots \rangle$ denotes an average over the equilibrium canonical ensemble, $\varrho_0 = \exp(-\beta H)/Z$. Notice that equation (2.1) is a generalized form of the Langevin equation familiar from the stochastic theory of Brownian motion[18]. However equation (2.1) is an exact equation for $\mathbf{a}(t)$, and hence is equivalent to the equations of motion for the many body system.

In the G.L.E. (2.1), the "frequency matrix" $\mathbf{\Omega}$ is defined by

$$i\mathbf{\Omega} \equiv \langle \dot{\mathbf{a}}\mathbf{a}^* \rangle \cdot \langle \mathbf{a}\mathbf{a}^* \rangle^{-1} \tag{2.3}$$

where $\mathbf{a}$ denotes $\mathbf{a}(0)$; the "damping matrix" $\boldsymbol{\phi}$ is given by

$$\boldsymbol{\phi}(\tau) \equiv \langle \mathbf{f}(\tau)\, \mathbf{f}^*(0) \rangle \cdot \langle \mathbf{a}\mathbf{a}^* \rangle^{-1}; \tag{2.4}$$

and the "random force" $\mathbf{f}(t)$ is given by

$$\mathbf{f}(t) \equiv \exp[it(1 - P)L]\, i(1 - P)\, L\mathbf{a}. \tag{2.5}$$

Here, $\mathbf{a}^*$ is the row vector adjoint to $\mathbf{a}$, P is a projection operator defined by its action on an dynamical variable G as

$$PG \equiv \langle G\mathbf{a}^* \rangle \cdot \langle \mathbf{a}\mathbf{a}^* \rangle^{-1} \cdot \mathbf{a}, \tag{2.6}$$

and L is the Liouville operator, $L = i\{H, \cdot\}$. The matrix $\langle \mathbf{a}\mathbf{a}^* \rangle^{-1}$ is the inverse of the static correlation matrix, $\langle \mathbf{a}\mathbf{a}^* \rangle \equiv [\langle a_i a_j^* \rangle]$.

The G.L.E. (2.1) can be used either to study correlations among the variables in the set $\mathbf{a}$, or to obtain equations of motion for the ensemble averaged components of $\mathbf{a}$. In the first instance, one proceeds by noting that

$$\langle \mathbf{f}(t)\, \mathbf{a}^* \rangle = 0, \quad t \geqq 0. \tag{2.7}$$

Hence by multiplying (2.1) by $\mathbf{a}^* \cdot \langle \mathbf{aa}^* \rangle^{-1}$ from the right, and averaging, one can derive an equation for the correlation matrix

$$\mathbf{R}(t) \equiv \langle \mathbf{a}(t)\, \mathbf{a}^* \rangle \cdot \langle \mathbf{aa}^* \rangle^{-1} \tag{2.8}$$

which takes the form

$$\frac{d\mathbf{R}}{dt} - i\boldsymbol{\Omega} \cdot \mathbf{R}(t) + \int_0^t d\tau\, \boldsymbol{\phi}(\tau) \cdot \mathbf{R}(t - \tau) = 0. \tag{2.9}$$

If one wishes, instead, to obtain an equation of motion for the ensemble average of $\mathbf{a}$, then Eq. (2.1) can be averaged over a constrained equilibrium ensemble[16]

$$\varrho(0) \equiv \exp\left[-\beta(H - \mathbf{a}^* \cdot \mathbf{b})\right]/Z_0 \tag{2.10}$$

to find an equation for $\bar{\mathbf{a}}(t) \equiv \langle \mathbf{a}(t) \rangle_{\varrho(0)}$:

$$\frac{d\bar{\mathbf{a}}}{dt} - i\boldsymbol{\Omega} \cdot \bar{\mathbf{a}}(t) + \int_0^t d\tau\, \boldsymbol{\phi}(\tau) \cdot \bar{\mathbf{a}}(t - \tau) = \bar{\mathbf{f}}(t). \tag{2.11}$$

In the linear approximation of small departures from equilibrium, $\bar{\mathbf{f}}(t) = 0$; but in the more general case, $\bar{\mathbf{f}}(t)$ will introduce nonlinear terms in $\mathbf{a}$ into the equation (for recent discussions of the derivation of such nonlinear equations using projection operators, see references 19 and 20).

Since Eqs. (2.9) and (2.11) are still exact, and hence only formal identities with the equations of motion, one must eventually resort to approximation in order to obtain useful results. The frequency matrix, $\boldsymbol{\Omega}$, can usually be calculated explicitly in terms of static quantities. However, the damping matrix requires the study of the modified propagator, $\exp\left[it(1 - P)L\right]$, which, in turn, would involve solving the many body problem directly. The attractive feature of equations such as (2.9) which are generated by projection operator techniques is that the "damping" or "memory" terms are quite susceptible to approximation or modeling. That is, the generalized Langevin equation is of value primarily because it re-expresses the quantities of interest (e.g., time correlation functions) in forms involving damping terms which can then be easily approximated.

In Mori's formalism[16], the choice of the set of dynamical variables $\mathbf{a}$ was essentially arbitrary. Different choices of $\mathbf{a}$ will lead to different—but exact—descriptions of the system under consideration. However a given approximation of the damping term will yield results which may vary considerably, depending upon the choice one makes for $\mathbf{a}$. Usually by increasing the number of components in the set **a**, one can improve the description of the system within the framework of a given approximation.

We shall examine two alternative choices of the set **a**: By choosing the set of dynamical variables as microscopic densities in configuration space (e.g., mass, current, and energy densities), one arrives at an exact hydrodynamic description of correlation functions in terms of frequency and wavelength dependent transport parameters. This procedure, which is formally equivalent to the hydrodynamic description of Kadanoff and Martin, has been developed in some detail by Akcasu and Daniels[21]. In particular, we will illustrate their method as applied to the calculation of the transverse current-current correlation function $R_\perp(k, \omega)$ for a liquid.

If, instead, one chooses the dynamical variables in **a** as the fluctuations in the microscopic phase density function, one obtains an exact kinetic equation description of the general correlation function

$$\mathscr{G}(\mathbf{x}, \mathbf{p}; \mathbf{x}', \mathbf{p}', t) \equiv \langle \delta g^*(\mathbf{x}', \mathbf{p}', 0)\, \delta g(\mathbf{x}, \mathbf{p}, t) \rangle. \tag{2.12}$$

We will develop this description in some detail, and study alternative approximate forms of this kinetic equation. In particular, by modeling the damping terms, we will obtain an approximate kinetic equation and then use this to calculate $R_\perp(k, \omega)$ and hence be able to compare these results with those obtained via the hydrodynamic description.

III THE CONFIGURATION SPACE OR "GENERALIZED HYDRODYNAMICS" DESCRIPTION

By choosing **a** to be the set of microscopic densities (appropriately Fourier transformed in configuration space)

$$\varrho(k) \equiv \sum_{\alpha=1}^{N} e^{i\mathbf{k}\cdot\mathbf{x}^\alpha} \tag{3.1}$$

$$J_i(k) \equiv \sum_{\alpha=1}^{N} (p_i^\alpha/m)\, e^{i\mathbf{k}\cdot\mathbf{x}^\alpha} \tag{3.2}$$

$$\Pi_{ij}(k) \equiv \sum_{\alpha=1}^{N} (p_i^\alpha p_j^\alpha/m)\, e^{i\mathbf{k}\cdot\mathbf{x}^\alpha} + \tfrac{1}{2} \sum_{\alpha=1}^{N} {\sum_{\beta=1}^{N}}' P_{ij}^{\alpha\beta} e^{i\mathbf{k}\cdot\mathbf{x}^\alpha} \tag{3.3}$$

$$E(k) \equiv \sum_{\alpha=1}^{N} (p^{\alpha 2}/2m)\, e^{i\mathbf{k}\cdot\mathbf{x}^\alpha} + \tfrac{1}{2} \sum_{\alpha=1}^{N} {\sum_{\beta=1}^{N}}' V^{\alpha\beta} e^{i\mathbf{k}\cdot\mathbf{x}^\alpha} \tag{3.4}$$

$$Q_i(k) \equiv \sum_{\alpha=1}^{N} (p_i^\alpha p^{\alpha 2}/2m^2)\, e^{i\mathbf{k}\cdot\mathbf{x}^\alpha} + \tfrac{1}{2} \sum_{\alpha=1}^{N} \sum_{\beta=1}^{N}{}' (p_i^\alpha/m)\, V^{\alpha\beta} e^{i\mathbf{k}\cdot\mathbf{x}^\alpha}$$

$$+ \tfrac{1}{2} \sum_{\alpha=1}^{N} \sum_{\beta=1}^{N}{}' [(p_j^\alpha + p_j^\beta)/m] P_{ji}^{\alpha\beta} e^{i\mathbf{k}\cdot\mathbf{x}^\alpha} \tag{3.5}$$

where

$$\mathbf{x}^{\alpha\beta} \equiv \mathbf{x}^\alpha - \mathbf{x}^\beta,\ V^{\alpha\beta} \equiv V(|\mathbf{x}^{\alpha\beta}|),\ P_{ij}^{\alpha\beta} \equiv x_i^{\alpha\beta} \frac{\partial V^{\alpha\beta}}{\partial x_j^\alpha} \frac{[e^{-i\mathbf{k}\cdot\mathbf{x}^{\alpha\beta}} - 1]}{i\mathbf{k}\cdot\mathbf{x}^{\alpha\beta}}, \tag{3.6}$$

Akcasu and Daniels[21] have utilized Mori's[16] technique to develop an exact generalization of the conventional hydrodynamics equations and used this scheme to study current-current correlations in liquids. However for the purposes of calculating transverse current-current correlations, a more restricted set of variables suffices. That is, we wish to calculate the cosine transform in time of

$$R_\perp(k, t) \equiv \langle J_{k_1}^*(0)\, J_{k_1}(t)\rangle \langle J_{k_1}^*(0)\, J_{k_1}(0)\rangle^{-1}, \tag{3.7}$$

where the coordinate system has been chosen such that $\mathbf{k}$ is parallel to the z-axis, while $\mathbf{k}$-subscripted quantities denote fluctuations from equilibrium values, e.g.

$$J_k \equiv J(\mathbf{k}) - \langle J(\mathbf{k})\rangle, \text{ etc.} \tag{3.8}$$

To this end, we choose the components of the state vector as

$$a_1 = J_{k_1}, \quad a_2 = \Pi_{k_{31}}. \tag{3.9}$$

Then straightforward calculation[21] yields the frequency matrix as

$$\mathbf{\Omega} = k \begin{bmatrix} 0 & 1 \\ \dfrac{C_{44}(k)}{mn} & 0 \end{bmatrix} \tag{3.10}$$

where we have identified the elastic modulus[22]

$$C_{44}(k) = n\left[\frac{1}{\beta} + n \int d^3R\, g(R) \frac{\partial^2 V}{\partial x_1^2} \left(\frac{1 - \cos kx_3}{k^2}\right)\right]. \tag{3.11}$$

Furthermore, since $(1 - P)\, a_1 = ik\,(1 - P)\, a_2 = 0$, we find

$$\boldsymbol{\phi}(t) = \begin{bmatrix} 0 & 0 \\ 0 & \varphi_{22}(t) \end{bmatrix} \tag{3.12}$$

where

$$\varphi_{22}(t) \equiv \langle (1 - P)\, \dot{a}_2^* \exp(it(1 - P)\, L]\, (1 - P)\, \dot{a}_2\rangle \langle a_2 a_2^*\rangle^{-1}. \tag{3.13}$$

Hence the generalized Langevin equation (2.1) for the set (3.9) becomes

$$\frac{da_1}{dt} - ika_2(t) = 0 \tag{3.14a}$$

$$\frac{da_2}{dt} - ik\frac{C_{44}(k)}{mn}a_1(t) + \int_0^t d\tau\, \varphi_{22}(\tau)\, a_2(t-\tau) = f_2(t). \tag{3.14b}$$

Now we see from (3.7) that

$$R_\perp(k,t) = \langle a_1(t)\, a_1^*\rangle \langle a_1 a_1^*\rangle^{-1}. \tag{3.15}$$

Hence by multiplying the set (3.14) by a_1^*, ensemble averaging, and then Laplace transforming in time, we can solve for

$$\tilde{R}_\perp(k,s) = \left[s + \frac{k^2 C_{44}(k)/mn}{s + \tilde{\varphi}_{22}(s)}\right]^{-1} \tag{3.16}$$

or

$$R_\perp(k,\omega) = Re\{[i\omega + k^2\eta_\perp(k, i\omega)/mn]^{-1}\}, \tag{3.17}$$

where we define

$$\eta_\perp(k, i\omega) \equiv \frac{C_{44}(k)}{i\omega + \tilde{\varphi}_{22}(k, i\omega)}. \tag{3.18}$$

As it stands, (3.17) is still exact expression for $R_\perp(k,\omega)$. But the actual calculation of $\tilde{\varphi}_{22}(k, i\omega)$ is as difficult as that involved in calculating $R_\perp(k,\omega)$ directly; hence we choose to introduce an approximation in (3.17) by replacing $\tilde{\varphi}_{22}$ by its $\omega \to 0$ limit

$$\lim_{\omega\to 0} \tilde{\varphi}_{22}(k, i\omega) \equiv \omega_\perp(k) \tag{3.19}$$

for small frequencies ω. This corresponds to a "Markovian" description[15] of the fluid in which we approximate

$$\int_0^t d\tau\, \varphi_{22}(\tau)\, a_2(t-\tau) \sim a_2(t) \int_0^\infty d\tau\, \varphi_{22}(\tau). \tag{3.20}$$

In this approximation, our form for $R_\perp(k,\omega)$ becomes

$$R_\perp(k,\omega) = \frac{\omega_\perp(k)\, k^2 C_{44}(k)/mn}{\omega^2\omega_\perp^2(k) + \left[\dfrac{k^2 C_{44}(k)}{mn} - \omega^2\right]^2}. \tag{3.21}$$

However, even $\omega_\perp(k)$ as given by (3.19) is extremely difficult to calculate. Hence instead of calculating $\omega_\perp(k)$ directly, Akcasu and Daniels[21] chose to model its k-dependence as follows: For small k, $\eta_{\perp}(k, 0)$ as defined in (3.18) reduces to the usual shear viscosity, η_s. Hence

$$\lim_{k\to 0} \omega_{\perp}(k) = C_{44}(0)/\eta_s. \tag{3.22}$$

Furthermore, for large k one expects to obtain the dispersion relation $\omega_m^{\perp}(k)$ [here, $\omega_m^{\perp}(k)$ is that frequency at which $R_{\perp}(k, \omega)$ assumes a maximum value for a given k] characteristic of an ideal gas*. Noting that for an ideal gas

$$R_{\perp}(k, \omega) = \left(\frac{\pi\beta m}{2k^2}\right)^{\frac{1}{2}} e^{-\frac{\omega^2}{k^2}\left(\frac{m\beta}{2}\right)}, \tag{3.23}$$

and hence $\omega_m^{\perp}(k) = 0$ for all k, Akcasu and Daniels[21] required $\omega_m^{\perp}(k)$ as calculated from (3.21) to approach zero as $k \to \infty$. This led to the requirement

$$\omega_{\perp}^2(k) \to 2k^2 C_{44}(k)/mn \quad \text{as} \quad k \to \infty. \tag{3.24}$$

For intermediate k, a simple interpolation formula was used

$$\omega_{\perp}^2(k) = \frac{2k^2 C_{44}(k)}{mn} + \left\{\frac{\omega_{\perp}^2(0) - 2k^2[C_{44}(k) - n/\beta]/mn}{1 + k^2/k_0^2}\right\} \tag{3.25}$$

where k_0 is an adjustable interpolation parameter. We will return to discuss this result in more in detail the next section, after we have developed the kinetic equation description.

IV THE PHASE SPACE OR "KINETIC EQUATION" DESCRIPTION

A Derivation of the kinetic equations

We now wish to use the G.L.E. to study fluctuations in the microscopic phase density function

$$g(\mathbf{x}, \mathbf{p}, t) \equiv \sum_{\alpha=1}^{N} \delta[\mathbf{x} - \mathbf{x}^{\alpha}(t)]\, \delta[\mathbf{p} - \mathbf{p}^{\alpha}(t)]. \tag{4.1}$$

To this end, choose $\mathbf{a}(t)$ to be a "vector" whose "components" $a_j \equiv a(\mathbf{p}, t)$ are indexed by a continuous parameter $\mathbf{p}$ and defined by

$$a(\mathbf{p}, t) \equiv \delta g(\mathbf{k}, \mathbf{p}, t) = \sum_{\alpha=1}^{N} e^{i\mathbf{k}\cdot\mathbf{x}^{\alpha}(t)}\, \delta[\mathbf{p} - \mathbf{p}^{\alpha}(t)] - n\,\delta(\mathbf{k})\, M(\mathbf{p}) \tag{4.2}$$

---

* This point is more apparent in the case of longitudinal current-current correlations[21], since **k** may then be interpreted as the momentum transferred between a scattered particle (e.g. neutron) and the scattering media (the system of particles).

where it has been noted that $\langle g(\mathbf{k}, \mathbf{p}, t)\rangle = n\,\delta(\mathbf{k})\,M(\mathbf{p})$, n being the equilibrium number density, and

$$M(\mathbf{p}) \equiv (\beta/2\pi m)^{3/2} \exp(-\beta p^2/2m). \tag{4.3}$$

The extension of the G.L.E. to such a continuous representation is straightforward:

$$\frac{\partial a}{\partial t} - i\int d^3p'\,\Omega(\mathbf{p}, \mathbf{p}')\,a(\mathbf{p}', t) + \int_0^t d\tau \int d^3p'\,\varphi(\mathbf{p}, \mathbf{p}', \tau)\,a(\mathbf{p}', t-\tau) = f(\mathbf{p}, t). \tag{4.4}$$

The appropriate projection operator becomes

$$PG(\mathbf{p}) \equiv \int d^3p' \int d^3p''\langle G(\mathbf{p})\,a^*(\mathbf{p}')\rangle\,\phi^{-1}(\mathbf{p}', \mathbf{p}'')\,a(\mathbf{p}''), \tag{4.5}$$

where $\phi^{-1}(\mathbf{p}', \mathbf{p}'')$ is the "inverse" of the static correlation function defined by

$$\phi(\mathbf{p}, \mathbf{p}') \equiv \langle a(\mathbf{p})\,a^*(\mathbf{p}')\rangle = n\,\delta(\mathbf{p} - \mathbf{p}')\,M(\mathbf{p}') + n^2M(\mathbf{p})\,M(\mathbf{p}')\,h(k), \tag{4.6}$$

where $h(k)$ is the Fourier transform of $[g(r) - 1]$, $g(r)$ being the static pair correlation function. By its definition, $\phi^{-1}(\mathbf{p}', \mathbf{p}'')$ satisfies

$$\int d^3p'\,\phi(\mathbf{p}, \mathbf{p}')\,\phi^{-1}(\mathbf{p}', \mathbf{p}'') = \delta(\mathbf{p} - \mathbf{p}''). \tag{4.7}$$

But substituting the explicit form (4.6) into (4.7), we arrive at an inhomogeneous integral equation for $\phi^{-1}(\mathbf{p}', \mathbf{p}'')$

$$nM(\mathbf{p})\,\phi^{-1}(\mathbf{p}, \mathbf{p}'') + n^2M(\mathbf{p})\,h(k) \int d^3p'M(\mathbf{p}')\,\phi^{-1}(\mathbf{p}', \mathbf{p}'') = \delta(\mathbf{p} - \mathbf{p}''). \tag{4.8}$$

It is straightforward to demonstrate that the unique solution to equation (4.8) is

$$\phi^{-1}(\mathbf{p}', \mathbf{p}'') = \frac{\delta(\mathbf{p}' - \mathbf{p}'')}{nM(\mathbf{p}')} - \frac{h(k)}{1 + nh(k)}. \tag{4.9}$$

Having thus obtained $\phi^{-1}(\mathbf{p}', \mathbf{p}'')$, we are now in a position to explicitly evaluate each of the terms in (4.4). First, notice that the "frequency kernel" becomes

$$i\Omega(\mathbf{p}, \mathbf{p}') = \int d^3p''\langle \dot{a}(\mathbf{p})\,a^*(\mathbf{p}'')\rangle\,\phi^{-1}(\mathbf{p}'', \mathbf{p}'). \tag{4.10}$$

But

$$\dot{a}(\mathbf{p}) = \sum_{\alpha=1}^{N} \left(\frac{i\mathbf{k}\cdot\mathbf{p}^\alpha}{m}\right) e^{i\mathbf{k}\cdot\mathbf{x}^\alpha}\,\delta(\mathbf{p} - \mathbf{p}^\alpha) + \sigma(\mathbf{p}) \tag{4.11}$$

where

$$\sigma(\mathbf{p}) \equiv \sum_{\alpha=1}^{N} e^{i\mathbf{k}\cdot\mathbf{x}^\alpha}\,\mathbf{F}^\alpha\cdot\frac{\partial}{\partial\mathbf{p}^\alpha}\,\delta(\mathbf{p} - \mathbf{p}^\alpha). \tag{4.12}$$

Hence

$$\langle \dot{a}(\mathbf{p})\, a^*(\mathbf{p}'')\rangle = \left(\frac{i\mathbf{k}\cdot\mathbf{p}}{m}\right)\phi(\mathbf{p}, \mathbf{p}'') + \langle \sigma(\mathbf{p})\, a^*(\mathbf{p}'')\rangle. \tag{4.13}$$

To calculate $\langle \sigma(\mathbf{p})\, a^*(\mathbf{p}'')\rangle$

$$\langle \sigma(\mathbf{p})\, a^*(\mathbf{p}'')\rangle = \sum_{\alpha=1}^{N}\sum_{\beta=1}^{N}\left\langle e^{i\mathbf{k}\cdot\mathbf{x}^\alpha}\,\mathbf{F}^\alpha \frac{\partial}{\partial \mathbf{p}^\alpha}\,\delta(\mathbf{p}-\mathbf{p}^\alpha)\, e^{i\mathbf{k}\cdot\mathbf{x}^\beta}\,\delta(\mathbf{p}''-\mathbf{p}^\beta)\right\rangle \tag{4.14}$$

Now noting $\dfrac{\partial \varrho_0}{\partial \mathbf{x}^\alpha} = \beta \mathbf{F}^\alpha \varrho_0$ and integrating by parts, one finds

$$\langle e^{i\mathbf{k}\cdot\mathbf{x}^\alpha}\,\mathbf{F}^\alpha\, e^{-i\mathbf{k}\cdot\mathbf{x}^\beta}\rangle = -\frac{1}{\beta}\left\langle \frac{\partial}{\partial \mathbf{x}^\alpha}\, e^{i\mathbf{k}\cdot\mathbf{x}^{\alpha\beta}}\right\rangle = -\frac{i\mathbf{k}}{\beta}\langle e^{i\mathbf{k}\cdot\mathbf{x}^{\alpha\beta}}\rangle_{\alpha\neq\beta} \tag{4.15}$$

Then (4.14) becomes

$$\begin{aligned}\langle \sigma(p)\, a^*(p'')\rangle &= \frac{\partial M}{\partial \mathbf{p}}\, M(\mathbf{p}'')\,\frac{ik}{\beta}\sum_{\alpha=1}^{N}\sum_{\beta=1}^{N}{}' \langle e^{i\mathbf{k}\cdot\mathbf{x}^{\alpha\beta}}\rangle \\ &= -\left(\frac{i\mathbf{k}\cdot\mathbf{p}}{m}\right) M(\mathbf{p})\, M(\mathbf{p}'')\, n^2 g(k). \end{aligned}\tag{4.16}$$

Hence using (4.13), and (4.16) in (4.10), we find

$$\begin{aligned} i\Omega(\mathbf{p}, \mathbf{p}') &= \left(\frac{i\mathbf{k}\cdot\mathbf{p}}{m}\right)\int d^3p''\,\phi(\mathbf{p}, \mathbf{p}'')\,\phi^{-1}(\mathbf{p}'', \mathbf{p}') \\ &\quad + \int d^3p''\left[-\left(\frac{i\mathbf{k}\cdot\mathbf{p}}{m}\right) M(\mathbf{p})\right] M(\mathbf{p}'')\, n^2 g(k)\,\phi^{-1}(\mathbf{p}'', \mathbf{p}') \\ &= \left(\frac{i\mathbf{k}\cdot\mathbf{p}}{m}\right)\delta(\mathbf{p}-\mathbf{p}') - \left(\frac{i\mathbf{k}\cdot\mathbf{p}}{m}\right) M(\mathbf{p})\,\frac{nh(k)}{1+nh(k)} \end{aligned}\tag{4.17}$$

In a similar manner, the "damping kernel" $\varphi(\mathbf{p}, \mathbf{p}', t)$ becomes

$$\begin{aligned}\varphi(\mathbf{p}, \mathbf{p}', t) &= \int d^3p''\langle f(\mathbf{p}, t)\, f^*(\mathbf{p}'', 0)\rangle\,\phi^{-1}(\mathbf{p}'', \mathbf{p}') \\ &= \frac{\langle f(\mathbf{p}, t)\, f(\mathbf{p}', 0)\rangle}{nM(\mathbf{p}')} - \frac{h(k)}{1+nh(k)}\langle f(\mathbf{p}, t)\int d^3p''\, f^*(\mathbf{p}'', 0)\rangle. \end{aligned}\tag{4.18}$$

Note

$$f(\mathbf{p}, 0) = i(1 - P)\, La(\mathbf{p}) = (1 - P)\, \dot{a}(\mathbf{p})$$
$$= (1 - P)\, \sigma(\mathbf{p}) + \left(\frac{i\mathbf{k} \cdot \mathbf{p}}{m}\right)(1 - P)\, a(\mathbf{p}) = (1 - P)\, \sigma(\mathbf{p}). \quad (4.19)$$

Hence the second term in (4.18) vanishes since $\int d^3p''\, \sigma(\mathbf{p}'') = 0$. Furthermore

$$P\sigma(\mathbf{p}) = \int d^3p' \int d^3p'' \langle \sigma(\mathbf{p})\, a^*(\mathbf{p}')\rangle\, \phi^{-1}(\mathbf{p}', \mathbf{p}'')\, a(\mathbf{p}'')$$
$$= -\left(\frac{i\mathbf{k} \cdot \mathbf{p}}{m}\right) M(\mathbf{p}) \frac{nh(k)}{1 + nh(k)} \int d^3p'' a(p'') \equiv \mu(\mathbf{p}) \int d^3p'' a(\mathbf{p}''). \quad (4.20)$$

but

$$\int d\mathbf{p}'' a(\mathbf{p}'') = \varrho(\mathbf{k}) - n\, \delta(\mathbf{k}) \equiv \varrho_k$$

so that (4.19) becomes

$$f(\mathbf{p}, 0) = \sigma(\mathbf{p}) - \mu(\mathbf{p})\, \varrho_k.$$

Noting

$$\langle f(\mathbf{p}, t)\, f^*(\mathbf{p}', 0)\rangle = \langle [e^{it(1-P)L}\, i(1 - P)\, La]\, [\sigma^*(\mathbf{p}') - \mu^*(\mathbf{p}')\, \varrho_{-k}]\rangle$$
$$= \langle [e^{itL(1-P)}\, i(1 - P)\, La]\, [(1 - P)\, \sigma^* - \mu^*(1 - P)\, \varrho_{-k}]\rangle,$$

and $(1 - P)\, \varrho_{-k} = 0$, our final form becomes

$$\varphi(\mathbf{p}, \mathbf{p}', t) = [nM(\mathbf{p}')]^{-1} \langle \sigma^*(\mathbf{p}') e^{it(1-P)L} [\sigma(\mathbf{p}) - \mu(\mathbf{p})\, \varrho_k]\rangle. \quad (4.21)$$

To proceed further requires a detailed investigation of the propagator $\exp [it(1 - P)\, L]$. This will be deferred until the next section.

In summary then, the generalized Langevin equation for $a(\mathbf{p}, t) = \delta g(\mathbf{k}, \mathbf{p}, t)$ can be written explicitly as

$$\frac{\partial}{\partial t} \delta g - i \frac{\mathbf{k} \cdot \mathbf{p}}{m} \delta g(\mathbf{k}, \mathbf{p}, t) + i \frac{\mathbf{k} \cdot \mathbf{p}}{\mathrm{m}} M(\mathbf{p}) \frac{nh(k)}{1 + nh(k)} \int d^3p'\, \delta g(\mathbf{k}, \mathbf{p}', t)$$
$$+ \int_0^t d\tau \int d^3p'\, \varphi(\mathbf{p}, \mathbf{p}', \tau)\, \delta g(\mathbf{k}, \mathbf{p}', t - \tau) = f(\mathbf{p}, t). \quad (4.22)$$

To obtain a kinetic equation for the singlet phase distribution function $f_1(\mathbf{k}, \mathbf{p}, t)$, one can follow Mori's prescription[16] of averaging (4.22) over the perturbed canonical ensemble (2.10). Then defining

$$f_1(\mathbf{k}, \mathbf{p}, t) \equiv \langle g(\mathbf{k}, \mathbf{p}, t)\rangle_{\varrho(0)}, \quad (4.23)$$

one obtains an exact kinetic equation for $f_1(\mathbf{k}, \mathbf{p}, t)$

$$\frac{\partial f_1}{\partial t} - i\frac{\mathbf{k}\cdot\mathbf{p}}{m} f_1(\mathbf{k}, \mathbf{p}, t) + i\frac{\mathbf{k}\cdot\mathbf{p}}{m} M(\mathbf{p}) \frac{nh(k)}{1 + nh(k)} \int d^3p' f_1(\mathbf{k}, \mathbf{p}', t) + \int_0^t d\tau \int d^3p' \, \varphi(\mathbf{p}, \mathbf{p}', \tau) f_1(\mathbf{k}, \mathbf{p}', t - \tau) = \langle f(\mathbf{p}, t)\rangle_{\varrho(0)}. \tag{4.24}$$

However our immediate concern is not with $f_1(\mathbf{k}, \mathbf{p}, t)$, but rather with the time correlation function

$$\mathscr{G}(\mathbf{k}, \mathbf{p}, \mathbf{p}'', t) \equiv \langle \delta g^*(\mathbf{k}, \mathbf{p}'', 0)\, \delta g(\mathbf{k}, \mathbf{p}, t)\rangle \tag{4.25}$$

If we multiply (4.22) by $\delta g^*(\mathbf{k}, \mathbf{p}'', 0)$ and then average, we obtain an exact kinetic equation for $\mathscr{G}$

$$\frac{\partial \mathscr{G}}{\partial t} - i\frac{\mathbf{k}\cdot\mathbf{p}}{m} \mathscr{G}(\mathbf{k}, \mathbf{p}, \mathbf{p}'', t) + i\frac{\mathbf{k}\cdot\mathbf{p}}{m} M(\mathbf{p}) \frac{nh(k)}{1 + nh(k)} \int d^3p' \, \mathscr{G}(\mathbf{k}, \mathbf{p}', \mathbf{p}'', t) = -\int_0^t d\tau \int d^3p' \, \varphi(\mathbf{p}, \mathbf{p}', t)\, \mathscr{G}(\mathbf{k}, \mathbf{p}', \mathbf{p}'', t - \tau) \tag{4.26}$$

which is just the explicit generalization of Eq. (2.9) to continuous parameters. Hence, as has been suggested by other authors[9], we find that the time correlation function $\mathscr{G}(\mathbf{k}, \mathbf{p}, \mathbf{p}'', t)$ obeys a kinetic equation identical to the linearized kinetic equation for the singlet distribution function $f_1(\mathbf{k}, \mathbf{p}, t)$. It should again be stressed, however, that (4.26) is an exact equation for $\mathscr{G}(\mathbf{k}, \mathbf{p}, \mathbf{p}'', t)$ in which no linearization assumption was necessary. Of course, since (4.26) is exact, its solution is tantamount to solving the equations of motion themselves. Hence (4.26) is of only formal significance until we introduce approximations sufficient to obtain an explicit and tractible form for the damping kernel $\varphi(\mathbf{p}, \mathbf{p}', t)$.

B Approximations of the damping Kernel $\varphi(\mathbf{p}, \mathbf{p}', t)$

One can generate approximate kinetic equations for $\mathscr{G}(\mathbf{k}, \mathbf{p}, \mathbf{p}'', t)$ in a consistent fashion by utilizing various approximations in the calculations of $\varphi(\mathbf{p}, \mathbf{p}', t)$. Of course, the crudest approximation would be to simply set

$$\varphi(\mathbf{p}, \mathbf{p}', \tau) \equiv 0. \tag{4.27}$$

Interestingly enough, the resulting approximate kinetic equation

$$\frac{\partial \mathscr{G}}{\partial t} - \left(\frac{i\mathbf{k}\cdot\mathbf{p}}{m}\right) \mathscr{G}(\mathbf{k}, \mathbf{p}, \mathbf{p}'', t) + \left(\frac{i\mathbf{k}\cdot\mathbf{p}}{m}\right) M(\mathbf{p}) \frac{nh(k)}{1 + nh(k)} \int d^3p' \mathscr{G}(\mathbf{k}, \mathbf{p}', \mathbf{p}'', t) = 0 \tag{4.28}$$

is just the Vlasov-like equation first derived by Zwanzig[23] and later applied to the study of neutron scattering in liquids by Nelkin and Ranganathan[13].

An alternative and perhaps more systematic approach to approximating $\varphi(\mathbf{p}, \mathbf{p}', t)$ consists of applying standard perturbative methods[24] directly to the form (4.21). That is, one can calculate $\varphi(\mathbf{p}, \mathbf{p}', t)$ to lowest order in a perturbation parameter (e.g., the density n or the interaction strength $\lambda = 0(V)$). To illustrate this procedure, we shall apply a perturbation expansion in the coupling parameter λ to obtain a Fokker–Planck-like form for the damping term. (A similar perturbation treatment in density should lead to a Boltzmann-like operator.)

We first remark that it can be demonstrated that

$$\langle f^*(0)\, e^{it(1-P)L} f(0)\rangle = \langle f^*(0)\, e^{itL_0} f(0)\rangle + 0(\lambda^3) \tag{4.29}$$

where

$$L_0 \equiv -i \sum_{\alpha=1}^{N} \frac{\mathbf{p}^\alpha}{m} \cdot \frac{\partial}{\partial \mathbf{x}^\alpha}.$$

Therefore from (4.21),

$$\varphi(\mathbf{p}, \mathbf{p}', t) = [nM(\mathbf{p}')]^{-1} \langle \sigma^*(\mathbf{p}')\, e^{itL_0}\, [\sigma(\mathbf{p}) - \mu(\mathbf{p})\, \varrho_k]\rangle + 0(\lambda^3) \tag{4.30}$$

If we now recognize that $\exp(itL_0)$ is an operator which merely displaces $\mathbf{x}^\alpha$ to $\mathbf{x}^\alpha + \mathbf{p}^\alpha t/m$, then we can calculate

$$\langle \sigma^*(\mathbf{p}')\, e^{itL_0} \mu(\mathbf{p})\, \varrho_k\rangle = \mu(\mathbf{p}) \sum_{\alpha=1}^{N} \sum_{\beta=1}^{N} \left\langle e^{-i\mathbf{k}\cdot\mathbf{x}^\alpha}\, \mathbf{F}^\alpha \cdot \frac{\partial}{\partial \mathbf{p}^\alpha} \delta(\mathbf{p}' - \mathbf{p}^\alpha)\, e^{i\mathbf{k}\cdot\mathbf{x}^\beta} e^{\frac{i\mathbf{k}\cdot\mathbf{p}^\beta}{m} t} \right\rangle$$

$$= -\left[\frac{i\mathbf{k}\cdot\mathbf{p}}{m} M(\mathbf{p})\right]\left[\frac{i\mathbf{k}\cdot\mathbf{p}'}{m} M(\mathbf{p}')\right] \frac{n^2 h^2(k)}{1 + nh(k)}\, e^{-k^2 t^2/2m\beta}. \tag{4.31}$$

Furthermore

$$\langle \sigma^*(\mathbf{p}')\, e^{itL_0} \sigma(\mathbf{p})\rangle$$

$$= \sum_{\alpha=1}^{N} \sum_{\beta=1}^{N} \left\langle e^{-i\mathbf{k}\cdot\mathbf{x}^\alpha}\, \mathbf{F}^\alpha \cdot \frac{\partial}{\partial \mathbf{p}^\alpha} \delta(\mathbf{p}' - \mathbf{p}^\alpha)\, e^{itL_0} e^{i\mathbf{k}\cdot\mathbf{x}^\beta} \mathbf{F}^\beta \cdot \frac{\partial}{\partial \mathbf{p}^\beta} \delta(\mathbf{p} - \mathbf{p}^\beta) \right\rangle$$

$$= \sum_{\alpha=1}^{N} \sum_{\beta=1}^{N} \sum_{\substack{\gamma=1 \\ \gamma\neq\beta}}^{N} \Bigg\langle e^{-i\mathbf{k}\cdot\mathbf{x}^\alpha}\, \mathbf{F}^\alpha \cdot \frac{\partial}{\partial \mathbf{p}^\alpha} \delta(\mathbf{p}' - \mathbf{p}^\alpha)\, e^{i\mathbf{k}\cdot\mathbf{x}^\beta}$$

$$\times\, e^{i\mathbf{k}\cdot\mathbf{p}t/m}\, \mathbf{F}^{\beta\gamma}(\mathbf{x}^\beta - \mathbf{x}^\gamma + \mathbf{p}^\beta t/m - \mathbf{p}^\gamma t/m) \cdot \frac{\partial}{\partial \mathbf{p}^\beta} \delta(\mathbf{p} - \mathbf{p}^\beta) \Bigg\rangle \tag{4.32}$$

or

$$\langle \sigma^*(\mathbf{p}') e^{itL_0} \sigma(\mathbf{p}) \rangle$$

$$= -\frac{\partial}{\partial \mathbf{p}'} M(\mathbf{p}') e^{i\mathbf{k}\cdot\mathbf{p}t/m} \cdot \left[\sum_{\alpha=1}^{N} \sum_{\beta=1}^{N}{}' \langle \mathbf{F}^\alpha \mathbf{F}^{\alpha\beta} (\mathbf{x}^\alpha - \mathbf{x}^\beta + \mathbf{p}'t/m - \mathbf{p}^\beta t/m) \rangle \right]$$

$$\cdot \frac{\partial}{\partial \mathbf{p}'} \delta(\mathbf{p} - \mathbf{p}') - \frac{\partial}{\partial \mathbf{p}'} M(\mathbf{p}') \frac{\partial}{\partial \mathbf{p}} M(\mathbf{p}) e^{i\mathbf{k}\cdot\mathbf{p}t/m}$$

$$: \left[\sum_{\alpha=1}^{N} \sum_{\beta=1}^{N}{}' \langle \mathbf{F}^\alpha \mathbf{F}^{\alpha\beta} (\mathbf{x}^\alpha - \mathbf{x}^\beta + \mathbf{p}'t/m - \mathbf{p}t/m) e^{i\mathbf{k}\cdot\mathbf{x}^{\alpha\beta}} \rangle + 0(\lambda^3) \right. \tag{4.33}$$

Here we have noted that the terms arising from $\alpha \neq \beta \neq \gamma$ in equation (4.33) are of $0(\lambda^3)$. Combining Eqs. (4.33) and (4.31), one arrives at an expression for the damping kernel $\varphi(\mathbf{p}, \mathbf{p}', t)$ which is correct to $0(\lambda^2)$.

The damping term becomes particularly transparent in the Markovian limit in which we let $\lambda \to 0$, $t \to \infty$, $\mathbf{x} \to \infty$ in such a way that $\lambda^2 t$ and $\lambda^2 \mathbf{x}$ remain constant. (Note that both t and $\mathbf{x}$ refer to displacements in time and space in a stationary, homogeneous medium.) Then

$$\int_0^t d\tau \int d^3p' \, \varphi(\mathbf{p}, \mathbf{p}', \tau) \, \mathscr{G}(\mathbf{k}, \mathbf{p}', \mathbf{p}'', t - \tau)$$

$$\to \int d^3p' \left[\int_0^\infty d\tau \langle \sigma^*(\mathbf{p}') e^{i\tau L_0} \sigma(\mathbf{p}) \rangle \bigg|_{\mathbf{k}=0} \right] \frac{\mathscr{G}(\mathbf{k}, \mathbf{p}', \mathbf{p}'', t)}{nM(\mathbf{p}')}$$

$$= -\int d^3p' \left\{ \frac{\partial}{\partial \mathbf{p}'} M(\mathbf{p}') \cdot \left[\int_0^\infty d\tau \sum_{\alpha,\beta}{}' \langle \mathbf{F}^\alpha \mathbf{F}^{\alpha\beta} (\mathbf{x}^\alpha - \mathbf{x}^\beta + \mathbf{p}'\tau/m - \mathbf{p}^\beta \tau/m) \rangle \right] \right.$$

$$\cdot \frac{\partial}{\partial \mathbf{p}'} \delta(\mathbf{p} - \mathbf{p}') + \frac{\partial}{\partial \mathbf{p}'} M(\mathbf{p}') \frac{\partial}{\partial \mathbf{p}}$$

$$\times M(\mathbf{p}) : \left[\int_0^\infty d\tau \sum_{\alpha,\beta}{}' \langle \mathbf{F}^\alpha \mathbf{F}^{\alpha\beta} (\mathbf{x}^\alpha - \mathbf{x}^\beta + \mathbf{p}'\tau/m - \mathbf{p}\tau/m) \rangle \right] \frac{\mathscr{G}(\mathbf{k}, \mathbf{p}', \mathbf{p}'', t)}{nM(\mathbf{p}')}$$

$$\equiv J_M[\mathscr{G}(\mathbf{k}, \mathbf{p}', \mathbf{p}'', t)] \tag{4.34}$$

Here we have assumed that the integration over all times does not change the magnitude of the $0(\lambda^3)$ terms [as it would if we had terms of the form $\lambda^3 \exp(-\lambda t)$ present]. If we integrate by parts twice and use the fact that

$$\langle \mathbf{F}^\alpha \mathbf{F}^{\alpha\beta} (\mathbf{x}^\alpha - \mathbf{x}^\beta + \mathbf{p}'\tau/m - \mathbf{p}^\beta \tau/m) \rangle$$

$$= \int d\mathbf{p}'' M(\mathbf{p}'') \langle \mathbf{F}^\alpha \mathbf{F}^{\alpha\beta} (\mathbf{x}^\alpha - \mathbf{x}^\beta + \mathbf{p}\tau/m - \mathbf{p}''\tau/m) \rangle \tag{4.35}$$

then we find

$$J_M[\psi(\mathbf{p}, t)] = \frac{1}{\beta} \int d^3\mathbf{p}' \frac{\partial}{\partial \mathbf{p}} \cdot \xi(\mathbf{p}', \mathbf{p}) \cdot \left\{ \left[\frac{\partial}{\partial \mathbf{p}'} + \frac{\beta}{m} \mathbf{p}' \right] \psi(\mathbf{p}', t)\, M(\mathbf{p}) - \left[\frac{\partial}{\partial \mathbf{p}} + \frac{\beta}{m} \mathbf{p} \right] \psi(\mathbf{p}, t)\, M(\mathbf{p}') \right\} \tag{4.36}$$

where

$$\xi(\mathbf{p}', \mathbf{p}) \equiv \beta \int_0^\infty d\tau \sum_{\alpha, \beta}{}' \langle \mathbf{F}^\alpha \mathbf{F}^{\alpha\beta}(\mathbf{x}^\alpha - \mathbf{x}^\beta + \mathbf{p}'\tau/m - \mathbf{p}\tau/m) \rangle$$
$$= \beta \int \frac{d^3k}{(2\pi)^3} \mathbf{k} V(k)\, nS(k)\, \mathbf{k} V(k)\, \pi\delta[\mathbf{k} \cdot (\mathbf{p}' - \mathbf{p})/m] \tag{4.37}$$

and $S(k)$ is the static structure factor, $S(k) = 1 + nh(k)$. Noting that

$$\int d^3p' \frac{\partial}{\partial \mathbf{p}} \cdot \xi(\mathbf{p}', \mathbf{p}) \cdot (\mathbf{p} - \mathbf{p}')\, [\psi(\mathbf{p}', t)\, M(\mathbf{p}) - \psi(\mathbf{p}, t)\, M(\mathbf{p}')] = 0 \tag{4.38}$$

one can reduce (4.36) to a more familiar form

$$J_M[\psi(\mathbf{p}, t)] = \int d^3p' \int \frac{d^3k}{(2\pi)^3} nS(k)\, V(k)\, \mathbf{k} \cdot \frac{\partial}{\partial \mathbf{p}} \pi\delta[\mathbf{k} \cdot (\mathbf{p} - \mathbf{p}')/m]\, V(k)\, \mathbf{k} \cdot \left(\frac{\partial}{\partial \mathbf{p}} - \frac{\partial}{\partial \mathbf{p}'} \right) [\psi(\mathbf{p}', t)\, M(\mathbf{p}) + \psi(\mathbf{p}, t)\, M(\mathbf{p}')] \tag{4.39}$$

In the absence of initial correlations, $S(k) = 1$, and (4.39) reduces to the Fokker–Planck operator derived in the standard weak-coupling theory[24,25] of many body systems (e.g. see Eq. (6.18) of reference 24).

Hence the damping term (4.34) to lowest order in the interaction strength λ can be regarded as a "generalized Fokker–Planck" operator which reduces to the more conventional form (4.39) in the Markovian limit.

In a similar manner, the damping term reduces to the linearized Boltzman operator following a density expansion and subsequent Markovian limit $n \to 0$, $t \to \infty$, $\mathbf{x} \to \infty$ such that nt and $n\mathbf{x}$ remain constant (refer to the paper of R. Zwanzig and M. Bixon[26] given at this Symposium).

C Calculation of transverse current-current correlations using a "Modeled" damping kernel

The above procedures for approximating $\varphi(\mathbf{p}, \mathbf{p}', t)$ yield integro-differential kinetic equations which, while very suggestive in appearance, are in fact rather difficult to solve. To circumvent these difficulties, we shall adopt a "modeled" description of the damping term as follows: Assume that the time behavior of the damping kernel $\varphi(\mathbf{p}, \mathbf{p}'', \tau)$ is exponential such that

$$\varphi(\mathbf{p}, \mathbf{p}', \tau) = \varphi(\mathbf{p}, \mathbf{p}', 0) \exp[-\alpha(k)\tau], \tag{4.40}$$

where the exact form for $\varphi(\mathbf{p}, \mathbf{p}', 0)$ will be used, while $\alpha(k)$ will be determined at a later stage of the calculation. This "single-relaxation-time" approximation is very similar to that introduced earlier by Berne, Boon, and Rice[27] in the calculation of the momentum autocorrelation function of a test particle in a liquid-except that we allow a k-dependent relaxation constant $\alpha(k)$. Such an approximation can be motivated by referring to the two-component kinetic description described in Reference 28 and recognizing that one expects the damping kernel to be a rapidly decaying function of time (at least on the time-scale of $\mathscr{G}$). However, we shall regard (4.10) as merely a model, and as such, justified only by agreement with experimental data.

It is a straightforward task to evaluate $\varphi(\mathbf{p}, \mathbf{p}', 0)$ from Eq. (4.21). Such a calculation yields

$$\int d^3p'\, \varphi(\mathbf{p}, \mathbf{p}', 0)\, \mathscr{G}(\mathbf{k}, \mathbf{p}', \mathbf{p}'', t) = -D(0)\left\{\frac{\partial}{\partial \mathbf{p}} \cdot \frac{\partial}{\partial \mathbf{p}} + \frac{\beta}{m}\frac{\partial}{\partial \mathbf{p}} \cdot \mathbf{p}\right\}\mathscr{G}$$

$$+ \mathbf{p}M(\mathbf{p}) \cdot \mathbf{D}'(k) \cdot \int d^3p'\mathbf{p}'\mathscr{G}(\mathbf{p}'), \tag{4.41}$$

where

$$\mathbf{D}'(k) \equiv \mathbf{kk}nc(k)/m^2 - (\beta^2/m^2)\,\mathbf{D}(k)$$

$$D_{ij}(k) \equiv (n/\beta)\int d^3R \frac{\partial^2 V}{\partial x_i^2} g(R) \cos kx_3\, \delta_{ij}$$

$$D(0) \equiv (n/\beta)\int d^3R \frac{\partial^2 V}{\partial x_i^2} g(R) \tag{4.42}$$

and $c(k) = h(k)/[1 + nh(k)]$ is the direct correlation function. Hence the modeled kinetic equation we will consider becomes, after a Laplace transform in time,

$$\left(s - \frac{i\mathbf{k}\cdot\mathbf{p}}{m}\right)\tilde{\mathscr{G}}(\mathbf{k}, \mathbf{p}, \mathbf{p}'', s) + \frac{i\mathbf{k}\cdot\mathbf{p}}{m} M(\mathbf{p})\, nc(k) \int d^3p'\, \tilde{\mathscr{G}}(\mathbf{k}, \mathbf{p}', \mathbf{p}'', s)$$

$$= \mathscr{G}(\mathbf{k}, \mathbf{p}, \mathbf{p}'', 0) + \frac{D(0)}{s + \alpha(k)}\left\{\frac{\partial}{\partial \mathbf{p}} \cdot \frac{\partial}{\partial \mathbf{p}} + \frac{\beta}{m} \cdot \frac{\partial}{\partial \mathbf{p}}\mathbf{p}\right\}\tilde{\mathscr{G}}(k, p, p'', s)$$

$$- \frac{\mathbf{p}M(\mathbf{p}) \cdot \mathbf{D}(k)}{s + \alpha(k)} \cdot \int d^3p'\, \mathbf{p}'\tilde{\mathscr{G}}(\mathbf{k}, \mathbf{p}', \mathbf{p}'', s), \tag{4.43}$$

where

$$\mathscr{G}(\mathbf{k}, \mathbf{p}, \mathbf{p}'', 0) = nM(\mathbf{p})\,\delta(\mathbf{p} - \mathbf{p}') + n^2 M(\mathbf{p})M(\mathbf{p}'')\,h(k)$$

To apply this equation to the calculation of the transverse current-current correlation function $R_\perp(k, \omega)$, we note from the definition (3.7) that

$$R_\perp(k, t) = (\beta/mn) \int d^3p\, p_1 \int d^3p''\, p_1'' \mathscr{G}(\mathbf{k}, \mathbf{p}, \mathbf{p}'', t) \tag{4.44}$$

Hence to calculate $R_\perp(k, t)$, we need only compute a moment of $\mathscr{G}(\mathbf{k}, \mathbf{p}, \mathbf{p}'', t)$.

Rather than attempting to solve the modeled kinetic equation (4.50) for $\tilde{\mathscr{G}}(\mathbf{k}, \mathbf{p}, \mathbf{p}'', s)$ directly, we will instead first multiply it by p_1 and p_1'' and the integrate over $\mathbf{p}''$ and p_1 and p_2 to find

$$\left(s - \frac{ikp}{m}\right) X(p, s) - \frac{D(0)}{s + \alpha(k)}\left[\frac{d^2}{dp^2} + \frac{\beta}{m} p \frac{d}{dp}\right] X(p, s)$$
$$- \frac{(\beta/m) D_{11}(k) M(p)}{s + \alpha(k)} \int_{-\infty}^{\infty} dp' X(p', s) = \left(\frac{mn}{\beta}\right) M(p), \tag{4.45}$$

where we have denoted $p_3 \to p$ and defined

$$X(p, s) \equiv \int_{-\infty}^{\infty} dp_1 \int_{-\infty}^{\infty} dp_2\, p_1 \int d^3p'' p'' \tilde{\mathscr{G}}(\mathbf{k}, \mathbf{p}, \mathbf{p}'', s). \tag{4.46}$$

Equation (4.45) is now an integro-differential equation in the single variable p. Recognize that we need only compute the integral of $X(p, s)$ over all p to arrive at $\tilde{R}_\perp(k, s)$. With this in mind, let us study Eq. (4.45) in two limiting cases: In the limit of large s it is evident that we can neglect the damping term in (4.45) to find

$$\tilde{R}_\perp(k, s) = \left(\frac{m\beta}{2\pi}\right)^{\frac{1}{2}} \int_{-\infty}^{\infty} dp \frac{e^{-p^2\beta/2m}}{s - ikp/m} \tag{4.47}$$

Of more interest is the cosine transform of $R_\perp(k, t)$ in time, given by

$$R_\perp(k, \omega) = Re\{\lim_{\varepsilon \to 0+} \tilde{R}_\perp(k, i\omega + \varepsilon)\} = \left(\frac{\pi m\beta}{2k^2}\right)^{\frac{1}{2}} \exp\left[-\frac{\omega^2}{k^2}\left(\frac{m\beta}{2}\right)\right]. \tag{4.48}$$

Of course, this is just the form of the transverse current correlation one would have obtained for an ideal gas—that is, a system in which there are no particle interactions. This is the same result which one obtains from the Vlasov kinetic equation for a liquid, hence verifying that the total neglect of the damping term is related to a large ω or short time approximation.

Having verified that our modeled kinetic equation yields the correct large ω behavior, let us now examine the other extreme of small k and ω. The most convenient approach is to multiply Eq. (4.45) by p^r and integrate over p, hence obtaining a recursion relation among the moments $X^{(r)}(s)$:

$$\left.\begin{aligned} &sX^{(r)} - \frac{ik}{m}X^{(r+1)} - \frac{D(0)}{s+\alpha(k)} \\ &\quad\times\left[r(r-1)X^{(r-2)} - \frac{\beta}{m}(r+1)X^{(r)}\right] = 0 \qquad r \text{ odd} \\ &sX^{(r)} - \frac{ik}{m}X^{(r+1)} - \frac{D(0)}{s+\alpha(k)} \\ &\quad\times\left[r(r-1)X^{(r-2)} - \frac{\beta}{m}(r+1)X^{(r)}\right] \\ &\quad - \frac{D_{11}(k)}{s+\alpha(k)}\left(\frac{m}{\beta}\right)^{\frac{r}{2}-1}\nu_r X^{(0)} = \left(\frac{m}{\beta}\right)^{\frac{r}{2}+1}\nu_r n, \qquad r \text{ even} \end{aligned}\right\} \tag{4.49}$$

where $\nu_r \equiv [1\cdot3\cdot5\cdots(r-1)]$ and

$$X^{(r)}(s) \equiv \int_{-\infty}^{\infty} dp\, p^r X(p, s) \tag{4.50}$$

In particular, note that

$$\tilde{R}_{\perp}(k, s) = (\beta/mn)\, X^{(0)}(s) \tag{4.51}$$

It is a straightforward task to compute the first few moments to find

$$X^{(0)}(s) = \left[s + \frac{\beta/m\,[D(0) - D_{11}(k)]}{s+\alpha(k)}\right]^{-1}\left[\frac{mn}{\beta} + \frac{ik}{m}X^{(1)}\right], \tag{4.52}$$

$$X^{(1)}(s) = \left[s + \frac{2\beta/m\, D(0)}{s+\alpha(k)}\right]^{-1}\left[\frac{ik}{m}X^{(2)}\right], \tag{4.53}$$

$$\begin{aligned} X^{(2)}(s) &= \left[s + \frac{3\beta/m\, D(0)}{s+\alpha(k)}\right]^{-1} \\ &\quad\times\left\{\left(\frac{m}{\beta}\right)^2 n + \left[\frac{2D(0) + D_{11}(k)}{s+\alpha(k)}\right]X^{(0)} + \frac{ik}{m}X^{(3)}\right\}. \end{aligned} \tag{4.54}$$

Comparing (3.11) and (3.42), we note

$$\frac{\beta}{m}[D(0) - D_{11}(k)] = \frac{k^2 C_{44}(k)}{mn}\left[1 - \frac{\eta}{\beta C_{44}(k)}\right] \tag{4.55}$$

Hence from (4.51) and (4.52) we find

$$\tilde{R}_{\perp}(k, s) = \left\{ s + \frac{k^2 C_{44}(k)}{mn} \frac{\left[1 - \dfrac{\eta}{\beta C_{44}(k)}\right]}{s + \alpha(k)} \right\}^{-1} [1 + \delta(k, s)] \quad (4.56)$$

where we have defined

$$\delta(k, s) \equiv \left(\frac{\beta}{mn}\right)\left(\frac{ik}{m}\right) X^{(1)}(s) \quad (4.57)$$

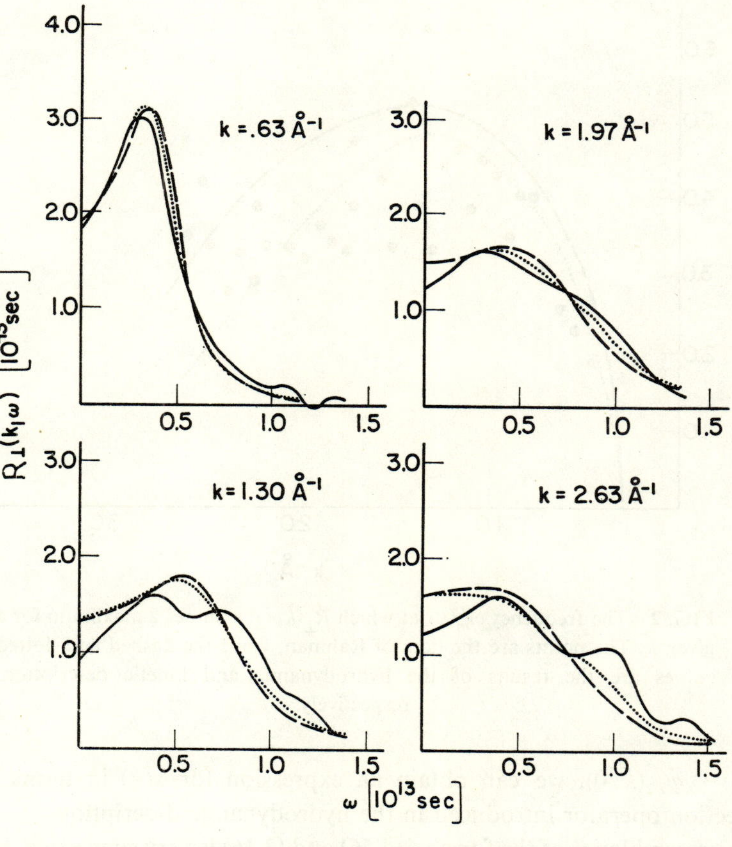

FIG. 1 The transverse current-current correlation function $R_{\perp}(k, \omega)$ for argon. The solid curves represent the computer data of Rahman[17], the dashed curves represent the hydrodynamics calculations, while the dotted curves represent a three moment kinetic description

In order to obtain an expression for $\alpha(k)$, we compare (4.56) and the exact hydrodynamic result (3.16) at $s = 0$ to find

$$\alpha(k) = \omega_\perp(k)\left[\frac{1 - \dfrac{\eta}{\beta C_{44}(k)}}{1 + \delta(k, s)}\right] \tag{4.58}$$

Since $\omega_\perp(k) = C_{44}(k)/\eta_\perp(k, 0)$, we find that $\alpha(k)$ is related to a transport parameter, the k-dependent shear viscosity. Furthermore, using the relation

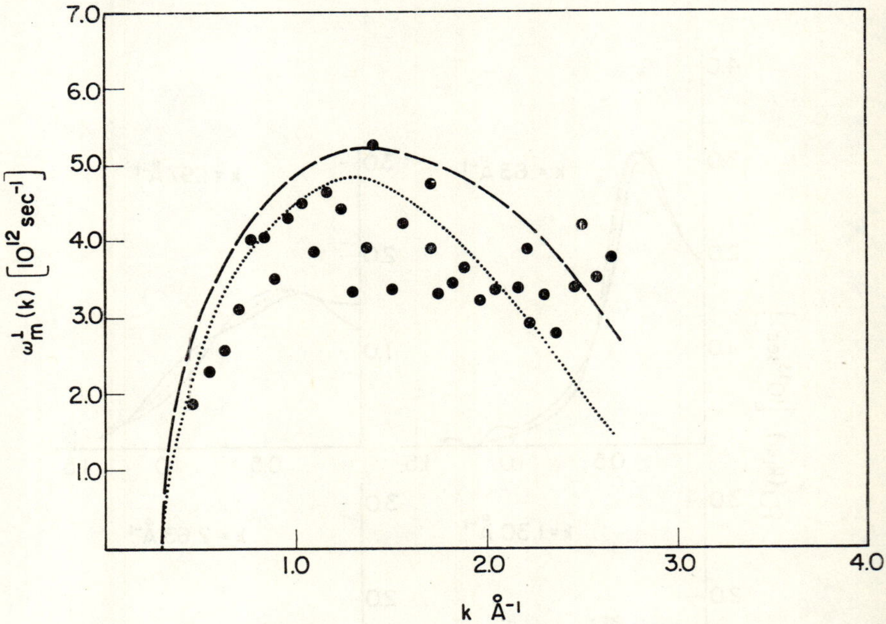

FIG. 2 The frequency $\omega_m^\perp(k)$ at which $R_\perp(k, \omega)$ assumes a maximum for a given k. The points are the data of Rahman, while the dashed and dotted curves are the results of the hydrodynamic and kinetic description, respectively

$\omega_\perp(k) = \tilde{\varphi}_{22}(k, 0)$, we can obtain an expression for $\alpha(k)$ in terms of the projection operator introduced in the hydrodynamic description.

The resemblance of the form of (4.56) and (3.16) is very suggestive. Indeed, since the additional term $\delta(k, s)$ in (4.56) is due to the streaming of the particles, represented by $\dfrac{ikp}{m} X(p, s)$ in (4.45), we may therefore argue that for dense fluids and small k, this term will be small compared to unity. Furthermore, the factor $[1 - n/\beta C_{44}(k)]$ in the denominator of (4.50) is very close to unity for small values of k. Hence for small k, the expressions (4.56) and (3.16) are equivalent, with $\alpha(k)$ also approaching $\omega_\perp(k)$ as seen from (4.58). Thus, the modeled kinetic description (4.43) yields the correct "hydrodynamic" results for small k with the choice of $\alpha(k)$ given in (4.58).

We have computed $R_\perp(k, \omega)$ for argon at $T = 76°\text{K}$, $mn = 1.407\ \text{g/cm}^3$ using the modeled hydrodynamic result (3.21) as well as the kinetic de-

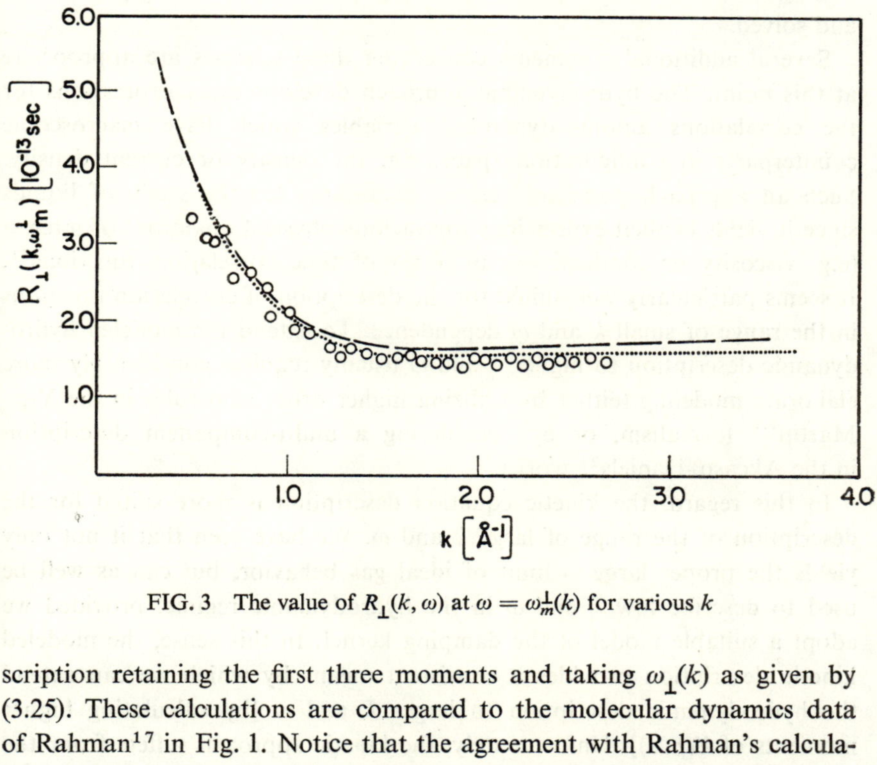

FIG. 3 The value of $R_\perp(k, \omega)$ at $\omega = \omega_m^\perp(k)$ for various k

scription retaining the first three moments and taking $\omega_\perp(k)$ as given by (3.25). These calculations are compared to the molecular dynamics data of Rahman[17] in Fig. 1. Notice that the agreement with Rahman's calculations is rather good in all cases. As one expects, the higher moments in the kinetic description are essentially negligible for $k < 1.5\ \text{Å}^{-1}$, and the hydrodynamic description yields satisfactory results. However for larger k, the kinetic description yields somewhat better results due to its more detailed treatment of the frequency dependence. Fig. 2 gives a comparison of the frequency $\omega_m^\perp(k)$ for which $R_\perp(k, \omega)$ assumes a maximum for a given value of k for the various theories, while Fig. 3 plots the magnitude of this maximum.

V CONCLUSIONS

In summary, we have used the generalized Langevin equation of Zwanzig and Mori to study correlations in simple, classical fluids. Two alternative procedures have been presented and compared: first, a generalized hydrodynamic approach which involves dynamical variables defined in configuration space was used to calculate the transverse current-current correlation function in a liquid; this same quantity was then calculated by obtaining an exact kinetic equation description which was then approximated and solved.

Several additional comments concerning these schemes are appropriate at this point. The hydrodynamic approach develops exact expressions for the correlations among dynamical variables which have macroscopic counterparts in configuration space, e.g. the density or current density. Such an approach possesses certain advantages for the study of liquids since it yields explicit expressions for various physical quantities of interest (e.g. viscosity or conductivity) in terms of time correlation functions[21]. It seems particularly well-suited for the description of correlation functions in the range of small k and ω dependence. To extend the modeled hydrodynamic description to higher k and ω usually requires considerably more elaborate modeling (either by utilizing higher order sum rules in the Yip–Martin[7,8] formalism, or by considering a multi-component description in the Akcasu–Daniels[21] work).

In this regard, the kinetic equation description is more suited for the description of the range of large k and ω. We have seen that it not only yields the proper large ω limit of ideal gas behavior, but can as well be used to describe low k and ω in the hydrodynamic regime, provided we adopt a suitable model of the damping kernel. In this sense, the modeled kinetic description provides a consistent means by which we can extend the hydrodynamic description to larger k and ω [by calculating higher moments of $X(p, s)$]. Unfortunately, the kinetic approach suffers from the complexities of modeling and obtaining a solution to the kinetic equation itself. It is, of course, much easier to model a function of k, as in the hydrodynamic description, than an operator, as we must in the kinetic approach. Moreover, there are very few "modeled collision operators" which will permit us to extract sufficient information about the behavior of $\mathscr{G}(\mathbf{k}, \mathbf{p}, \mathbf{p}'', t)$ to allow a practical calculation of time correlation functions.

These considerations, in addition to the recognition that the generalized hydrodynamic formalism seems perfectly adequate for describing time correlation functions for those ranges of k and ω ecountered in neutron or light scattering, leads us to conclude that the hydrodynamic approach seems the more effective method for studying the dynamics of simple fluids.

Acknowledgments

This work has benefited rather substantially from numerous discussions with Professors Richard Osborn and Noel Corngold. The authors are also extremely grateful to Eugene Daniels who allowed them to describe a portion of his work with the first author prior to its actual publication.

References

1. For a comprehensive review of this subject, see R. Zwanzig, in *Annual Review of Physical Chemistry*, Vol. **16** (Annual Reviews, Inc., Palo Alto, 1965).
2. H. B. Callen and T. A. Welton, *Phys. Rev.* **83**, 34 (1951).
3. L. D. Landau and E. M. Lifschitz, *Statistical Physics* (Pergamon Press, London, 1959).
4. S. M. Rytov, *Soviet Phys. JETP* **6**, 130 (1958).
5. R. Mountain, *Rev. of Mod. Phys.* **38**, 205 (1966).
6. L. P. Kadanoff and P. C. Martin, *Ann. Phys.* **24**, 419 (1963).
7. P. C. Martin and S. Yip, *Phys. Rev.* **170**, 151 (1968).
8. C. H. Chung and S. Yip, *Phys. Rev.* **182**, 323 (1969).
9. M. Nelkin, J. Van Leeuwen, and S. Yip, in *Neutron Inelastic Scattering* (International Atomic Energy Agency, Vienna, 1964) Vol. II, p. 36.
10. J. M. J. Van Leeuwen and S. Yip, *Phys. Rev.* **139**, A1138 (1965).
11. M. Nelkin and S. Yip, *Phys. Fluids* **9**, 380 (1966).
12. J. J. Greytak and G. B. Benedek, *Phys. Rev. Letters* **17**, 179 (1966).
13. M. Nelkin and S. Ranganathan, *Phys. Rev.* **164**, 222 (1967).
14. M. Nelkin and P. J. Ortoleva, in *Neutron Inelastic Scattering* (International Atomic Energy Agency, Vienna, 1968).
15. R. Zwanzig, "Statistical Mechanics of Irreversibility, "in *Lectures in Theoretical Physics*, ed. by W. E. Britten (John Wiley, New York, 1961).
16. H. Mori, *Progr. Theoret. Phys.* (Kyoto) **33**, 423 (1965).
17. A. Rahman, in *Neutron Inelastic Scattering* (IAEA, Vienna, 1968).
18. S. Chandrasekhar, *Rev. Mod. Phys.* **15**, 1 (1943).
19. B. Robertson, *Phys. Rev.* **144**, 151 (1966).
20. R. A. Piccirelli, *Phys. Rev.* **175**, 77 (1968).
21. A. Z. Akcasu and E. Daniels, *Phys. Rev.* A, **2** (1970).
22. P. Schofield, *Proc. Phys. Soc.*, (British) **88**, 149 (1966).
23. R. Zwanzig, *Phys. Rev.* **144**, 170 (1966).
24. P. Résibois, in *N Particle Physics*, ed. by E. Meeron (Gordon and Breach, New York, 1967).

25. R. Brout and I. Prigogine, *Physica* **22**, 621 (1956).
26. R. Zwanzig and M. Bixon, private communication.
27. B. Berne, J. Boon, and S. A. Rice, *J. Chem. Phys.* **45**, 1086 (1966).
28. A. Z. Akcasu and J. J. Duderstadt, *Phys. Rev.* **188**, 479 (1969).
29. J. J. Duderstadt and A. Z. Akcasu, *Phys. Rev.* **A, 1**, 905 (1970); *Phys. Rev.* **A, 2** (1970).

Note added in proof: In a subsequent publication[29] the authors have recalculated the transverse and longitudinal current correlation functions using the exact solution of Eqn. (4.43). In this later work a more satisfactory determination of the relaxation parameter $\alpha(k)$ is given.

PAPER A–5

Kinetic equations with radiation effects

R. HAKIM*

Département de Physique Mathématique†
Montpellier, France

and

A. MANGENEY

Laboratoire de Radio-Astronomie Spatiale
Observatoire de Meudon, France

In this talk, I would like to report on a work effected in collaboration with *André Mangeney* about relativistic kinetic equations involving radiation effects.

The problem is the following: How to describe a radiating electron plasma? Of course the methods given hereafter also apply to more general plasmas such as many component plasmas etc.

In order to be consistent I shall consider a relativistic electron plasma. I said "to be consistent" essentially because of the fact that electromagnetic phenomena are of relativistic nature: the non relativistic approximation of Maxwell's equations is obtained by suppressing the displacement current and hence no wave can propagate. Therefore a consistent treatment of a plasma involving radiation should be relativistic.

Furthermore relativistic electron plasmas do exist in astrophysics. Think, for instance, to the old component of quasars, composed of relativistic electrons, as indicated by synchroton emission.

As you know, the system (electron + radiation) has been described for the first time—in a statistical sense, of course—by Brittin and Klimontovich who used methods close to the one considered in turbulence theory.

---

* Postal Address: Département de Physique Mathématique, Faculté des Sciences, 34 – MONTPELLIER, FRANCE.

† Physique Mathématique et Théorique, Equipe de recherche associée au C.N.R.S.

For instance, they considered moments of the electromagnetic field, averages of field and particles quantities etc. ..., and gave equations for them.

What amounts to the same thing, is the equivalent treatment of the electromagnetic field considered as an assembly of oscillators. The relativistic treatment has been given by Prigogine and Mangeney.

Here, I would like to present a completely different approach and various applications.

For the sake of simplicity, let us first consider the simpler case of only one charged particle embedded in an external force field K^μ and submitted to the action of its self electromagnetic field $F^{\mu\nu}_{\text{self}}(\mu = 0, ..., 3)$.

In the *field approach* the statistical description of such a system is fully contained in the following equations:

$$\begin{cases} u^\mu \partial_\mu N + \left[\dfrac{e}{m_0} F^{\mu\nu}_{\text{self}} u_\nu + \dfrac{K^\mu}{m_0}\right] \dfrac{\partial}{\partial u^\mu} N = 0 \\ \partial_\mu F^{\mu\nu}_{\text{self}} = e \displaystyle\int_{-\infty}^{+\infty} d\tau\, u^\nu(\tau)\, \delta[x^\varrho - x^\varrho(\tau)],\ \partial_\mu F^{\mu\nu *}_{\text{self}} = 0 \end{cases}$$

where m_0 is the (*infinite*) bare mass of the particle. $F^{\mu\nu}_{\text{self}}$ is the electromagnetic field due to the current of the particle and thus is its self-field (we only consider the retarded field).

It follows that we have to deal with a system which possesses an *infinity* of degrees of freedom. Next we must obtain equations for quantities such as:

$$\langle F^{\mu\nu}_{\text{self}} \rangle, ..., \langle F^{\mu\nu}_{\text{self}} N \rangle \text{ etc.}$$

Our approach is completely different. Instead of considering the self-field of the electron we rather consider derived simpler quantities. Indeed, it may be shown (see e.g. Barut, Rohrlich), that the problem of an electron embedded in an external force field *and* its self-field amounts to a change in the equations of motion for the electron, the self electromagnetic field being eliminated. The new equations are the Lorentz-Dirac equations:

$$m\gamma^\mu = K^\mu + \tau_0 [g^{\mu\nu} - u^\mu u^\nu]\, \dot{\gamma}_\nu$$

with

τ_0 = non causality time $= \dfrac{2}{3} \dfrac{e^2}{mc^3}$, ($10^{-23}$ s for an electron)

m = (*finite*) renormalized mass

γ^μ = 4-acceleration, $\dot{\gamma}^\mu = \dfrac{d}{d\tau} \gamma^\mu$.

Therefore we have to face the problem of the statistical description of a particle obeying this equation.

What is gained by proceeding in such a way? First, we have to deal with a finite system. Secondly, our equations will involve only the *finite observable* mass and hence no renormalization procedure will be needed (R. Hakim). Thirdly, the effects of radiation will be included in our equations through the radiation reaction term:

$$\tau_0[g^{\mu\nu} - u^\mu u^\nu]\,\dot{\gamma}_\nu .$$

Of course, this term is generally extremely weak. However a weak loss of energy (for a particle) often represents quite an amount of radiation.

The statistical description we are looking for may be obtained by considering an *enlarged* μ-space involving the *acceleration variables* or, equivalently, by considering the one-particle phase space spanned by the variables $(x^\mu, u^\mu, \dot{\gamma}^\mu)$. This equivalence stems from the fact that γ^μ may be expressed as a function of $(x^\mu, u^\mu, \dot{\gamma}^\mu)$ and conversely $\dot{\gamma}^\mu$ may be expressed as a function of $(x^\mu, u^\mu, \gamma^\mu)$.

Our goal will be reached when we know the generalized distribution function $\hat{N}(x^\mu, u^\mu, \dot{\gamma}^\mu)$ on this phase space. I do not insist on the normalization of this distribution function: it just has to take the constraints

$$u^\mu u_\mu = 1$$

$$u^\mu \gamma_\mu = 0$$

into account.

At this stage it is worth to indicate a nice feature of the formalism when dealing with radiation quantities. Indeed, all these latter quantities are functions of $(x^\mu, u^\mu, \dot{\gamma}^\mu)$ so that the data of $\hat{N}$, allows us to compute their averages. For instance, the invariant power radiated is something like:

$$P \sim \gamma^\mu \gamma_\mu$$

while the radiation field emitted is something like:

$$F^{\mu\nu}_{\text{rad.}} \sim u^\mu \dot{\gamma}^\nu - u^\nu \dot{\gamma}^\mu .$$

Let us now write an equation for $\hat{N}$. We first write the continuity equation in our enlarged μ-space:

$$u^\mu \partial_\mu \hat{N} + \frac{\partial}{\partial u^\mu}[\gamma^\mu \hat{N}] + \frac{\partial}{\partial \dot{\gamma}^\mu}[\ddot{\gamma}^\mu \hat{N}] = 0.$$

Next, γ^μ and $\ddot{\gamma}^\mu$, expressed as functions of $(x^\mu, u^\mu, \dot{\gamma}^\mu)$ by using the equations of motion, are inserted in the continuity equation and an equation for $\hat{N}$ is obtained. Of course, since none of the equations:

$$\frac{\partial}{\partial u^\mu}\gamma^\mu = 0$$

or

$$\frac{\partial}{\partial \dot{\gamma}^\mu}\ddot{\gamma}^\mu = 0$$

is true, we have no Liouville theorem. However, this circumstance, although ennoying if we want to define equilibrium states, is by no means troublesome in order to obtain a non equilibrium description of the system.

Suppose now that we are interested in the more conventional distribution function:

$$N(x^\nu, u^\nu) = \int d_4\dot{\gamma}\; N(x^\nu, u^\nu, \dot{\gamma}^\nu).$$

Then, it is sufficient to integrate the equation verified by $\hat{N}$ over $\dot{\gamma}^\mu$ to obtain:

$$u^\mu\,\partial_\mu N + \frac{K^\mu}{m}\frac{\partial}{\partial u^\mu}N = -\tau_0\frac{\partial}{\partial u^\mu}\{\;\},$$

with

$$\{\;\} = [g^{\mu\nu} - u^\mu u^\nu]\int d_4\dot{\gamma}\;\dot{\gamma}_\nu\hat{N}.$$

The left handside of this equation is nothing but the usual one-particle Liouville equation while the right handside contains the effects of radiation as indicated by the occurence of τ_0. Note that τ_0 is actually a new parameter which occurs because a new constant, the velocity of light, is at our disposal. τ_0 characterizes the effects of radiation emission and hence will be used as an expansion parameter to obtain kinetic equations.

THE MODIFIED VLASOV EQUATION

Suppose now that the external force field K^μ is due to the *collective* electromagnetic field of the electrons of the plasma, i.e. suppose that:

$$K^\mu = eF^{\mu\nu}u_\nu$$

$$\partial_\mu F^{\mu\nu} = e\int d_4u\,u^\nu N,\quad \partial_\mu F^{\mu\nu *} = 0.$$

Then we get a self-consistent kinetic equation which is a modification of the usual relativistic Vlasov equation. Let us also consider only the first order in radiation effects, i.e. in τ_0; then we get the following kinetic equation:

$$u^{\mu}\partial_{\mu}N + \left(\frac{4\pi e^2}{m}\right)u_{\nu}\int d_4x'\,d_4u'[u'^{\nu}\partial^{\mu} - u'^{\mu}\partial^{\nu}]\,D_{\text{ret.}}(x^{\varrho} - x^{\varrho\prime}) \times N' \otimes \frac{\partial}{\partial u^{\mu}}N$$

$$= -\left(\frac{4\pi e^2\tau_0}{m}\right)\frac{\partial}{\partial u^{\mu}}\{[g^{\mu\nu} - u^{\mu}u^{\nu}]\,u^{\varrho}u^{\alpha}\} \times \int d_4x'\,d_4u'[u'_{\alpha}\,\partial_{\nu} - u'_{\nu}\,\partial_{\alpha}]$$

$$\times\, \partial_{\varrho}D_{\text{ret.}}(x_{\varrho} - x'_{\varrho}) \times N' \otimes N.$$

Of course, all this can be performed for a system of N electrons and a BBGKY hierarchy can be obtained, using Klimontovich method. As a consequence the same modified Vlasov equation is obtained, once we have assumed that correlations are negligible.

An interesting feature of this equation is that an H-theorem can be proved, which expresses nothing but the *irreversible* emission of radiation.

APPLICATIONS

Several applications of this equation have been studied—or are presently investigated. In particular we have studied the dispersion relations of waves propagating in the plasma. We studied two cases: (1) supraluminous waves, (2) suprathermal waves.

The longitudinal dispersion relation writes

$$\Omega_p^2 + k^2 = 2\omega_p^2\omega I_0 + \omega_p^2(\omega^2 - k^2)\frac{\partial}{\partial\omega}I_0 + i\tau_0\omega_p^2\frac{\omega}{k} \times$$

$$\times \left\{\omega + (\omega^2 - k^2)\frac{\partial}{\partial\alpha}I_0 + (\omega^2 - k^2)\left(G + \frac{3}{\alpha}\right)I_0\right\},$$

where $I_0(\alpha, \omega, k)$ is a complicate integral and where

$$\Omega_p^2 = \omega_p^2\frac{K_1(\alpha)}{K_2(\alpha)}, \quad \alpha = \frac{mc^2}{kT}.$$

The dispersion relation for transverse modes reads

$$\frac{\Omega_p^2 + k^2}{\omega^2 - k^2} = 2I_0\frac{\omega_p^2\omega}{\omega^2 - k^2} + \omega_p^2\frac{\partial}{\partial\omega}I_0 + i\tau_0\omega_p^2\omega\frac{\omega J_3^0 + kJ_0^0}{\omega^2 - k^2}$$

where J_3^0 and J_0^0 are integrals which can be expressed in terms of I_0. Essentially, we found that the "radiation term" leads to a damping of these waves: this is by no means surprising. This effect can also be seen on the form of the radiation term occurring in the modified Vlasov equation. Indeed, this term may by rewritten as:

$$\text{radiating term} = A^{\mu} \frac{\partial}{\partial u^{\mu}} N + \frac{\partial}{\partial u^{\mu}} A^{\mu} \cdot N.$$

The first term is a contribution to the 4-force while the second is of the form $B \cdot N$ and hence (after due analysis of the sign of B) is something like the simplification of a collision term and thus leads to a damping of the plasma waves.

In case (1) the transverse modes are negligibly damped, which is not the case for the longitudinal modes which are damped.

In case (2) there are no transverse modes and the longitudinal modes are damped; the radiation damping is of the order of the Landau damping in some cases.

A second kind of effects is a typically relativistic effect which consists of a plasma frequency shift. If we put numbers in the formula obtained for that shift, we can expect a change of plasma frequency of 10% for an electronic temperature of the order of 4×10^8 °K. However I do not think this effect to be observable.

Finally, it seems that the stream instability may under certain circumstances transform into a stream stability due to radiation damping.

References

R. Hakim and A. Mangeney, *J. Math. Phys.* **9**, 116 (1968).
R. Hakim and A. Mangeney, to appear in *Phys. Fluids.*
W. E. Brittin, *Phys. Rev.* **106**, 843 (1957).
Yu. L. Klimontovich, *Soviet Phys. J. E. T. P.* **34** (7), 119 (1958).
I. Prigogine and B. Leaf, *Physica* **25**, 1067 (1959).
A. Mangeney, *Ann. Phys.* **10**, 191 (1965).
A. O. Barut, *Electrodynamics and Classical Theoty of Fields and Particles* (McMillan Co., New York 1964).
F. Rohrlich, *Classical Charged Particles* (Addison-Wesley, Reaging 1965).
R. Hakim, *J. Math. Phys.* **8**, 1378 (1967).

PAPER A–6

The Boltzmann-Langevin equation

R. ZWANZIG, M. BIXON

University of Maryland
Institute for Fluid Dynamics
College Park, Maryland

Abstract

The one particle distribution function which satisfies the Boltzmann equation is interpreted as the average of the phase space function $\sum_{i=1}^{N} \delta(r - r_i)\,\delta(v - v_i)$. The equation of motion for this function is a generalized Langevin equation. This equation is the linear Boltzmann equation to which a fluctuating force term is added.

Expression for the second moment of this force in terms of the Boltzmann kernel and the equilibrium second moment of the distribution function is derived in analogy with the known procedure with the Langevin equation. The second moments of the fluctuating pressure tensor and the heat flow vector are evaluated by using the first Chapmann-Enskog approximation. They are equal to the expressions derived by Landau and Lifshitz using thermodynamic fluctuation theory in relation to the linearized hydrodynamic equations.

PART B

QUANTUM AND RELATIVISTICS

Contributors

W. B. Thompson
Department of Physics
University of California
San Diego, California

R. Résibois
Université Libre de Bruxelles
Bruxelles, Belgique

W. R. Chappell
Department of Physics and Astrophysics
University of Colorado
Joint Institute for Laboratory Astrophysics
Boulder, Colorado

J. Cooper
Department of Physics and Astrophysics
University of Colorado
Joint Institute for Laboratory Astrophysics
Boulder, Colorado

E. W. Smith
National Bureau of Standards
Boulder, Colorado

F. Henin
Faculté des Sciences
Université Libre de Bruxelles
Belgium

H. Smith
Laboratory of Atomic and Solid State Physics
Cornell University
Ithaca, New York

J. W. Wilkins
Laboratory of Atomic and Solid State Physics
Cornell University
Ithaca, New York

R. F. Snider
Department of Chemistry
University of British Columbia
Vancouver, B. C. Canada

PAPER B–1

The self-consistent test particle approach to relativistic kinetic theory

W. B. THOMPSON
Department of Physics
University of California, San Diego

INTRODUCTION

Experiments with relativistic electron beams and observations of extremely energetic plasmas in astronomical objects arouse some interest in the theoretical study of plasmas at relativistic temperatures. Such plasmas have been studied by use of the relativistic Vlasov Equation (Clemmow & Wilson, Buti, Rosenbluth & Watson), but usually these discussions have been couched in a three dimensional language. There may be some advantages in using a manifestly Lorentz invariant four-vector description and Synge has shown how kinetic theory may be so formulated. Using these methods for classical kinetic theory runs into the difficulty that the molecular interaction is represented by a potential, a non-relativistic object. For the plasma, however, the interaction is strictly electromagnetic, and if magnetic as well as Coulomb interactions are included, is Lorentz invariant.

Of course, plasma kinetic theory suffers from the long range nature of the particle interactions, and the consequent development of correlations in position and velocity. These, however, may conveniently be allowed for by using the dressed-particle approach (Thompson & Hubbard, Thompson, Rostoker) which as we shall show readily adapts itself to the relativistic case.

What follows is a preliminary and rather specialized treatment of the problem, which considers only a plasma free of any large electromagnetic fields, and which neglects important quantum phenomena such as the possible presence, creation and annihilation of antiparticles. Moreover, in order to derive a kinetic equation of the usual type, we have restricted our discussion to stable plasmas, in which the electromagnetic field remains small, and in which particles do not produce Cerenkov radiation. In order to include these effects, we would need to consider at least part of the field spectrum as a variable additional to the distribution function, as has been done by Rogister and Oberman. By avoiding these difficulties, we easily obtain a kinetic equation of a simple and elegant form; although there may be no interesting systems to which it can be applied. As a further artificiality, we will have to restrict ourselves to a single (neutralized) component.

FORMAL DEVELOPMENT

We will make use of a Lorentz invariant distribution function

$$f(x^\mu, u^\mu); \quad x^\mu = (ct, \mathbf{x}); \quad u^\mu = \frac{dx^\mu}{d\tau}$$

The distribution f is related to the local 4-flux of particles by

$$J^\mu = \int f(u)\, u^\mu d^4u = (nc, n\mathbf{v}) \tag{1.1}$$

We also find it useful to introduce a scalar $\tilde{n} = \int f d^4u$. The variables u^μ are not independent, but are related by the orthogonality condition $u^\mu u_\mu = c^2$; $u_\mu = g_{\mu\nu}u^\nu$

$$g = \begin{bmatrix} 1 & & & 0 \\ & -1 & & \\ & & -1 & \\ 0 & & & -1 \end{bmatrix}$$

the Lorentz metric. This relation is preserved by equations of motion and we will find it convenient to introduce this into the distribution function by a δ-function thus

$$f(u^\mu) = g(u_0, \mathbf{u})\, \delta(u^\mu u_\mu - c^2) \tag{1.2}$$

The relation between the value of f in any given rest frame and the associated classical distribution function F is determined by (1.1), thus

$$\int f u^0\, d^4u = nc = c \int F\, d^3v$$
$$\int f \mathbf{u}\, d^4u = n\mathbf{v} = \int F\mathbf{v}\, d^3v$$

There are several possible analogues to the Maxwell distribution. We shall fix on $f = K\exp[-S_\mu u^\mu]$. To determine the normalizing constant K, we note that in the rest frame of the vector S, $J^\mu = (nc, 0, 0, 0)$ and

$$n_0 c = K \int e^{-S_0 u_0} u_0\, \delta(u_0^2 - u^2 - c^2)\, dn_0\, d^3\mathbf{u}$$

$$= \frac{K}{2}\int d^3\mathbf{u}\, e^{-S_0\sqrt{u^2+c^2}} = 2\pi K \int du\, u^2 e^{-S_0\sqrt{u^2+c^2}} = 2\pi c^2 K\left[\frac{K_2(S_0 c)}{S_0 c}\right]$$

$$\therefore K = \frac{n_0}{2\pi c^2}\frac{S_0 c^2}{K_2(S_0 c)} : K_2 \text{ being a Bessel function.}$$

S_0 can be related to the temperature T as $S_0 c = mc^2/kT$ and

$$f_M = \frac{n_0 S_0}{2\pi K_2(S_0 c)} e^{-S_\mu u^\mu}\, \delta(u^\mu u_\mu - c^2) \tag{1.3}$$

Moreover

$$\tilde{n} = \int f\, d^4u = n_0 \frac{K_1(S_0 c)}{K_2(S_0 c)}$$

Note that in this form of the Maxwell distribution we are restricted to positive time like S, and u. In general, we might use negative u_0 to describe antiparticles in the same general formalism.

The Vlasov equation

Since the equations of motion can be written in Hamiltonian form, $f(x, p)$ is a constant of the motion, $\left[\frac{d}{d\tau}(dx^\mu\, dp^\mu) = 0\right]$. A direct calculation of the Jacobian between x, p and x, u, shows that $f(x^\mu, u^\mu)$ is also a constant. In external fields, $F_{\mu\nu}$ we can then write the Vlasov equation for particles of mass m, and charge e as

$$u^\mu \frac{\partial f}{\partial x^\mu} + \frac{e}{mc} u^\nu F_\nu^\sigma \frac{\partial f}{\partial u^\sigma} = 0 \tag{1.4}$$

It is probably advisable to express the fields in terms of the potentials, A_ν thus $F_\nu^\sigma = g^{\sigma\mu}(A_{\mu,\nu} - A_{\nu,\mu})$.

At this point we make the assumption that the fields are weak, and can be treated by perturbation theory. This immediately casts out any zero order electric or magnetic field, an enormous simplification (and an enormous limitation). Further, we assume that on the scale of interest, the plasma is initially uniform, except as perturbed by the fields. We then may write

$f = f_0(u^\mu) + \delta f(x^\mu, u^\mu)$ and consider the potentials A and the distribution function δf as Fourier analysed.

$$A, \delta f \sim A(k), f(k)\, e^{ik_\mu x^\mu}$$

(1.4) then yields δf as

$$ik_\mu u^\mu \, \delta f = \frac{e}{mc} i[k_\mu A_\nu - k_\nu A_\mu]\, g^{\mu\sigma} u^\nu \, \partial f^0/\partial u^\sigma \tag{1.5}$$

From this we may construct the 4-current [at least, the contribution due to a single species!] as

$$\begin{aligned} j^\lambda &= \frac{e^2}{mc^2}\int d^4u \frac{[k_\mu A_\nu - k_\nu A_\mu]\, u^\lambda u^\nu}{k_\lambda u^\lambda} g^{\sigma\mu} \frac{\partial f}{\partial u^\sigma} \\ &= \frac{e^2}{mc^2}\int d^4u \left\{ \frac{u^\lambda u^\nu}{k_\lambda u^\lambda} k^\sigma \frac{\partial f}{\partial u^\sigma} A_\nu - u^\lambda A^\sigma \frac{\partial f}{\partial u^\sigma} \right\} \\ &= \frac{e^2}{mc^2}\int d^4u\, f A^\lambda + \frac{e^2}{mc^2}\int d^4u \frac{u^\lambda u^\nu}{k_\lambda u^\lambda} k^\sigma \frac{\partial f}{\partial u^\sigma} A_\nu \end{aligned} \tag{1.6}$$

Maxwell's equation have the form

$$k^\mu k_\mu A^\lambda = 4\pi j^\lambda, \tag{1.7}$$

provided that A satisfies the Lorentz condition

$$k_\lambda A^\lambda = 0$$

$$\therefore \left[k^\mu k_\mu - \frac{4\pi e^2}{mc^2} \tilde{n} \right] A^\lambda - \frac{4\pi e^2}{mc^2} \int d^4u \frac{u^\lambda u^\nu}{k_\lambda u^\lambda} k^\sigma \frac{\partial f}{\partial u^\sigma} g_{\nu\tau} A^\tau = 4\pi j^\lambda_\varphi \tag{1.8}$$

In the remaining integral occurring here, the singularity may be handled in the usual Landau way [$k_0 = \lim_{\varepsilon \to 0} k_0 - i\varepsilon$, for a causal solution!], and leads to the usual Landau complex part (Buti, Romig).

$$\frac{4\pi e^2}{mc^2} \int d^4u\, u^\lambda u_\tau k^\sigma \frac{\partial f}{\partial u^\sigma} i\pi\, \delta(k_\nu u^\nu)\, A^\tau \tag{1.9}$$

The set of algebraic eqs. (1.7) can clearly be inverted to yield

$$A^\lambda = 4\pi T^\lambda_{\ \sigma} j^\sigma \tag{1.10}$$

provided T^{-1} is non-singular.

For the special case of an isotropic distribution important simplifications occur. Suppose there is a frame in which $f(u)$ takes the special form $f(u^0, u_k^2)$, then, that frame can be rotated so that

$$k^\mu = (k^0, k^1, 0, 0) = (\omega/c, k^1, 0, 0)$$

The components of Maxwells equations then become

$$\left[\frac{\omega^2}{c^2} - k^2 - \frac{4\pi e^2}{mc^2}\tilde{n}\right]$$

$$\times \begin{Bmatrix} A_0 \\ A_1 \\ A_2 \\ A_3 \end{Bmatrix} - \frac{4\pi e^2}{mc^2}\int d^4u k^\sigma \frac{\dfrac{\partial f}{\partial u^\sigma}}{k_\lambda u^\lambda} \begin{Bmatrix} u_0[u_0A^0 + u_1A^1] \\ u^1[u_0A^0 + u_2A^2] \\ -(u_2)^2 A^2 \\ -(u_3)^2 A^3 \end{Bmatrix} = 4\pi \begin{Bmatrix} j_0 \\ j^1 \\ j^2 \\ j^3 \end{Bmatrix}$$

The two perpendicular components A^2 and A^3 clearly separate, yielding

$$\left\{\left[k^\mu k_\mu - \frac{4\pi e^2}{mc^2}\tilde{n}\right] + \frac{4\pi e^2}{mc^2}\int d^4u k^\sigma \frac{\dfrac{\partial f}{\partial u^\sigma}}{k_\lambda u^\lambda}\frac{1}{2}u_\perp^2\right\} A^{2,3} = 4\pi j^{2,3} \tag{1.11}$$

To handle the remaining two, which are coupled by the gauge condition and the equation of continuity, it is advantageous to shift to the Coulomb gauge. This leaves the components of $A_\perp$ unchanged, but replaces A^0 and A^1 by

$$\Phi = A^0 - \frac{k^0}{k^1}A^1$$

Note then, that

$$k^\mu k_\mu A^0 = -k^2\Phi$$

and

$$-\frac{4\pi e^2}{mc^2}\int d^4u\left[f + \frac{k^\sigma \dfrac{\partial f}{\partial u^\sigma}}{k^\lambda u_\lambda}u^0u_0\right]A^0 + \left[\frac{k^\sigma \dfrac{\partial f}{\partial u^\sigma}}{k^\lambda u_\lambda}u^0u_1A^1\right]$$

$$= -\frac{4\pi e^2}{mc^2}\int d^4u\left[f + \frac{k^\sigma \dfrac{\partial f}{\partial u^\sigma}}{k^\lambda u_\lambda}u^0u_0\right]\left[A^0 - \frac{k^0}{k^1}A^1\right]$$

$$-\frac{4\pi e^2}{mc^2}\left\{\int d^4uf\frac{k^0}{k^1}A^1 + \frac{k^\sigma \dfrac{\partial f}{\partial u^\sigma}}{k^\lambda u_\lambda}\left[u^0u_0\frac{k^0}{k^1}A^1 + u^0u_1A^1\right]\right.$$

$$\left. = \int d^4uf\frac{k^0}{k^1}A^1 + k^\sigma\frac{\partial f}{\partial u^\sigma}\frac{u^0}{k^1}\left[\frac{k^\lambda u_\lambda}{k^\lambda u_\lambda}\right]A^1 = \int d^4uf\frac{k^0}{k^1}A^1 - \frac{k^0}{k^1}A^1 = 0\right\}$$

where a partial integration has been used.

In the Coulomb gauge we find thus

$$-k^2\Phi - \frac{4\pi e^2}{mc^2}\tilde{n}\Phi - \frac{4\pi e^2}{mc^2}\int \frac{k^\sigma \dfrac{\partial f}{\partial u^\sigma}}{k^\lambda u_\lambda} u^0 u_0 \Phi = 4\pi j^0 = 4\pi g \tag{1.12}$$

which is again diagonal.

An alternative procedure, is to retain the Lorentz gauge and use the Lorentz condition to eliminate the unwanted component.

$$\begin{aligned} u^0 u_0 A^0 + u^0 u_1 A^1 &= \left[u^0 u_0 - u_1 u^0 \frac{k_0}{k^1}\right] A^0 \\ &= (u^0 u_0 + u^1 u_1)\, A^0 - \frac{u_1}{k_1}(k^\lambda u_\lambda)\, A^0 \end{aligned} \tag{1.13}$$

$$\therefore \left[k^\mu k_\mu - \frac{4\pi e^2}{mc^2}\tilde{n}\right] A^0 - \frac{4\pi e^2}{mc^2}\int du^4 \frac{k^\sigma \dfrac{\partial f}{\partial u^\sigma}}{k^\lambda u_\lambda} u^0 [u_0 A^0 + u_1 A^1]$$

$$= \left[k^\mu k_\mu - \frac{8\pi e^2}{mc^2}\tilde{n}\right] A^0 - \frac{4\pi e^2}{mc^2}\int du^4 \frac{k^\sigma \dfrac{\partial f}{\partial u^\sigma}}{k^\lambda u_\lambda} [u^0 u_0 + u^1 u_1]\, A^0 = 4\pi j^0$$

The test particle equation

We turn next to a consideration of the effect on the distribution function of the discreteness of the particles. This will necessarily induce fluctuations in the distribution. Indeed, since the fine grained distribution function vanishes except on the world lines of particles, it must have the value

$$f = \sum_i \int_{-\infty}^{\infty} d\tau\, \delta(u^\mu - u_i^\mu)\, \delta(x^\lambda - x_i^\lambda(\tau)) \tag{1.14}$$

If there are no external forces, and we consider particle interactions as switched off, this may be Fourier analysed to yield

$$\delta f_k^0 = 2\pi \sum e^{ik_\lambda x_i{}^\lambda}\, \delta(k_\mu u^\mu)\, \delta(u^\mu - u_i^\mu) \tag{1.15}$$

To treat the particle interaction we observe that these fluctuations yield a current of the form

$$j_k^\mu = 2\pi \sum e \frac{u_i^\mu}{c} e^{ik_\lambda x_i{}^\lambda}\, \delta(k_\mu u^\mu) = \sum \frac{e}{c} u_i^\mu\, \delta f \tag{1.16}$$

The fluctuating current will produce in turn, a fluctuating field of the form (using 1.10)

$$A_k^\lambda = 8\pi^2 \sum \frac{e}{c} T_\mu^\lambda u_i^\mu e^{ik_\lambda x_i^\lambda} \delta(k_\mu u^\mu) = 4\pi T_\mu^\lambda \sum \frac{e}{c} u^\mu \, \delta f_k \tag{1.17}$$

and a further contribution to the fluctuations in f

$$\delta f_k^1 = \frac{e}{mc} \frac{[k_\mu A_\nu - k_\nu A_\mu]}{k_\tau u^\tau} g^{\mu\sigma} u^\nu \frac{\partial f}{\partial u^\sigma} \tag{1.18}$$

Now we may write down an expression determining the averaged distribution function, including the effects of the fluctuations thus

$$u^\mu \frac{\partial f}{\partial x^\mu} + \left\langle \frac{e}{mc} u^\nu \, \delta F_\nu^\sigma \frac{\partial}{\partial u^\sigma} \delta f \right\rangle = 0 \tag{1.19}$$

The bracketted term here may be written

$$\left\langle \frac{1}{(2\pi)^4} \int d^4k \frac{e}{mc} u^\nu \, \delta F_\nu^\sigma(-k) \frac{\partial}{\partial u^\sigma} \delta f(k) \right\rangle$$

For a single species (neutralized) plasma

$$\begin{aligned} \langle \; \rangle &= \left\langle \frac{1}{(2\pi)^4} \int d^4k \frac{e}{mc} u^\nu \, \delta F_\nu^\sigma(-k) \frac{\partial}{\partial u^\sigma} \left[\delta f^0(k) - \frac{e}{mc} \frac{\delta F_\lambda^\sigma u^\lambda}{k^\tau u_\tau} \frac{\partial f}{\partial u^\sigma} \right] \right. \\ &= \left\langle -\frac{4\pi}{(2\pi)^4} \frac{\partial}{\partial u^\sigma} \int d^4k u^\nu g^{\sigma\mu} \, i[k_\nu T_{\mu\lambda} - k_\mu T_{\nu\lambda}] j^\lambda \right. \\ &\quad \times \left\{ \delta f^0(k) + \frac{4\pi e}{mc} g^{\alpha\beta} u^\gamma \frac{[k_\gamma T_{\beta\delta} - k_\beta T_{\gamma\delta}]}{k^\tau u_\tau} \delta j^\delta(k) \frac{\partial}{\partial u^\alpha} f_0 \right\} \end{aligned} \tag{1.20}$$

Now, we use the following results

$$\left. \begin{aligned} \langle j^\lambda \, \delta f^o \rangle &= \frac{e}{c} u^\lambda f(u) \, \delta(k^\tau u_\tau) \\ \text{and} \qquad & \\ \langle j^\lambda j^\gamma \rangle &= \frac{e^2}{c^2} \int d^4u \, u^\lambda u^\delta f(u) \, \delta(k^\tau u_\tau) \end{aligned} \right\} \tag{1.21}$$

Further observe that

$$T_{\lambda\nu} u^\nu u^\lambda k^\sigma \frac{\partial}{\partial u^\sigma} = k^\sigma \frac{\partial}{\partial u^\sigma} [T_{\nu\lambda} u^\nu u^\lambda]$$

and that $T(-k) = T^*(k)$. The bracketted expression then becomes

$$-\left\langle \frac{4\pi}{(2\pi)^4} \int d^4k \frac{4\pi e^2}{mc^2} k^\sigma \frac{\partial}{\partial u^\sigma} \left\{ \operatorname{Im} T^*_{\nu\lambda} u^\nu u^\lambda \, \delta(k^\tau u_\tau) f(u) \right. \right.$$

$$\left. - \frac{4\pi^2 e^2}{mc^2} k^\alpha T^*_{\nu\lambda} T_{\gamma\delta} u^\nu u^\gamma \frac{\partial f}{\partial u^\alpha} \delta(k^\tau u_\tau) \int d^4\bar{u}\, f(\bar{u})\, \bar{u}^\lambda \bar{u}^\delta\, \delta(k^\tau \bar{u}_\tau) \right\} \tag{1.22}$$

Finally we observe that

$$T^*_{\nu\lambda} = T_{\nu\gamma} \tilde{T}^{\gamma\delta} T^*_{\delta\lambda}$$

Where $\tilde{T} = T^{-1}$, and that

$$\operatorname{Im} \tilde{T}^{\gamma\delta} = \frac{4\pi e^2}{mc^2} \int d^4u\, \delta(k_\tau u^\tau)\, k^\sigma \frac{\partial f}{\partial u^\sigma} \tag{1.23}$$

hence, that (1.19) becomes

$$u^\mu \frac{\partial f^0}{\partial x^\mu} + \frac{4e^4}{m^2c^4} \int d^4k\, k^\alpha k^\sigma \frac{\partial}{\partial u^\sigma} \int d^4\bar{u}\, T^*_{\nu\lambda} T_{\gamma\delta}\, u^\nu u^\gamma \bar{u}^\lambda \bar{u}^\delta$$

$$\times\, \delta(k^\tau u_\tau)\, \delta(k^\tau \bar{u}_\tau) \left[\frac{\partial f(\bar{u})}{\partial \bar{u}^\alpha} f(u) - f(\bar{u}) \frac{\partial f(u)}{\partial u^\alpha} \right] = 0 \tag{1.24}$$

or

$$u^\mu \frac{\partial f}{\partial x^\mu} + \frac{\partial}{\partial u^\alpha} \int d^4\bar{u}\, D^{\alpha\beta} \left[f(u) \frac{\partial f(\bar{u})}{\partial \bar{u}^\beta} - f(\bar{u}) \frac{\partial f(u)}{\partial u^\beta} \right] \tag{1.25}$$

where

$$D^{\alpha\beta} = \frac{4e^4}{m^2c^4} \int d^4k\, T^*_{\nu\lambda} T_{\gamma\delta}\, u^\nu u^\gamma \bar{u}^\lambda \bar{u}^\delta\, \delta(k^\tau \bar{u}_\tau)\, \delta(k^\tau u_\tau) \tag{1.26}$$

The coefficient of $k^\alpha k^\beta$, in this expression has a simple physical interpretation. Consider the contribution due to a particle at $\bar{x}$, having velocity $\bar{u}$ to the field at x. This is $A_\gamma(x) = \dfrac{e}{c} T_{\gamma\delta}\, \bar{u}^\delta\, e^{ik^\tau(x-\bar{x})_\tau}$. The scalar product of this with the current at x due to a particle with velocity u is $\left(\dfrac{e}{c}\right)^2 u^\gamma A_\gamma$. The product of this and its complex conjugate $\left(\dfrac{e}{c}\right)^2 u^\gamma A_\gamma u^\nu A^*_\nu = \left(\dfrac{e}{c}\right)^4 u^\gamma u^\nu T_{\gamma\delta} \times T^*_{\nu\lambda}\, \bar{u}^\delta \bar{u}^\lambda$ is necessarily real and positive, and $4m^2$ times the coefficient of $k^\alpha k^\beta$ in D, and 1.26 may be written

$$D^{\alpha\beta} = \frac{4}{m^2} \int d^4k\, k^\alpha k^\beta\, |\delta(u)\, A(\bar{u})|^2\, \delta(k \cdot u)\, \delta(k \cdot \bar{u})$$

Formal properties of the kinetic equation

We find it useful to exploit the following symmetry property, for any F

$$\int F(u, \bar{u})\left[f(u)\frac{\partial f(\bar{u})}{\partial \bar{u}} - f(\bar{u})\frac{\partial f}{\partial \bar{u}}\right] d^4u\, d^4\bar{u}$$

$$= -\int F(\bar{u}, u)\left[f(u)\frac{\partial f}{\partial \bar{u}} - f(\bar{u})\frac{\partial f}{\partial u}\right] d^4u\, d^4\bar{u}$$

$$= \frac{1}{2}\int [F(u, \bar{u}) - F(\bar{u}, u)]\left[f(u)\frac{\partial f}{\partial \bar{u}} - f(\bar{u})\frac{\partial f}{\partial u}\right] \tag{1.27}$$

Further $D^{\beta\alpha}$ is a symmetric function of $u, \bar{u}$, and $D^{\alpha\beta} = \int d^4k\, R\, k^\alpha k^\beta$, R being positive definite.

We now display the basic properties of our kinetic equation

i) It conserves matter $J^\mu_{,\mu} = 0$

$$\int u^\mu \frac{\partial f}{\partial x^\mu} = J^\mu_{,\mu} = -\int \frac{\partial}{\partial u^\mu} D^{\alpha\beta}[\ ] = 0$$

ii) It conserves the energy momentum tensor $T^{\nu\mu}_{,\mu} = 0$

$$T^{\nu\mu} = m_0 \int u^\nu u^\mu f$$

For

$$\int u^\nu u^\mu \frac{\partial f}{\partial x^\mu} = \frac{1}{m_0} T^{\nu\mu}_{,\mu} = \int u^\nu \frac{\partial}{\partial u^\alpha} D^{\alpha\beta}[\ ] = -\int D^{\nu\beta}\left[f\frac{\partial \bar{f}}{\partial \bar{u}} - \bar{f}\frac{\partial f}{\partial u}\right] = 0$$

because of the symmetry of $D_{\alpha\beta}$ in $u, \bar{u}$.

iii) If f is initially everywhere positive it remains positive, (A. Lenard). If f becomes negative, then it must first be tangent to the $f = 0$ plane, in u space, hence $f = 0$, $\dfrac{\partial f}{\partial u} = 0$ and $\dfrac{\partial^2 f}{\partial u^\alpha \partial u^\beta}$ must be a non-negative quadratic form and at this point $\dfrac{df}{d\tau} < 0$

$$\frac{\partial f}{\partial \tau} = \int d^4\bar{u}\, f(\bar{u})\, D^{\alpha\beta} \frac{\partial^2 f}{\partial u^\alpha \partial u^\beta}$$

In the coordinate system that diagonalizes $\dfrac{\partial^2 f}{\partial u^\alpha \partial u^\beta}$ $(= \Lambda^\alpha \delta^{\alpha\beta})$

this becomes $\int d^4\bar{u} f(\bar{u}) \int d^4k\, R(k)\, k^\alpha k^\beta \Lambda^\alpha \delta_{\alpha\beta} \geqq 0$

since $\Lambda \geqq 0$

iv) An H theorem holds.

If we define

$$H^\alpha = \int u^\alpha f \log f\, d^4u$$

then

$$H^\alpha_{,\alpha} \lesssim 0$$

for

$$\frac{\partial}{\partial x^\alpha} \int u^\alpha f \log f = \int [1 + \log f]\, u^\alpha \frac{\partial f}{\partial x^\alpha}$$

$$= -\int [1 + \log f] \frac{\partial}{\partial u^\alpha} D^{\alpha\beta} [\;]$$

$$= \int \frac{1}{f} \frac{\partial f}{\partial u^\alpha} D^{\alpha\beta} [\;]$$

$$= \frac{1}{2} \int \left[\frac{1}{f} \frac{\partial f}{\partial u^\alpha} - \frac{1}{\bar{f}} \frac{\partial \bar{f}}{\partial \bar{u}^\alpha} \right] D^{\alpha\beta} \left[f \frac{\partial \bar{f}}{\partial \bar{u}^\beta} - \bar{f} \frac{\partial f}{\partial u^\beta} \right]$$

$$= -\frac{1}{2} \int du^4 \int d\bar{u}^4 \int d^4k \frac{R}{f\bar{f}} k^\alpha \left[f \frac{\partial \bar{f}}{\partial \bar{u}^\alpha} - \bar{f} \frac{\partial f}{\partial u^\alpha} \right] k^\beta \left[f \frac{\partial \bar{f}}{\partial \bar{u}^\beta} - \bar{f} \frac{\partial f}{\partial u^\beta} \right]$$

$$= -\frac{1}{2} \int du^4 \int d\bar{u}^4 \int d^4k \frac{R}{f\bar{f}} \left\{ k^\alpha \left[f \frac{\partial \bar{f}}{\partial \bar{u}^\alpha} - \bar{f} \frac{\partial f}{\partial u^\alpha} \right] \right\}^2 \leqq 0$$

v) $\dfrac{\partial H^\alpha}{\partial x^\alpha} = 0$ for the "Maxwellian" distribution, $f_\mu = -Ke^{-S_\mu u^\mu}$

for

$$\frac{\partial f}{\partial u^\alpha} = -S_\alpha f$$

hence

$$\left[\bar{f} \frac{\partial f}{\partial u^\alpha} - f \frac{\partial \bar{f}}{\partial \bar{u}^\alpha} \right] = -S_\alpha [\bar{f} f - f \bar{f}] = 0$$

CONCLUSIONS

The dressed particle approach leads rather simply to a Lorentz invariant kinetic equation, which satisfies all the conditions that we expect of a kinetic equation. It is limited, however, in that it does not take into account undamped radiation. Indeed, the derivation remains valid only if the coarse graining assumption holds good; hence radiation from any electron must be damped within a mean free path, and oscillation must die out within a mean free time. This limitation does not affect calculation of bremsstrahlung losses, which depend on e^6, hence is of higher order than our kinetic equation, but if T^{-1} is singular for any k satisfying $k.\,u = 0$, as may happen for an anisotropic velocity distribution, then particles can produce long range Cerenkov radiation, which would invalidate the coarse graining assumption. At the same time, the plasma would become at best marginally stable, and a different procedure required for attacking Kinetic Theory of the sort developed by Rogister and Oberman. If there is radiation in a relativistic plasma, however, pair production could occur and the particle number would no longer be constant. This introduces further complications into kinetic theory, and probably requires a quantum treatment. Since synchrotron radiation occurs in the same order as particle collisios, these complications are necessarily always present for a relativistic magnetoplasma.

References

B. Buti, *Phys. Fluids* **5**, 1 (1962).
P. C. Clemmow and A. Wilson, *Proc. Roy Soc. A*, **237**, 117 (1956).
A. Lenard, *Ann. Phys.* **3**, 390 (1960).
A. Rogister and C. Oberman, *J. Plasma Physics* **2**, 33 (1968).
J. Romig, M. A. Thesis: Oxford University 1962.
N. Rostoker, *Nuclear Fusion* **1**, 101 (1961).
W. B. Thompson and J. Hubbard, *Rev. Mod. Phys.* **32**, 714 (1960).
W. B. Thompson, in "*Advanced Plasma Theory*", Academic Press (1964).
K. M. Watson, S. A. Bludman, and M. N. Rosenbluth, *Phys. Fluids.* **3**, 741.
J. L. Synge, *The Relativistic Gas*, Interscience 1957.

PAPER B–2

Kinetic equations in spin systems

P. RÉSIBOIS
Université Libre de Bruxelles
Bruxelles, Belgique

1 THE PROBLEM

In this talk, we want to discuss some recent work, mostly done in collaboration with M. De Leener, on the dynamics of many spin systems[1]. As we shall be interested mainly in isotropic systems above their critical temperature, we shall discuss the following spin correlation function:

$$\tilde{\Gamma}_{ab}(t) = \langle \mathbf{S}_a(t) \cdot \mathbf{S}_b(0) \rangle \tag{1}$$

where $\mathbf{S}_a(t)$ is the Heisenberg representation of spin operator $\mathbf{S}$ at lattice site $\mathbf{a}$ and the bracket denotes an average over the canonical equilibrium distribution.

Moreover, for definitess, we take a Heisenberg spin system with Hamiltonian:

$$H = -\sum_{i,j} J_{ij} \mathbf{S}_i \cdot \mathbf{S}_j \tag{2}$$

and spin $|s| = \frac{1}{2}$; extensions to more complicated cases are discussed in Ref. 2 and 3.

The autocorrelation (1) is of great interest to both the experimentalist and the theoretician.

a) from the experimental point of view, inelastic neutron scattering experiments on magnetic substances give a direct measure of the double Fourier transform of (1). More precisely, if $\sigma_q(\omega)$ denotes the neutron inelastic cross-section for energy transfer ω and momentum transfer q, we have:

$$\sigma_q(\omega) \propto Re\,\tilde{\Gamma}_q(\omega) \tag{3}$$

$$\tilde{\Gamma}_q(\omega) = \int_{-\infty}^{+\infty} dt \exp[-i\omega t]\tilde{\Gamma}_q(t) \tag{4}$$

$$\tilde{\Gamma}_q(t) = \sum_b \exp[iq(ab)]\tilde{\Gamma}_{ab}(t) \tag{5}$$

Similarly, the output signal in N.M.R. experiments is often directly related to quantities analogous to (1);

b) from the theoretical point of view, the analysis of (2) offers a challenging difficulty. Indeed, the usual perturbation methods developed for time-dependent phenomena fail because, as is clear from (2), there is no umperturbed Hamiltonian compared to which the interaction between the spins would be small! The traditional method to overcome this difficulty is the moment analysis[4]: one calculates the so-called moments:

$$\langle\omega^{2n}\rangle_q = \int_{-\infty}^{+\infty} d\omega\, \omega^{2n}\tilde{\Gamma}_q(\omega) = (-1)^n \frac{d^{2n}}{dt^{2n}} \tilde{\Gamma}_q(t)\bigg|_{t=0} \tag{6}$$

for n small ($n = 2, 4$) and these short-time results are used by ad-hoc extrapolation procedures to guess the behavior of $\tilde{\Gamma}_q$ for long times (or low frequencies).

Although this method often gives reasonable answers, there are known cases where it definitely fails, even qualitatively. For instance, it predicts the following behavior for the diffusion coefficient of magnetization, D_q, in a ferromagnet close to the Curie point:

$$D_q q^2 \underset{\substack{q\to 0\\ T\to T_c}}{\propto} q^2(q^2 + \varkappa^2) \tag{7}$$

(here k is the inverse correlation lenght), a result which is not observed experimentally.

2 THE METHOD

Using a suitable projection operator technique (see for instance Ref. 5), there is no difficulty in deriving a formal kinetic equation for $\Gamma_q(t)$ in the form*:

$$\partial_t \Gamma_q(t) = \int_0^t G_q(\tau)\, \Gamma_q(t-\tau)\, d\tau \tag{8}$$

---

* From now on, we find it convenient to consider the quantity $\Gamma_q(t) = \tilde{\Gamma}_q(t)/\tilde{\Gamma}_b(0)$, in such a way that $\Gamma_q(0) = 1$, eliminating thereby the initial equilibrium correlations.

where $G_q(\tau)$ is a non-Markoffian kernel, which can be *formally* defined in terms of H.

However, the whole problem remains of finding a suitable *explicit* approximation for this kernel $G_q(\tau)$; otherwise Eq. (8) remains useless, being little more than a definition for $G_q(\tau)$!

In order to understand the difficulties connected with the spin problem let us naïvely expand $G_q(\tau)$ in power of the exchange integral J. The lowest order non vanishing contribution is the Born approximation; at infinite temperature, an explicit calculation leads to:

$$G_q^{(2)} \propto \int d\mathbf{q}_1 (J_{q+q_1} - J_{q_1})\, \tau_{q_1} \tag{9}$$

a time-independent quantity!

This result is very different from the well-known Born approximation for a weakly interacting quantum gas, where the transition probability analogous to (9) behaves like[6]:

$$G_{\text{gas}}^{(2)} \propto \int d\mathbf{q}_1\, |V_{q_1}|^2 \exp i(\varepsilon_{k+q_1} + \varepsilon_{k'-q_1} - \varepsilon_k - \varepsilon_{k'})\, \tau \tag{10}$$

In this formula, which describes the scattering of two particles with incident energies ε_k and $\varepsilon_{k'}$, one recognizes the familiar free-particle oscillating factors, expressing the propagation of the particles between the two interactions V_{q_1}. These factors play an essential role in Eq. (10) because after the q_1-integration is performed, they lead to a kernel $G_{\text{gas}}^{(2)}(\tau)$ which decreases in time, with a cut-off at the characteristic collision time τ_c (see Fig. 1a).

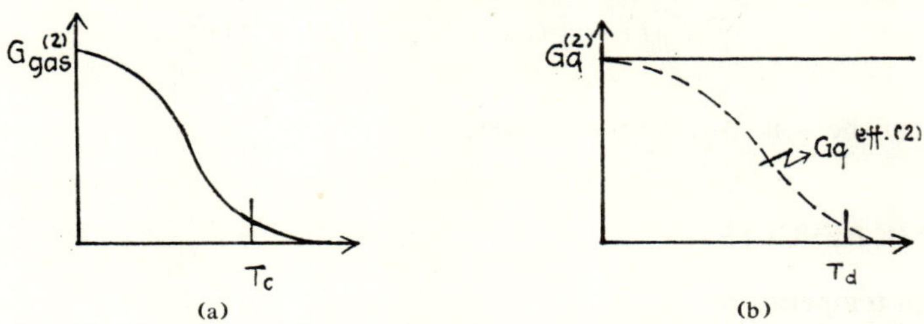

FIG. 1 Born approximation to the kernel: (a) Interacting gas (b) Spins.

This contrasts sharply with the spin case (Fig. 1b) where the "collision" between two spins has an arbitrary long duration: this merely expresses that spins are fixed on the lattice and have thus no free-propagation.

However it is clear that, if the "collision" between two given spins (a) and (b) is arbitrarily long, we have no right to consider spins (a) and (b) as isolated. If τ is large, these spins will also interact with the "bath" made of the remaining spins; after a certain characteristic time τ_d, the magnetization initially put on these spins will have diffused over the whole bath and the interaction between (a) and (b), although always present, will become ineffective: we expect thus that, by taking the "bath" into account, it will be possible to define an *effective* transition probability which will decay to zero after a certain time τ_d. As the function which describes the propagation of the magnetization is precisely the correlation function $\Gamma_q(\tau)$ itself, it is not surprising that this effective transition probability takes the form (see 9)

$$G_q^{\text{eff}(2)} \propto \int d\mathbf{q}_1 (J_{q+q_1} - J_{q_1})\, J_{q_1} \Gamma_{q+q_1}(\tau)\, \Gamma_{-q_1}(\tau) \tag{11}$$

Notice the quadratic dependence of Eq. (11) on $\Gamma_{q'}$, which appears because the magnetization diffuses from each of the spins (a) and (b).

The argument developed here qualitatively can be formalized in theorems, which are valid in the *Weiss limit* where the number of neighbours Z go to infinity. In this limit, it has been shown that the whole perturbation series defining $G_q(\tau)$ can be rewritten as a non-linear functional of $\Gamma_{q'}$, itself:

$$G_q(\tau) \underset{Z\to\infty}{\equiv} \tilde{G}_q(\tau \mid \{\Gamma_{q'}\}) \tag{12}$$

The simplest approximation to (12) leads again to Eq. (11), but higher order corrections can also be systematically computed. When introduced in Eq. (8), the kernel (12) gives a closed non-linear kinetic equation for $\Gamma_q(t)$:

$$\partial_t \Gamma_q(t) = \int_0^t \tilde{G}_q(\tau \mid \{\Gamma_{q'}\})\, \Gamma_q(t-\tau)\, d\tau \tag{13}$$

which can be solved in various cases.

3 SOME RESULTS

a High temperature

i) *Large q*: At high temperature, it is expected that, for large q, $\Gamma_q(t)$ is not very sensitive to the exact q value. In particular, it is reasonable to replace in the kernel (12) $\Gamma_q(t)$ by its average value:

$$\Gamma_q(t) \simeq \frac{1}{N} \sum_q \Gamma_q(t) = \Gamma(t) \tag{14}$$

where $\Gamma(t)$ denotes the *direct* correlation function. We get thus:

$$\tilde{G}_q(\tau \mid \{\Gamma_{q.}\}) \simeq \tilde{G}_q(\tau \mid \Gamma) \tag{15}$$

Using (15) and taking the average of (13) over all q values, it is then a simple matter to obtain a closed equation for the direct correlation $\Gamma(t)$:

$$\partial_t \Gamma(t) = \int_0^t \tilde{G}(\tau \mid \Gamma) \Gamma(t-\tau)\, d\tau \tag{16}$$

In the simplest approximation, one gets:

$$\tilde{G}(\tau \mid \Gamma) \simeq \tilde{G}^{(2)}(\tau \mid \Gamma) = -2ZJ^2\Gamma^2(\tau) \tag{17}$$

and the solution of (16), (17) can be obtained either numerically or by approximate analytical methods. One shows that $\Gamma(t)$ starts like a Gaussian and approaches then its equilibrium value through damped oscillations with a characteristic time $\tau_d = \sqrt{2Z}\,J$. In the case of an isotropic Heisenberg Hamiltonian, these oscillations are rather weak and seem to be suppressed when higher order correction are considered.

However, an analysis similar to the one given here shows that such oscillations play an important role in the interpretation of some N.M.R. experiments, as for instance the transverse relaxation in CaF_2 studied first experimentally by Lowe and Norberg[7]: the details can be found in References 2.

ii) *Small q*: For small q, it is easily verified that the approximation leading to (15) remains valid, and the kernel $\tilde{G}_q(\tau \mid \{\Gamma_{q'}\})$ decays thus to zero after times of the order τ_d. On the other hand, from (11) it is easily verified that:

$$\frac{\partial_t \Gamma_q}{\Gamma_q} \propto \frac{q^2}{\tau_d} \tag{18}$$

and the correlation $\Gamma_q(t)$ decays thus very slowly to zero for small q. We get in this case a clearcut separation between the characteristic time scales of Γ_q and of $\tilde{G}_q$, which allows us to replace Eq. (8) by a Markoffian diffusion equation:

$$\partial t\, \Gamma_q(t) = Dq^2\Gamma_q(t) \tag{19}$$

with

$$D = \frac{\partial^2}{\partial q^2} \int_0^\infty G_q(\tau \mid \Gamma)\, d\tau \tag{20}$$

From the knowledge of $\Gamma(\tau)$ obtained from (16), one can also get an explicit expression for the diffusion coefficient D, which will not be reprodduced here.

b Critical temperature

The above arguments have been generalized to finite temperature. Without entering a detailed analysis of this case, let us stress that the main modification of the theory is the replacement, in some contributions to the kernel $\tilde{G}_q(\tau \mid \{\Gamma_{q'}\})$, of the exchange interaction J_q by an *effective exchange interaction* $\tilde{J}_q$:

$$\tilde{J}_q = J_q/(1 - \beta J_q/2) \tag{21}$$

which is temperature dependent and involves the well-known spin pair correlation function $(1 - \beta J_q/2)$, which is exact in the Weiss limit.

This modification has a drastic effect close to the critical temperature where[8]:

$$\tilde{J}_q \propto 1/(q^2 + \varkappa^2) \tag{22}$$

($\varkappa$ being the inverse correlation length). Indeed the Ornstein–Zernicke factor (22) tends give to a very large weight to small q values in the finite temperature generalization of (11) which is:

$$\tilde{G}_q^{\text{eff}(2)} \propto \int d\mathbf{q}_1 (J_{q+q_1} - J_{q_1}) \frac{J_{q_1}}{(1 - \beta J_{q_1}/2)} \Gamma_{q+q_1}(\tau)\, \Gamma_{-q_1}(\tau)\, (1 - \beta J_q/2) \tag{23}$$

As a result, the argument leading to (15) becomes invalid; on the contrary, it can be shown that the dominant contributions in the kernel (23) come from value of q_1 such that:

$$q_1 \lesssim O(\max(q, \varkappa)) \tag{24}$$

Thus, if $q \gtrsim \varkappa$, there is no more any separation between the characteristic time scales of $\Gamma_q(t)$ and of $\tilde{G}_q(\tau \mid \{\Gamma_{q'}\})$; the step leading to the diffusion equation (19) is no more valid and $\Gamma_q(t)$ *never reaches a diffusion regime*, even for small q. In other words, $\Gamma_q(t)$ has now to be self-consistently determined from the full non-Markoffian equation (23). The consequences of this are not trivial. Let us mention a few of them:

i) Line width at T_c

It has been shown that the characteristic relaxation time τ_q of our non Markoffian kinetic equation at T_c is given by:

$$\begin{aligned} \tau_q^{-1} &\propto q^{5/2} \quad \text{(ferromagnets)} \\ \tau_q^{-1} &\propto q^{3/2} \quad \text{(antiferromagnets)} \end{aligned} \tag{25}$$

These results, which are also consequences of the dynamical scaling law of Halperin and Hohenberg[9], compare surprisingly well with the experimental data[10].

ii) *Temperature dependence of the line width*

Here also, we recover the homogeneity property of dynamical scaling:

$$\frac{\tau_q^{-1}(\varkappa)}{\tau_q^{-1}(\varkappa = 0)} = \Omega(q/\varkappa) \tag{26}$$

However, from the solution of our kinetic equation, the explicit form of the homogeneous function $\Omega(x)$ can be obtained. It shows a non trivial minimum at $x \neq 0$, which has been observed experimentally in $RbMnF_3$ (see Ref. 10).

iii) *Line shape*

Once more, the scaling property:

$$\Gamma_q(\omega) = \frac{1}{q^{5/2}} \Gamma(\omega/q^{5/2}, q/\varkappa) \quad \text{(Ferromagnet)} \tag{27}$$

is an exact consequence of this theory. However, the explicit form of the universal function $\Gamma(x, y)$ is difficult to obtain, for it requires the detailed solution of our kinetic equation. Preliminary investigation indicates that the qualitative features of the observed spectral function are correctly reproduced by the theory; in particular, $\Gamma_q(t)$ at T_c decays to zero through damped oscillations which probably correspond to the "pseudo-spin wave" behavior experimentally observed.

References

1. P. Résibois and M. De Leener, *Phys. Rev.* **152**, 305, 318 (1966); **178**, 806, 819 (1969); *Phys. Letters* **25A**, 65 (1967).
2. P. Résibois, P. Borckmans, and D. Walgraef, *Phys. Rev. Letters* **17**, 1290 (1966); P. Borckmans and D. Walgraef, *Physica* **35**, 1280 (1967); *Phys. Rev.* **167**, 282 (1968); *Phys. Rev. Lett.* **21**, 1516 (1968); *Proceedings of the XVth Colloque*; AMPERE (North Holl. Pub., Amsterdam, 1969), p. 418.
3. P. Résibois, *Bull. Ac. Sc. Belgique* **54**, 1493 (1969).
4. See for instance: W. Marshall and Lowde: *Reports on Progress in Physics* **31**, Part 2 (1968).
5. P. Mazur and R. Terwiel: *Physica* **32**, 1813 (1966); **36**, 299 (1967).
6. See for instance: I. Prigogine: "*Non Equilibrium Statistical Mechanics*" (Interscience Pub., New York, 1963).

7. I. Lowe and R. Norberg: *Phys. Rev.* **107**, 46 (1957).
8. R. Brout: "*Phase Transition*" (Benjamin, New York, 1965).
9. B. Halperin and P. Hohenberg: *Phys. Rev.* **177**, 952 (1969).
10. R. Nathans, F. Menzinger, and J. Pickart: *J. Appl. Phys.* **39**, 1237 (1968); M. Collins, V. Minkiewicz, R. Nathans, L. Passel, and G. Shirane: *Phys. Rev.* **179**, 2, 241 (1969).

PAPER B–3

*The role of plasma kinetic theory in spectral line shapes**

W. R. CHAPPELL AND J. COOPER

Department of Physics and Astrophysics and Joint Institute for Laboratory Astrophysics†
University of Colorado, Boulder, Colorado

and

E. W. SMITH

National Bureau of Standards, Boulder, Colorado

The purpose of this talk is to relate some of our work involving applications of plasma kinetic theory to atomic spectroscopy. Several authors[1-3] have pointed out that spectral line shapes are phenomena which, in many ways, can best be treated by the methods of kinetic theory. We will confine ourselves here to the case of Stark broadening which results from the interaction of the radiating atom with the electric fields in a plasma. We will show that spectroscopy offers the possibility of measuring a quantity of importance to plasma kinetic theory—namely the electric field autocorrelation function

$$\langle \mathbf{E}(\mathbf{r}, t) \cdot \mathbf{E}(\mathbf{r}, o)\rangle.$$

The last statement should become obvious once one realizes what the basic problem involves. An atom immersed in a plasma can be perturbed in several ways. It interacts with the surrounding radiation field, giving rise to natural line broadening which in cases of interest is negligible. It also interacts with the plasma. This interaction in lowest order is represented by

$$V = -\mathbf{d} \cdot \mathbf{E}(\mathbf{r}), \tag{1}$$

---

* This research was supported by the Advanced Research Projects Agency of the Department of Defense and was monitored by Army Research Office-Durham under Contract No. DA-31-124-ARO-D-139.

† of the National Bureau of Standards and University of Colorado.

where **d** is the atomic dipole operator and **E(r)** is the instantaneous electric field at the position, **r**, of the atom. A third broadening mechanism is due to the motion of the atom (Doppler broadening). Doppler broadening is rather easily handled separately from pressure or Stark broadening, so we will henceforth assume that the atom is stationary.

If the electrons and ions in the plasma were fixed in known positions we could, conceivably, calculate the effect of the interaction in Eq. (1) in the usual manner and a Stark shift of the levels would ensue. However, the electrons move significant distances during times of interest (the lifetime or inverse half-width) and the positions of the ions, which are relatively static during this time, are not known. Consequently, we expect the statistical nature of the system to give rise to broadening and shifts of the corresponding spectral lines. It is obvious that the shapes of these lines will contain information about the fluctuating electric fields in the plasma.

Since we are trying to give an idea of the role that plasme kinetic theory plays in these problems, we will avoid as many specific details as possible in order to convey the general picture.

LINE SHAPE

The power radiated by dipole radiation per unit frequency, per unit volume by a group of atoms of density n_a immersed in a plasma is given by[4]

$$P(\omega) = \frac{4\omega^4 n_a}{3c^2} \sum_{m,\,n} \langle m\,|\mathbf{d}|\,n\rangle \cdot \langle n\,|\mathbf{d}|\,m\rangle\, \varrho_m\, \delta(\omega - \omega_{mn}), \tag{2}$$

where $|m\rangle$ and $|n\rangle$ are eigenstates of the total Hamiltonian, $\omega_{mn} = (E_m - E_n)/\hbar$, and ϱ_m is the probability that the system was initially in state m. If we make use of the time translation operator

$$U(t, o)\,|m\rangle = e^{-iHt/\hbar}\,|m\rangle = e^{-iE_m t/\hbar}\,|m\rangle \tag{3}$$

we can rewrite Eq. (2) in the form

$$P(\omega) = \frac{4n_a\omega^4}{3c^2}\, I(\omega), \tag{4}$$

where $I(\omega)$, the line shape function containing all the interesting frequency dependence, is given by

$$I(\omega) = \pi^{-1}\,\mathrm{Re} \int_0^\infty e^{i\omega t}\, Tr\,\{\mathbf{d} \cdot U(t, o)\, \mathbf{d}\varrho U^{+}(t, o)\} \tag{5}$$

$$= \pi^{-1}\,\mathrm{Re} \int_0^\infty e^{i\omega t}\, \langle \mathbf{d}(t) \cdot \mathbf{d}(o)\rangle, \tag{6}$$

where $\langle \rangle$ denotes an average of the entire system and $\mathbf{d}(t)$ is the dipole operator in the Heisenberg picture.

Several approximations are worth making in order to simplify the discussion. The fact that the line shape is given by a Laplace transform allows us to use the Fourier uncertainty principle to relate frequency differences of interest to times of interest. The basic time of interest is the inverse half-width. We noted earlier that the ions are fairly stationary over this time period (which for Lyman alpha is about 10^{-12} sec.). We can therefore consider them as stationary, calculate the corresponding Stark shift, and then average over all possible configurations of the ions. This is the "quasi-static" approximation[4]. On the other hand, the electrons can not be considered static over these time periods and must be treated dynamically.

This separation of the electron and ion effects results in several unresolved problems that we will not go into here. Among these problems is the fact that the ion and electron dynamics are coupled together through the shielding clouds.

If the atomic levels consist of widely-separated groups of degenerate or nearly degenerate levels, the interaction with the plasma will lead primarily to radiationless transitions (or level mixing) within the groups. Moreover, the transitions within the final state group $\{|b\rangle\}$ will lead to a much smaller effect than those in the initial state group $\{|a\rangle\}$. We can then make the following approximation

$$\langle b|\, U(t)\, |b'\rangle = e^{-i\varepsilon_b t} \quad \langle b|b'\rangle, \tag{7}$$

where $|b\rangle$ and $|b'\rangle$ are final *atomic* states and ε_b is the energy of the atom (appropriately Stark shifted by the static ion field) in state $|b\rangle$. This approximation allows us to write the line shape as[5]

$$I(\omega) = \pi^{-1} \operatorname{Re} \sum_{a,\, a',\, b} \mathbf{d}_{a'b} \cdot \mathbf{d}_{ba} \langle a|\, C(\Delta\omega_{ab})\, |a'\rangle\, \varrho_{a'}, \tag{8}$$

where

$$C(\Delta\omega_{ab}) = \int_0^\infty e^{it\Delta\omega_{ab}}\, F(t)\, dt, \tag{9}$$

and

$$F(t) = \langle U_I(t, o)\rangle_p. \tag{10}$$

The quantity $\Delta\omega_{ab} = \omega - \omega_{ab}$ and $\langle \rangle_p$ denotes an average over the plasma states. The quantity $U_I(t, o)$ is the time translation operator in the interaction picture

$$U_I(t, o) = e^{-itH_0/\hbar}\, U(t, o), \tag{11}$$

where

$$H = H_a + H_p - \mathbf{d} \cdot \mathbf{E} = H_0 - \mathbf{d} \cdot \mathbf{E} \tag{12}$$

and $\mathbf{E}$ is the instantaneous electric field due to the *electrons* at the position of the atom, H_a is the Hamiltonian for the atom in the presence of the static ion field, and H_p is the plasma Hamiltonian.

We know that the quantity $U_I(t, o)$ obeys the equation

$$i\hbar \frac{\partial U_I(t, o)}{\partial t} = V_I(t)\, U_I(t, o), \tag{13}$$

where

$$V_I(t) = e^{itH_0/\hbar}\, V e^{-itH_0/\hbar}. \tag{14}$$

At this point we can apply Zwanzig's[5] technique to obtain an equation for $F(t)$ by choosing a projection operator P appropriate to averaging over the plasma states. We then obtain the equation

$$\frac{\partial F(t)}{\partial t} = -\int_0^t dt' \langle V_I(t)\, G(t, t')\, V_I(t') \rangle\, F(t'), \tag{15}$$

where

$$G(t, t') = 0 \exp \left[-\hbar^{-1}{}_i \int_{t'}^{t} (1 - P)\, V_I(s)\, ds\right] \tag{16}$$

and 0 is the time-ordering operator, where $PA = \varrho_e\, Tr\{\varrho_e A\}$ and ϱ_e is the normalized density matrix for the electrons.

Since the average is only over plasma states the kernel in Eq. (15) is not a function of $t - t'$ alone. However, it can be shown that[5]

$$\frac{\partial F(t)}{\partial t} = -\int_0^t e^{it'H_a/\hbar} \langle V_I(t - t')\, G(t - t', o)\, V_I(o) \rangle\, e^{-it'H_a/\hbar} F(t')\, dt'. \tag{17}$$

We then obtain the following resolvant operator expression for the quantity

$$C(\Delta\omega_{ab}) = i[\omega + E_b - H_a + L(\omega + E_b - H_a)]^{-1}, \tag{18}$$

where

$$L(\omega + E_b - H_a) = +i \int_0^\infty e^{it(\omega + E_b - H_a)/\hbar} \langle V_I(t)\, G(t, o)\, V_I(o) \rangle\, dt. \tag{19}$$

We see that generally the width of the line is related to the imaginary part of the operator L and the shift to the real part of L. We will confine our discussion to the imaginary part, L_I. The correlation time for L_I is the inverse plasma frequency $\omega_p^{-1} = (4\pi ne^2/m)^{-1/2}$. The Laplace transform variable $(\omega + E_b - H_a)\hbar^{-1} = \Delta\omega$ corresponds to the separation from the unshifted line center (in the case of a forbidden line—the separation from the forbidden line center). For $\Delta\omega \ll \omega_p$ we are interested in times that are long compared to the correlation time. Consequently, we expect a Markovian behavior. For $\Delta\omega \gg \omega_p$, corresponding to short times, we expect a non-Markovian behavior. Except for $\Delta\omega$ very large compared to ω_p the shape is dominated by the long range collisions which are weak but numerous. The short range collisions are accounted for by a minimum impact parameter cutoff.

In many cases ω_p is large compared to the halfwidth, which means we can use an expansion in powers of the electric field over a large portion of the line. Thus, to lowest order we can replace $G(t, o)$ by the unit operator. We then see that in this approximation the line width is proportional to the real part of the Laplace transform of $\langle \mathbf{E}(\mathbf{r}, t) \cdot \mathbf{E}(\mathbf{r}, o)\rangle$, where $\mathbf{r}$ is the position of the atom.

ELECTRIC FIELD AUTOCORRELATION FUNCTION

This autocorrelation function is of great interest in plasma kinetic theory. It is a measure of the density fluctuations. It is important to ion-electron bremsstrahlung calculations. It is related to the conductivity and it gives a measure of the importance of collective modes. As a consequence, it has been calculated by several authors and its behavior, at least for the equilibrium, no-field case, is well known. In the absence of magnetic fields it is given by, in the random phase approximation[6],

$$L_I(\Delta\omega) = \mathrm{Re}\int_0^\infty dt\, e^{i\Delta\omega t}\,\langle \mathbf{E}(\mathbf{r}, t) \cdot \mathbf{E}(\mathbf{r}, o)\rangle$$

$$= \frac{(4\pi e)^2 n}{2(2\pi)^2}\int \frac{d\mathbf{k}}{k^2\,|\varepsilon(k, \Delta\omega)|^2}\int d\mathbf{v}\, f(\mathbf{v})\, \delta(\Delta\omega - \mathbf{k}\cdot\mathbf{v}), \qquad (20)$$

where

$$\varepsilon(\mathbf{k}, \Delta\omega) = 1 + \frac{\omega_p^2}{k^2}\int d\mathbf{v}\, \frac{\mathbf{k}\cdot\dfrac{\partial f(\mathbf{v})}{\partial v}}{\Delta\omega - \mathbf{k}\cdot\mathbf{v} + i\eta}. \qquad (21)$$

The real part $L_R(\omega)$ can be related to $L_I(\omega)$ as follows

$$L_R(\omega) = +\frac{1}{\pi}\mathscr{P}\int_{-\infty}^{\infty} d\omega' \frac{L_I(\omega')}{\omega - \omega'}. \tag{22}$$

The quantity $f(\mathbf{v})$ is the one-particle distribution function, η is the positive infinitesimal, and $\mathscr{P}$ denotes the Cauchy principle part of the integral.

It is relatively easy to obtain the high ($\Delta\omega \gg \omega_p$) and low ($\Delta\omega \ll \omega_p$) frequency limits for $L_I(\Delta\omega)$[7]. The low frequency limit is obtained by letting $\Delta\omega \to 0$ in which case the static dielectric constant leads to convergence at the lower limit of the k integration. The divergence at the upper limit is eliminated by the strong collision cutoff mentioned previously[8]. We have shown[9] that the low frequency contribution leads to a 10–20% correction to the impact theory[10] which evaluated the autocorrelation function incorrectly.

On the other hand, the high frequency ($\Delta\omega \gg \omega_p$) limit is obtained by replacing the dielectric function by unity. The resulting expression gives the results derived by Lewis[11], who proposed a frequency-dependent cutoff at low k in this region. The same high k cutoff is used here as before.

In the case of an equilibrium plasma the quantity $L_I(\Delta\omega)$ can be integrated numerically. This has been done by Dawson and Oberman[12] in their work on plasma conductivity. We have repeated this integration for values of the parameters which are of interest to spectroscopists.

The results of this numerical integration are shown in Fig. 1 for the case in which the minimum impact parameter is one-tenth the Debye length.

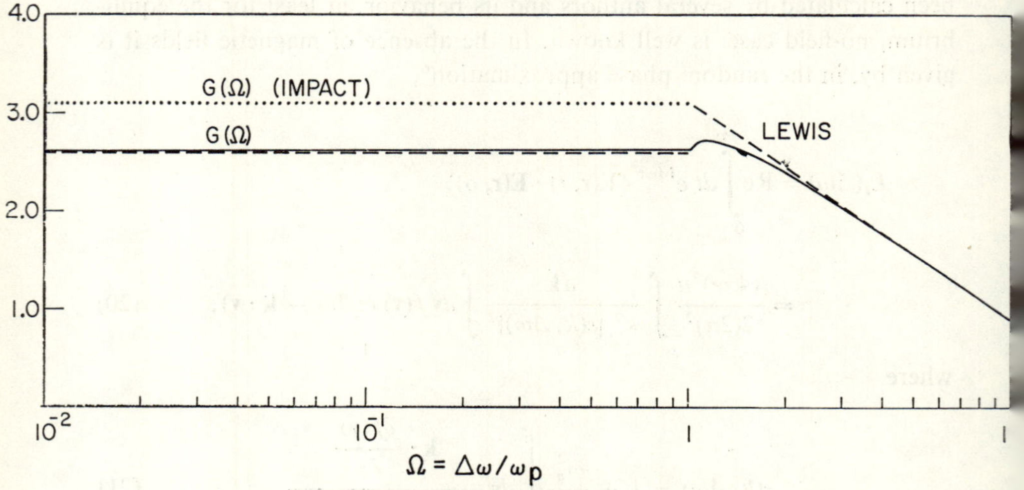

FIG. 1 Equilibrium electron gas

The quantity $G(\Omega)$ (which is proportional to $L_I(\Delta\omega)$) shown in Fig. 1 is

$$G(\Omega) = \int_{z_0}^{\infty} \frac{dz\, e^{-\Omega^2 z^2}}{z\,|\varepsilon_1(z,\Omega)|^2}, \tag{23}$$

where $z = k_D/\sqrt{2}\,k$ and $\Omega = \Delta\omega/\omega_p$. We see exactly the same type of behavior as that found by Dawson and Oberman. The solid lines represent the numerical results the dashed lines indicate the high and low frequency asymptotes, and the dotted line represents the old impact theory[10]. We note that the low frequency, or Markovian approximation, is amazingly accurate even at $\Delta\omega = \omega_p$. The high frequency approximation is quite good for frequencies as low as ω_p. Actually a quite good approximation over the entire range can be obtained simply by replacing $\varepsilon(z, \Omega)$ by $\varepsilon(z, o)$ in Eq. (23)[9].

The small bump at $\omega = \omega_p$ is the result of plasma waves. This contribution arises from a resonance in the integrand caused by a zero of $\varepsilon(z, \Omega)$ which occurs near $\Delta\omega = \omega_p$. The integration over wave numbers tends to smear out this contribution giving rise to a negligible contribution in the equilibrium case.

COLLECTIVE MODES

Although the collective modes do not make a significant contribution in the equilibrium case, their effect can be greatly enhanced in some nonequilibrium plasmas. In such systems the bump at ω_p can be significantly enhanced giving rise to structure on the line at approximately ω_p on either side of the line center (for allowed lines, e.g. hydrogen).

In 1961, Baranger and Mozer[13] considered the effect of an electric field oscillating at the plasma frequency on a spectrum in the region of a first-order forbidden line. They looked at forbidden lines because the effect to be described is less likely to be swamped by the allowed lines. The use of second-order perturbation theory then shows that satellites appear on either side of the forbidden line center. The height of the satellites is proportional to the square of the field amplitude. They then indicated that these satellites would be unobservable in an equilibrium plasma because of insufficient energy in the plasma waves.

By a straightforward calculation, Baranger and Mozer's work can be extended to account for a more accurate treatment of the electric field. To second-order in the interaction we can write Eq. (10) as

$$\langle a|\, F(t)\, |a'\rangle = -\hbar^{-2} \sum_{i} \int_{0}^{t} dt_1 \int_{0}^{t_1} dt_2\, e^{i(\omega_l - \omega_l')(t_1 - t_2)} \times [\langle lm\, | V(\mathbf{r}, t_1) |\, i\rangle \langle i\, | V(\mathbf{r}, t_2) |\, lm'\rangle]_{\text{ave}}$$

$$= -\hbar^{-2} \sum_{m''} \int_{0}^{t} dt_1 \int_{0}^{t_1} dt_2\, e^{i(\omega_l - \omega_l')(t_1 - t_2)} \times \langle lm\, |\mathbf{d}|\, l'm''\rangle \langle l'm''\, |\mathbf{d}|\, lm'\rangle : \langle \mathbf{E}(\mathbf{r}, t_1)\, \mathbf{E}(\mathbf{r}, t_2)\rangle_p . \tag{24}$$

This equation can also be obtained from Eq. (18) by an expansion in inverse powers of the separation from the allowed line (a wing expansion). We note that the above quantity is proportional to the electric field autocorrelation function. The line shape will then reflect any resonances contained in the autocorrelation function. The frequency difference, $\Delta\omega$, is the separation from the forbidden line center.

The test particle picture[14] shows that we can separate the microscopic electric field $\mathbf{E}(\mathbf{r}, t)$ into a part due to individual particle motion and a part due to the poles in $\varepsilon(\mathbf{k}, \omega)^{-1}$ near $\omega = \pm\omega_p$. The latter contribution is the part Baranger and Mozer had in mind.

Several models have been proposed in which the collective contribution is significantly enhanced over its equilibrium value. These models have mainly been of interest in the calculation of electron-ion bremsstrahlung[15, 16]. One such model proposed by Tidman and Dupree[15], consists of a two-temperature electron gas. This system is composed of a main body of electrons of density n_1 described by a Boltzmann distribution with temperature T_1 and a second body of electrons of density n_2 and temperature T_2 such that $n_1 \gg n_2$ and $T_2 \gg T_1$. The superthermal electrons are efficient producers of plasma waves by Cerenkov radiation. This is reflected mathematically by the fact that the quantity in the numerator of Eq. (20)

$$F(\omega/k) = k \int d\mathbf{v}\, f(\mathbf{v})\, \delta(\omega - \mathbf{k}\cdot\mathbf{v}) \tag{25}$$

remains relatively large at small k while the imaginary part of $\varepsilon(\mathbf{k}, \omega)$ is relatively small (compared to equilibrium values). Moreover, this occurs over a wide range of k so the effect is not washed out in the integration. (It is not sufficient to build up the plasma waves at a single k value, this being a set of measure zero). The integration has been carried out numerically and the results are shown in Fig. 2.

The results in Fig. 2 are for the case $n_1/n_2 = 10^8$, $T_1/T_2 = 10^{-2}$ and $r_D/r_{\min} = 10^7/\sqrt{2}$, where $r_{\min}$ is the strong impact cutoff (this is the value of

$r_{\min}$ used by Dawson and Oberman; the plasmon contribution is quite insensitive to this cutoff, however). We see that the plasmon contribution is 50% above the low frequency value. For $r_{\min}$ near 0.1 r_D the height of the bump is three times the low frequency value.

Although the two-temperature plasma is a model it does indicate that the collective mode contribution can become quite significant. In fact, the plasma

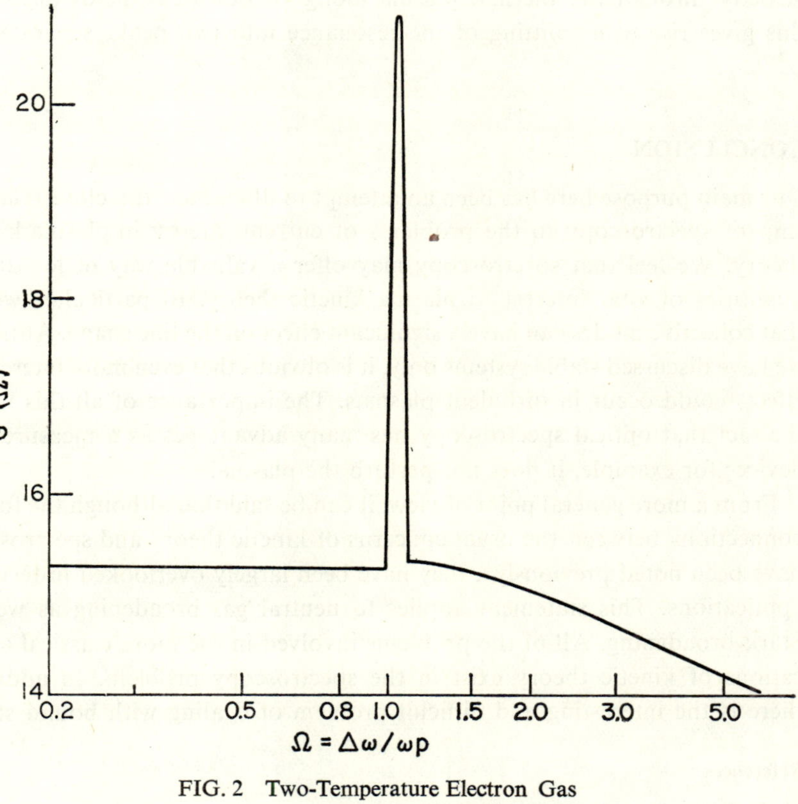

FIG. 2 Two-Temperature Electron Gas

satellites have been observed in two independent experiments within the past year. It was first observed by Kunze and Griem[17] who looked at helium in a Theta pinch. More recently, Cooper[18] observed it (plus another satellite separated from the first by the electron cyclotron frequency) by looking at a region of instability in a beam of electrons passing through neutral helium in the presence of a magnetic field.

Tidman and Dupree[15] also discussed two other models which enhance the effect of the plasma waves. There are undoubtedly many interesting effects associated with plasmas in magnetic fields, however, very little work has been done on calculating the autocorrelation function $\langle \mathbf{E}(\mathbf{r}, t) \cdot \mathbf{E}(\mathbf{r}, o) \rangle$ when magnetic fields are present (although formal expressions do exist)[7]. One of the few cases which has been considered in detail[16] involves a system consisting of a superthermal electron distribution with a kinetic temperature perpendicular to **B** much larger than that parallel to **B** and having a drift velocity through the thermal plasma along **B**. For weak fields ($\omega_c \ll \omega_p$) this gives rise to a splitting of the resonance into two peaks separated by $\omega_c^2/2\omega_p$.

CONCLUSION

Our main purpose here has been an attempt to illuminate the close relationship of spectroscopy to the problems of current interest in plasma kinetic theory. We feel that spectroscopy may offer a valuable way of looking at quantities of vital interest to plasma kinetic theory. In particular, we see that collective modes can have a significant effect on the line shape. Although we have discussed stable systems only, it is obvious that even more interesting effects could occur in turbulent plasmas. The importance of all this lies in the fact that optical spectroscopy has many advantages as a measurement device; for example, it does not perturb the plasma.

From a more general point of view, it can be said that although the formal connections between the usual concerns of kinetic theory and spectroscopy have been noted previously[1], they have been largely overlooked in terms of applications. This statement applies to neutral gas broadening as well as Stark broadening. All of the problems involved in the more classical applications of kinetic theory exist in the spectroscopy problem. In addition, there is the interesting and difficult problem of dealing with bound states.

References

1. U. Fano, *Phys. Rev.* **131**, 259 (1963).
2. E. W. Smith and C. F. Hooper, Jr., *Phys. Rev.* **157**, 126 (1967).
3. B. Bezzerides, *Phys. Rev.* **181**, 379 (1969).
4. M. Baranger, *Atomic and Molecular Processes*, D. Bates, Ed. (Academic Press, Inc., N. Y., 1962), Chap. 13.
5. E. Smith, J. Cooper, and R. Vidal, *Phys. Rev.* (in press).
6. G. Bekefi, *Radiation Processes in Plasmas* (John Wiley and Sons, Inc., N. Y., 1966).
7. W. Chappell, *J. Math. Phys.* **8**, 298 (1967).
8. H. Griem, A. Kolb, and K. Y. Shen, *Phys. Rev.* **116**, 4 (1959).
9. W. Chappell, J. Cooper, and E. Smith, *J. Quant. Spectrosc. Radiat. Transfer* **9**, 149 (1969).

10. H. Griem, M. Baranger, A. Kolb, and G. Oertel, *Phys. Rev.* **125**, 177 (1962).
11. M. Lewis, *Phys. Rev.* **121**, 501 (1960).
12. J. Dawson and C. Oberman, *Phys. Fluids* **5**, 517 (1962).
13. M. Baranger and B. Mozer, *Phys. Rev.* **123**, 25 (1961).
14. N. Rostoker, *Phys. Fluids* **7**, 479, 491 (1964).
15. D. A. Tidman and T. H. Dupree, *Phys. Fluids* **8**, 1860 (1965).
16. D. A. Tidman, T. J. Birmingham, and H. M. Stainer, *Ap. J.* **146**, 207 (1966).
17. H. J. Kunze and H. R. Griem, *Phys. Rev. Letters* **21**, 1048 (1968).
18. W. Cooper (private communication). Cooper and Ringler have also verified Baranger and Mozer's idea by applying an external A.C. electric field of known strength and frequency, W. S. Cooper III and H. Ringler, *Phys. Rev.* **179**, 226 (1969).

PAPER B-4

Interaction between atom and radiation field

F. HENIN
Faculté des Sciences, Université Libre de Bruxelles
Belgium

INTRODUCTION

The aim of this lecture initially was to discuss the interaction between an electron bound in an atom and the radiation field, using the methods developed in Brussels in non equilibrium statistical mechanics. Unfortunately, Professor Prigogine has not been able to attend this conference and, therefore, I cannot rely on his talk for the general concepts. I shall thus try to introduce them and, as far as time permits, show the kind of results one obtains in the discussion of the light-matter interaction.

This lecture will be divided into two parts. In the first one, we shall introduce the concept of subdynamics. In the second one, we shall show how quasiparticles can be introduced.

As well known, there are now various methods to derive kinetic equations to describe the long time behavior of a large class of N body systems. In the first part of this lecture, we shall see that the kinetic equations are exact equations, not for the density matrix of course, but for its projection on a well defined subspace. The most interesting feature is that this subspace is the space of thermodynamics.

While the first part of this talk will be concerned with the derivation of kinetic equations, the second part will deal with the interpretation of these equations. This is of course very easy in the lowest order approximation (Boltzmann equation, Pauli equation ...). Then, the evolution is due to real, energy conserving collision processes. At equilibrium, we deal with independent quantities. However, at higher orders, the situation is no longer so simple. In order to have an intuitive picture, it is necessary to use new entities, the quasiparticles.

2 HAMILTONIAN AND VON-NEUMANN–LIOUVILLE EQUATION

We shall only consider systems such that their hamiltonian H can be split into an unperturbed part H_0 and a perturbation λV:

$$H = H_0 + \lambda V \tag{2.1}$$

The unperturbed part is a sum of contributions of the individual degrees of freedom while V couples together these degrees of freedom.

For an electron bound in an atom and interacting with the radiation field, H_0 is the sum of the unperturbed atom hamiltonian (μ: energy levels) and the free field hamiltonian:

$$H = H_0^{at} + H_0^{f} = \sum_{\mu} n_\mu \omega_\mu + \sum_k n_k \omega_k \tag{2.2}$$

The interaction induces transitions in the atom by means of emission or absorption of one photon:

$$V = \sum_{\mu\nu k} \{V_{\mu\nu k}\, a_\mu^+ a_\nu a_k + V_{\mu\nu k} a_\mu a_\nu^+ a_k^+\} \tag{2.3}$$

The n, a^+, a are respectively occupation number, creation and annihilation operators.

For the matrix elements of any operator A, instead of the variables n, n', we shall use ν and N:

$$\nu = n - n' \qquad 2N = n + n' \tag{2.4}$$

and the notation:

$$A_{\{\nu\}}(\{N\}) \equiv \langle\{n\}\, |A|\, \{n'\}\rangle \tag{2.5}$$

Then, the $\varrho_{\{\nu\}}(\{N\})$, with $\{\nu\} \neq \{0\}$, correspond to off diagonal elements of the density matrix and describe correlations among the various degrees of freedom. $\varrho_{\{0\}}(\{N\})$ corresponds to the vacuum of correlations, i.e. the set of diagonal elements of the density matrix. The Von Neumann equation is then:

$$\frac{\partial \varrho_{\{\nu\}}(\{N\}, t)}{\partial t} = \sum_{\{\nu'\}} \langle\{\nu\}\, |L|\, \{\nu'\}\rangle\, \varrho_{\{\nu'\}}(\{N\}, t) \tag{2.6}$$

The matrix elements $\langle\{\nu\}\, |L|\, \{\nu'\}\rangle$ can be easily found. The unperturbed part is diagonal:

$$\langle\{\nu\}\, |L_0|\, \{\nu'\}\rangle = \{\delta_{\nu,\nu'}\} \sum_i \nu_i \omega_i \tag{2.7}$$

The perturbed part δL is non diagonal and its matrix elements contain displacement operators acting on the N variables.

In this representation, the Von Neumann equation (2.6) is formally identical to the classical Liouville equation. As a consequence, the general results presented here are valid for both classical and quantum systems.

3 KINETIC EQUATIONS

Let us now briefly review one method of deriving kinetic equations (see for instance[2]). The Liouville–Von Neumann equation is formally solved by means of the resolvent operator:

$$R(z) = (z - L)^{-1} \tag{3.1}$$

For instance, for the diagonal elements, we have:

$$\varrho_0(t) = -\frac{1}{2\pi i}\int_C dz\, e^{-izt}\, \{\langle O|R(z)|\,O\rangle\, \varrho_0(O) + \sum_{\{\nu\}} \langle O|R(z)|\,\nu\rangle\, \varrho_\nu(O)\} \tag{3.2}$$

where C is a parallel to the real axis above all singularities of the integrand.

The next step is then to isolate intermediate states identical to the vacuum of correlations. This is easily achieved if one defines operators irreducible with respect to the vacuum of correlations. For instance, the diagonal irreducible operator:

$$\psi(z) = \left\langle O \,\middle|\, \delta L \sum_{n=1}^{\infty} \left(\frac{1}{z - L_0}\,\delta L\right)^n \middle|\, O \right\rangle_{irr} \tag{3.3}$$

where "irr" means: no intermediate state identical to the vacuum of correlations. Then, one can easily write (3.2) in the form:

$$\varrho_0(t) = -\frac{1}{2\pi i}\int_C dz\, e^{-izt} \sum_{n=0} \frac{1}{z^{n+1}}\, [\psi(z)]^n\, \{\varrho_0(O) + \sum_\nu \mathscr{D}_\nu(z)\, \varrho_\nu(O)\} \tag{3.4}$$

where $\mathscr{D}_\nu$ is the irreducible operator describing the destruction of a correlation ν.

If all singularities of the irreducible operators $\psi(z)$ etc. ... are in the lower half plane at a distance of the real axis much larger than the inverse relaxation time, the only singularity which must be kept for times of the order of this relaxation time is the pole at $z = 0$. Straightforward integration and rearrangement of the series so obtained leads to the kinetic equation:

$$\frac{\partial \bar{\varrho}_0(t)}{\partial t} = -i\Omega\psi\bar{\varrho}_0(t) \tag{3.5}$$

for the asymptotic limit $\bar{\varrho}_0(t)$ of $\varrho_0(t) \cdot \psi$ is $\psi(z \to 0)$ and Ω can be expressed in terms of ψ and its derivatives for $z \to 0$. The equation (3.5) must be solved with the initial condition:

$$\bar{\varrho}_0(O) = A\{\varrho_0(O) + \sum_\nu D_\nu \varrho_\nu(o)\} \tag{3.6}$$

where A, D are expressed in terms of ψ, $\mathscr{D}$ and their derivatives for $z \to 0$.

Similarly, one finds that the asymptotic correlations $\bar{\varrho}_\nu(t)$ are functionals of $\bar{\varrho}_0(t)$:

$$\bar{\varrho}_\nu(t) = C_\nu \bar{\varrho}_0(t) \tag{3.7}$$

where the operator $C\nu$ can be expressed in terms of the irreducible operator $\mathscr{C}_\nu$ describing the creation of a correlation ν, the operator ψ and the derivatives of $\mathscr{C}$ and ψ for $z \to 0$.

Mathematically, these equations are valid if the operators $\psi(z)$ etc. ... have no singularities on or close to the real axis. If we consider a gas, this is easily verified for the lowest order terms for simple potentials[3], if we have short range forces and short range initial correlations. We must of course assume that it will be true at any order, but this seems quite a reasonable assumption.

However, there are cases where this is not true. For instance, in the light-atom interaction, correlations between discrete states have a life time of the order of the relaxation time[4]. Nevertheless, in the Born approximation, the kinetic equation (3.5) (i.e. the Pauli equation) correctly describes the evolution of the populations, but even asymptotically, we must add an equation which describes the decay of correlations between atomic levels. As we shall see now, this is a special case of a more general description of the evolution of many body systems.

4 SUBDYNAMICS

This description can be briefly summarized in the following way[5,6,7].

1) We start with the formal solution (3.4), but now, we only assume that all the operators $\psi(z)$ etc. ... are regular at $z = 0$.

2) We split $\varrho(t)$ into two parts:

$$\varrho(t) = \bar{\varrho}(t) + \hat{\varrho}(t) \tag{4.1}$$

where $\bar{\varrho}(t)$ is the contribution of the pole at $z = 0$ and $\hat{\varrho}(t)$ that of all remaining singularities.

3) $\bar{\varrho}(t)$ of course obeys the kinetic equations (3.5) and (3.7).

4) One can show that

$$\bar{\varrho}(t) = \Pi \varrho(t) \tag{4.2}$$

or, more explicitly, using an obvious matrix notation:

$$\begin{pmatrix} \bar{\varrho}_0(t) \\ \{\bar{\varrho}_\nu(t)\} \end{pmatrix} = \begin{pmatrix} A & \{AD_\nu\} \\ \{C_\nu A\} & \{C_\nu A D_{\nu'}\} \end{pmatrix} \begin{pmatrix} \varrho_0(t) \\ \{\varrho_\nu(t)\} \end{pmatrix} \tag{4.3}$$

The interesting feature is that Π is a projection operator:

$$\Pi^2 = \Pi \tag{4.4}$$

Thus, $\bar{\varrho}(t)$ is the projection of the density matrix on the Π-subspace and the kinetic equations (3.5), (3.7) are exact evolution equations for this projection of the density matrix.

5) As a consequence, $\hat{\varrho}(z)$ is the projection of the density matrix on the complementary subspace $\hat{\Pi}$:

$$\hat{\varrho}(t) = \hat{\Pi} \varrho(t) = (1 - \Pi)\, \varrho(t) \tag{4.5}$$

with

$$\hat{\Pi}^2 = \hat{\Pi} \tag{4.6}$$

and the orthogonality condition:

$$\Pi \hat{\Pi} = \hat{\Pi} \Pi = 0 \tag{4.7}$$

The projection $\hat{\varrho}(t)$ of the density matrix on the $\hat{\Pi}$ subspace satisfies independent evolution equations. In contrast with what happens for $\bar{\varrho}(t)$, the correlation part plays the important role. It obeys the equation:

$$\frac{\partial \hat{\varrho}_\nu(t)}{\partial t} = \sum_{\nu'} \{\langle \nu | L | \nu' \rangle - \langle \nu | L | o \rangle\, D_{\nu'}\}\, \hat{\varrho}_{\nu'}(t) \tag{4.8}$$

while the diagonal part is completely known as soon as the correlation part is known. Indeed, one has:

$$\hat{\varrho}_0(t) = -\sum_\nu D_\nu \hat{\varrho}_\nu(t) \tag{4.9}$$

We have thus achieved a decomposition of the density matrix into components which evolve independently. This definition of independent subdynamics has been obtained by an analysis of the structure of the resolvent operator. Now, the main question is, of course; what is the physical meaning of this decomposition? This is easily answered if one takes into account the fact that all invariants of motion are in the Π-subspace[5]. For instance, for the energy; we have

$$\Pi H = H \tag{4.10}$$

for the equilibrium distribution, we have

$$\bar{\varrho}_{eq} = \Pi \varrho_{eq} \sim \Pi e^{-\beta H} = e^{-\beta H} \sim \varrho_{eq} \tag{4.11}$$

The Π subspace is thus the space of equilibrium statistical mechanics and thermodynamics. The usual kinetic equations are exact equations for the evolution of the projection of the density matrix on the subspace of thermodynamics. They are such that this function evolves towards the equilibrium distribution. On the other hand, the evolution in the complementary subspace $\hat{\Pi}$ describes the progressive randomnization of phases through the disappearance of that part of initial correlations which does not exist in thermodynamical equilibrium.

Another interesting property, if we think of quantum field theory, is the following: all stable states lie in the Π subspace. By this, we mean an eigenstate of the total hamiltonian, corresponding to a real eigenvalue (energy). (The density matrix for a system in such an eigenstate can be obtained very easily from the expansion of the wave function in terms of the eigenstates of the unperturbed hamiltonian; this expansion can be obtained, using Rayleigh-Schrodinger perturbation expansion). In the light-atom interaction, we have such a state: the atom ground state in the absence of photons[8].

As an example of the two independent subdynamics, let us again consider the light-atom interaction and let us restrict ourselves to the weak coupling limit. The diagonal elements of the density matrix are then in the thermodynamical subspace:

$$\varrho_0(t) = \bar{\varrho}_0(t) \tag{4.12}$$

They satisfy the lowest order kinetic equation, the Pauli equation:

$$\begin{aligned}\frac{\partial \bar{\varrho}_0(t)}{\partial t} &= -2\pi \sum_{\mu\nu k} |V_{\mu\nu k}|^2 \, \delta(\omega_\mu - \omega_\nu - \omega_k) \\ &\quad \times \{(N_k + 1)\,[\bar{\varrho}_0(N, t) - \bar{\varrho}_0(N_\mu - 1, N_\nu + 1, N_k + 1, \{N\}', t)] \\ &\quad + N_k[\bar{\varrho}_0(N, t) - \bar{\varrho}_0(N_\mu + 1, N_\nu - 1, N_k - 1, \{N\}', t)]\}\end{aligned} \tag{4.13}$$

We easily recognize the usual loss and gain terms due to absorption or emission of one photon. The delta function means that the only processes which must be kept are those such that the energy is conserved in the transition. An $\mathscr{H}$-theorem can be easily derived from this equation.

On the other hand, asymptotically, the only non vanishing correlations are those between atomic levels; in the weak coupling limit, they are in the complementary subspace:

$$\varrho_{1_\mu, -1_\lambda}(t) = \hat{\varrho}_{1_\mu, -1_\lambda}(t) \tag{4.14}$$

and satisfy the following equation[9]:

$$\frac{\partial \hat{\varrho}_{1_\mu, -1_\lambda}}{\partial t} + i(\tilde{\omega}_\mu - \tilde{\omega}_\chi)\, \hat{\varrho}_{1_\mu, -1_\lambda} = \frac{1}{2}(\Gamma_\mu + \Gamma_\lambda)\, \hat{\varrho}_{1_\mu, -1_\lambda} \tag{4.15}$$

where the $\tilde{\omega}_i$'s are the unperturbed frequencies plus the second order Lamb shift and the Γ_i's are the life times of the ith excited state. This equation describes the disappearance of correlations and is the asymptotic limit of the equation (4.8) for $\hat{\varrho}_\nu$.

So far, we have seen that, provided all irreducible operators are regular at $z = 0$, the Von Neumann–Liouville equation can be exactly replaced by a set of equations corresponding to independent subdynamics. We have also seen that, in the weak coupling limit, the evolution equations in both subspaces are easily interpreted. The main point is now to understand them beyond the weak coupling limit. So far, the main results concern the evolution in the thermodynamical subspace. A large effort is now being done on the discussion of the randomnization process in the complementary subspace $\hat{\Pi}$ and we hope to report on it very soon. Anyway, because of lack of time, we shall, from now on, restrict ourselves to the discussion of processes in the thermodynamical subspace Π.

5 QUASIPARTICLES

In the weak coupling limit, the situation is simple. At equilibrium, the particles are independent. Out of equilibrium, the interaction only appears in the collision process.

At higher orders, the situation is no longer so simple, due to the mixing of real and virtual processes. However, for all processes which may be described in the Π-subspace, it is possible to perform a transition to a new representation in which an intuitive picture is restored:

$$\tilde{\varrho}(t) = \chi^{-1} \bar{\varrho}_0(t) \tag{5.1}$$

This quasiparticle distribution function will satisfy the kinetic equation:

$$\frac{\partial \tilde{\varrho}}{\partial t} = -i\varphi\tilde{\varrho} \tag{5.2}$$

with

$$\varphi = \chi^{-1}\Omega\psi\chi \tag{5.3}$$

Also, the mean value of any observable α will be given by:

$$\langle\alpha\rangle = \sum_N \alpha\bar{\varrho} = \sum_N \alpha_R\tilde{\varrho} \tag{5.4}$$

with the renormalized operator α_R given by:

$$\alpha_R = \chi^+\{\alpha_0 + \sum_\nu D_\nu\alpha_\nu\} \tag{5.5}$$

To find the operator χ^{-1}, we impose the following conditions[10,11]:

1) The energy will be the sum of the energies of the different quasiparticles:

$$H_R = \sum_i N_i\tilde{\omega}_i \tag{5.6}$$

2) The partition function can be written:

$$Z = \sum_N \exp(-\beta H) = \sum_N \exp(-\beta H_R) \tag{5.7}$$

and the equilibrium distribution will be of the form:

$$\tilde{\varrho}_{eq} = Z^{-1}\exp(-\beta H_R) \tag{5.8}$$

This implies that:

$$\exp(-\beta H_R) = \chi^{-1}[\exp(-\beta H)]_0 \tag{5.9}$$

3) The collision operator φ will be hermitian and will describe real processes, i.e. such that the energy of the quasiparticles taking part in the collision is conserved.

In this way, we have a description of strongly coupled systems in terms of quasiparticles which closely parallels that of weakly coupled systems in terms of particles.

Now, one can show[10] that, if χ satisfies the condition

$$A = \chi\chi^+ \tag{5.10}$$

the corresponding collision operator φ will be hermitian. However, this equation does not determine χ uniquely. Therefore, we have a whole class of transformations such that the first part of condition 3) is satisfied, and we shall look for the dressing operator χ^{-1} among this class. Then, the renormalized energy operator will take the simple form:

$$H_R = \chi^{-1} H_0 \tag{5.11}$$

Moreover, if χ^{-1} satisfies the equation:

$$\frac{\partial \chi^{-1}}{\partial \lambda} = \chi^{-1} \sum_{\nu} A D_{\nu} \frac{\partial C_{\nu}}{\partial \lambda} \tag{5.12}$$

(5.10) and (5.7) will be satisfied at all orders (Mandel, George). Turner has also verified that, for non dissipative systems, (5.12) is such that one has:

$$\{\chi^{-1} \langle N | H_0 | N \rangle\}^n = \chi^{-1} \langle N | H^n | N \rangle \tag{5.13}$$

at all orders in λ. For dissipative systems, he has verified this property up to order λ^4 and has strong indications that the proof can be extended to all orders in λ. This would give us a proof that, if χ^{-1} is solution of (5.12), then (5.9) is indeed satisfied. Conditions 1) and the second part of condition 3) have been verified up to order λ^4 and λ^6 respectively for several models. (In the case of the light-atom interaction, see reference [12]).

Another interesting property is the following. As we have seen, all stable states, i.e. all states with an infinite lifetime, are in the Π subspace. Now, for all such states, our definition of quasiparticles agrees with the results of ordinary Rayleigh–Schrodinger perturbation expansion[8]. In this case, the transition from the bare particle to the quasiparticle picture can be obtained through a canonical transformation. However, for unstable states, this is no longer the case; the present formalism shows us how to define perturbed unstable states.

To make this a little more intuitive, let us again consider the problem of interaction between light and atom.

6 QUASIPARTICLE PICTURE FOR LIGHT-ATOM INTERACTION

In the weak coupling limit, the evolution equation is the Pauli equation (4.13). It describes one photon emission or absorption processes. The transition probability for emission (induced and spontaneous) of a photon k with a transition between levels μ and ν in the atom is then:

$$\Gamma_{\mu\nu k} = |V_{\mu\nu k}|^2 \, \delta(\omega_\mu - \omega_\nu - \omega_k)\,(N_k + 1) \tag{6.1}$$

At the next order in λ, we obtain, for the quasiparticle distribution function, an equation of the form[1]:

$$\frac{\partial \tilde{\varrho}}{\partial t} = -i(\tilde{\varphi}_2 + \tilde{\varphi}_4)\tilde{\varrho} \tag{6.2}$$

where $\tilde{\varphi}_2$ describes one photon emission and absorption processes inducing transitions between the perturbed levels of the atom. The transition probability is given by an expression analogous to (6.1). For instance, for emission, we have:

$$\tilde{\Gamma}_{\mu\nu k} = \widetilde{|V_{\mu\nu k}|^2}\, \delta(\tilde{\omega}_\mu - \tilde{\omega}_\nu - \omega_k)\,(N_k + 1) \tag{6.3}$$

where the $\tilde{\omega}$ are the energies of the perturbed levels and $\widetilde{|V|^2}$ is a modified cross section. The energies of the perturbed levels are those which one obtains from (5.11). In vacuum, up to this order, we recover the usual expression for the Lamb shift, in the non relativistic limit.

The remaining part, $\tilde{\varphi}_4$, of the collision operator in (6.2) describes the influence of two photon processes on the evolution. It has the same structure as $\tilde{\varphi}_2$, except that now, two photons are involved. Processes are most easily described by the corresponding energy conservation condition. One finds all the two photon processes which one intuitively expects: double emission or absorption: $\delta(\tilde{\omega}_\mu - \tilde{\omega}_\nu - \omega_k - \omega_{k'})$, Raman scattering: $\delta(\tilde{\omega}_\mu - \tilde{\omega}_\nu - \omega_k + \omega_k)$, resonance scattering of light: $\delta(\omega_k - \omega_{k'})$.

The λ^6 contributions have been investigated for the two level Wigner–Weisskopf model[12]. Essentially, the same features are found: renormalization of the one and two photon operators and a part $\tilde{\varphi}_6$ describing the role of three photon processes on the evolution. One point however should be mentioned: this is the first order where our definition of perturbed levels for unstable states is not identical to the definition one obtains from the discussion of the phase of the wave function in simple problems like spontaneous emission[13].

Equation (6.2) and the corresponding equation at order λ^6 have been the starting point for the discussion of problems such as spontaneous emission, resonant light-scattering[1,12]. In both cases, one can see how multiple photon processes are responsible of the line shape.

7 CONCLUDING REMARKS

Let us first stress the main results obtained so far.

First, we have shown how the complete dynamical problem can be split into subdynamics in orthogonal subspaces. One of these subspace, Π, play a specially important role, as it is the subspace of thermodynamics and equilibrium statistical mechanics, and also of all states with an infinite lifetime. The evolution in the complementary $\hat{\Pi}$ subspace corresponds to a randomnization of phases. This randomnization is necessary to reach thermodynamical equilibrium.

For processes which can be described in the Π-subspace, we have seen how we can understand what happens in terms of new entities, the quasiparticles. These entities are independent at equilibrium. The interaction only appears in collision processes. This is a picture of strongly coupled systems which parallels completely the description of weakly coupled systems. In general, the transformation to quasiparticle is not a canonical transformation; this is only so for non dissipative systems.

At this point, of course, many questions could be raised. For instance, we have mentioned that the decomposition of the complete problem into independent subdynamics can be achieved provided that all the irreducible operators $\psi(z)$ etc. ... are regular at $z = 0$. This seems to be true for many systems but there are known exceptions. One of them is met in the light-matter interaction problem; indeed, there, the irreducible operators have poles on the real axis at the differences of energies between the atomic levels. If two of the levels are degenerate, the corresponding pole is at $z = 0$; some work is being done on this problem now, which indicates how the difficulty can be removed. One first has to perform partial summations and redefine the operators. Physically, it amounts to take into account the level separation due to the interaction and to introduce the proper base for the perturbation expansion (linear combination of the unperturbed states). Other examples of problems where the irreducible operators are singular at $z = 0$ are systems interacting with long range electrostatic or gravitational forces. In the first case, we again know that there is, in fact, a screening effect. To take it into account, one again has to perform partial summations and define new irreducible operators, which are no longer singular at $z = 0$. As to gravitational systems, so far it is not clear whether this formalism can be applied to them.

Another, more interesting, question is the following: so far, we have introduced the quasiparticle picture only for processes which can be entirely described in the thermodynamical subspace. In other words, we know how to define the population of quasiparticles, but not their correlations. For non dissipative systems, the complete quasiparticle picture can be easily obtained from the knowledge of the eigenstates of the total hamiltonian. In this case, correlations among the perturbed states are obtained by a transformation similar to (5.1) but on the $\hat{\varrho}_\nu$ part of the correlations. The main question is now to extend this picture to the case of dissipative systems. This seems to be quite feasible and we hope to report the results very soon.

Acknowledgements

This work has been done in Brussels in constant collaboration with Professor Prigogine. Among the people who have participated in the recent developments, I would like to specially mention C. George, R. Balescu, P. Mandel, W. Turner and P. Clavin.

I also gratefully acknowledge contributions to my travel expenses by the Fonds National Belge de la Recherche Scientifique, the organizers of the Symposium and the Air Force Office of Scientific Research through the European Office of Aerospace Research OAR, United States Air Force under grant AF EOAR 69–0058.

References

1. F. Henin, *Physica* **39**, 599 (1968).
2. C. George, *Physica* **30**, 1513 (1964).
3. P. Résibois, in *Physics of Many Particle Systems*, vol. I, ed. E. Meeron, Gordon and Breach (1966).
4. C. Cohen-Tannoudji, These, *Ann. de Phys.*
5. W. Turner, *Bull. Acad. Roy. Belg. Cl. Sci.* **54**, 1526 (1968).
6. I. Prigogine, C. George, and F. Henin, *Physica* **45**, 418 (1969).
7. I. Prigogine, C. George, and F. Henin, *Proc. Nat. Acad. Sc. USA* **59**, 7 (1968).
8. F. Henin, *Bull. Acad. Roy. Belg. Cl., Sci.* **54**, 1439 (1968).
9. I. Prigogine and F. Henin, in Problemi Theoretitsheskoi Physiki, Nauka Moscow (1969).
10. C. George, *Bull. Acad. Roy. Belg. Cl. Sci.* **53**, 623 (1967).
11. C. George, *Physica* **37**, 182 (1967).
12. F. Henin and M. De Haan, *Physica* **40**, 399 (1968).

PAPER B–5

*Upper and lower bounds on transport coefficients as applied to the Boltzmann equation of a Fermi liquid**

H. SMITH† AND J. W. WILKINS‡

Laboratory of Atomic and Solid State Physics
Cornell University
Ithaca, New York 14850.

Abstract

Various upper and lower bounds on the transport coefficients arising from a linearized Boltzmann equation are discussed, and a systematic approach to their construction is indicated. The usefulness of the bounds is demonstrated by considering the particular Boltzmann equation of a Fermi liquid with (a) particle-particle scattering and (b) particle-particle scattering *and* impurity scattering as the relaxation mechanisms. In the latter case the two scattering rates are shown to be non-additive to a significant extent.

Since the bulk of this work has recently appeared elsewhere[1,2] we give only a brief account here and refer the reader to references 1 and 2 for details. A major part of the work was performed in collaboration with H. Højgaard Jensen.

One of the favorite methods of attacking the Boltzmann equation is by the application of a variational principle which was first used in transport theory by Kohler. This variational principle determines a lower bound on transport coefficients like electrical or thermal conductivity.

* Work supported in part by the National Science Foundation under Grant GP-7198 and in part by the Advanced Research Projects Agency through the Materials Science Center at Cornell University, MSC Report 1228.

† Permanent Address: Physical Laboratory I, H. C. Ørsted Institute, University of Copenhagen, Denmark.

‡ Alfred P. Sloan Foundation Fellow.

The fact that there exist equivalent variational principles leading to upper bounds on the same transport coefficients has not received a comparable attention. A variety of bounds, both upper and lower, may be derived in a systematic fashion[1] starting from the usual lower bound of the Kohler-type. By suitable choices of trial functions these upper and lower bounds may be brought close to each other and hence close to the exact result. We have applied the methods to the Boltzmann equation of a Fermi liquid at low temperatures where the collision integral is determined by the scattering between the quasiparticles. The combined effect of particle-particle scattering and scattering against impurities has also been treated[2] by these methods.

The linearized Boltzmann equation is symbolically written as

$$X = HQ \tag{1}$$

where Q is the unknown deviation of the distribution function from equilibrium, H is an integral-operator characterizing the collision processes, and the inhomogeneous part X originates from the applied external fields. The transport coefficient T is given by a scalar product

$$T \equiv (X, Q).$$

The integral-operator H is symmetric and positive

$$(A, HB) = (B, HA) \text{ for all } A \text{ and } B.$$

The conventional lower bound on T may be derived in the manner of a Schwarz inequality. Since H is positive

$$((Q - \lambda U), \quad H(Q - \lambda U)) \geqq 0$$

where U is a trial function and λ some number. Using $HQ = X$ and symmetry we get

$$(X, Q) + \lambda^2(U, HU) - 2\lambda(U, X) \geqq 0.$$

The choice $\lambda = \dfrac{(U,X)}{(U, HU)}$ minimizes the left-hand side and gives the bound:

$$T = (X, Q) \geqq \frac{(U, X)^2}{(U, HU)}. \tag{2}$$

Other bounds, upper as well as lower, can be obtained on the assumption that a decomposition

$$H = J + L$$

is possible, where J and L are both positive and symmetric and J^{-1} exists. Then

$$H'Q \equiv (HJ^{-1} - 1)\, HQ = (HJ^{-1} - 1)\, X \equiv X'$$

where H' is positive and symmetric. Consequently

$$(X', Q) \geqq \frac{(U, X')^2}{(U, H'U)}$$

which results in an upper bound on T

$$T = (X, Q) \leqq (X, J^{-1}X) - \frac{(U, (HJ^{-1} - 1)\, X)^2}{(U, (HJ^{-1} - 1)\, HU)}. \tag{3}$$

Such decompositions may often be performed in several different ways, leading to a variety of upper bounds. If L is negative and J positive, then H' is negative. In such cases the inequality (3) is therefore reversed to give new lower bounds on T in addition to (2).

We now turn to the discussion of an example, that of the Boltzmann equation of a Fermi liquid at low temperatures. We first consider the case where the relaxation mechanism arises from interactions between the quasi-particles themselves. To be specific we discuss the thermal conductivity; similar results have been obtained for the spin-diffusion and viscosity coefficients.

The appropriate Boltzmann equation may be reduced to the form

$$X(t) = f(t)\, Q(t) - \alpha \int_{-\infty}^{\infty} dt'\, F(t - t')\, Q(t'). \tag{4}$$

Here $t = (\varepsilon - \mu)/k_B T$ is a reduced energy variable, μ being the Fermi energy. The right hand side of (4) comes from the collision integral, whereas the left hand side arises from the driving term.

The unknown deviation from equilibrium of the distribution function is

$$n(\mathbf{p}) - n_0(\mathbf{p}) \propto \frac{Q(t)}{\cosh t/2}.$$

Furthermore

$$X(t) = \frac{t}{\cosh t/2},\ f(t) = \pi^2 + t^2,\ F(t) = \frac{t/2}{\sinh t/2}.$$

The parameter α is

$$\alpha = 2\, \frac{\left\langle \dfrac{w(\theta, \varphi)}{\cos \theta/2} (1 + 2 \cos \theta) \right\rangle}{\left\langle \dfrac{w(\theta, \varphi)}{\cos \theta/2} \right\rangle}$$

where $w(\theta, \varphi)$ is the collision probability for the scattering between two quasi particles (θ and φ are scattering angles).

The thermal conductivity $\varkappa$ can finally be shown to be

$$\varkappa = \frac{1}{3} C_V V_F^2 \tau_0 \left(\frac{3}{2\pi^2} \right) \int_{-\infty}^{\infty} dt\, X(t)\, Q(t)$$

where

$$\frac{1}{\tau_0} = \frac{(m^*)^3 (k_B T)^2}{8\pi^4 \hbar^6} \left\langle \frac{w(\theta, \varphi)}{\cos \theta/2} \right\rangle$$

is the inverse relaxation time for a quasi-particle at the Fermi surface. The specific heat C_V and the Fermi velocity V_F have their usual meanings in Fermi liquid theory.

The Boltzmann equation (4) has the desired form (1), since H is positive and symmetric and the (reduced) transport coefficient is

$$K = \int_{-\infty}^{\infty} dt\, X(t)\, Q(t) \equiv (X, Q).$$

We have decomposed H $(= J + L)$ in the following three ways: i) $J =$ smallest eigenvalue of H, ii) $J = f(t)$, and iii) $J = f(t) - \alpha \int_{-\infty}^{\infty} dt'\, F(t - t')$. In each case we have calculated the upper or lower bounds with a variety of trial functions. The bounds on the thermal conductivity are exhibited as functions of the parameter α, the results showing that the upper and lower bounds may be brought very close to each other and hence close to the exact transport coefficient.

When elastic scattering against impurities is included in the collision integral in addition to the particle-particle scattering we have considered previously, the only change in the Boltzmann equation (4) occurs through the addition of a positive constant to $f(t)$:

$$f(t) = \pi^2(1 + \beta) + t^2 \tag{5}$$

where β is essentially the ratio of the previously introduced relaxation time τ_0 to the relaxation time τ_{imp} for elastic scattering against impurities.

With two different scattering mechanisms in the collision integral it is now possible to ask whether they interfere. According to Matthiessen's rule the total thermal resistance is simply the sum of the resistance due to quasi-particle scattering alone and that due to impurity scattering alone. The rule, then, is a statement that the individual scattering rates are additive. How deviations from additivity occur can be roughly understood by considering the $\alpha = 0$ case of (4) where now $f(t)$ includes impurity scattering according to the prescription (5). Then $Q(t) = [\pi^2(1 + \beta) + t^2]^{-1} X(t)$ and the resulting $\varkappa^{-1}$ is clearly not the sum of the thermal resistances due to the individual scattering processes acting alone. Generally speaking, if the individual collision operators have solutions for $Q(t)$ with different energy dependences, then there will be deviation from additivity.

We find from the upper and lower bounds that the scattering rates are non-additive to a significant degree. If this interference effect were absent, the temperature-dependence of $T/\varkappa$ and the electrical resistance ϱ would be of the form $\{aT^2 + b\}$, with a independent of temperature as well as of the impurity-concentration, and b dependent only on the impurity-concentration. The presence of interference leading to non-additivity of scattering rates is demonstrated in the figure. Instead of being strictly constant $a(T)$

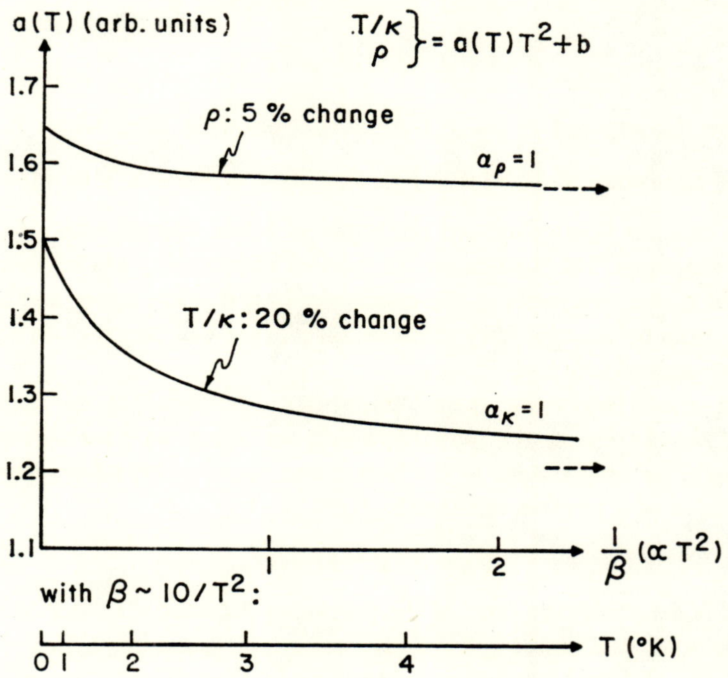

FIG. 1 The effective temperature-dependent coefficients $a(T)$ for the "T^2-term" in the electrical resistance (ϱ) and thermal resistance (actually $T/\varkappa$) are plotted as a function of $\beta^{-1} = \pi^2\tau_{\text{imp}}/2\tau_0$, a quantity which is proportional to T^2. Deviations of $a(T)$ from a constant are a measure of the breakdown in Matthiessen's rule. The lower of the two abscissa scales is obtained from an estimate for Indium metal in which there is 1 ppm of impurity present

shows a significant variation with temperature and impurity-concentration through the parameter β. The particular values of the parameters α_ϱ and $\alpha_\varkappa$ are initial estimates for a model[3] of Indium metal which allows for Umklapp-scattering between the quasi-particles. At present there have not been sufficiently accurate experiments to observe these effects.

References

1. H. Højgaard Jensen, Henrik Smith, and John W. Wilkins, *Phys. Rev.* **185**, 323 (1969).
2. H. Smith and J. W. Wilkins, *Phys. Rev.* **183**, 624 (1969).
3. W. E. Lawrence (to be published).

PAPER B–6

The Boltzmann equation and gaseous NMR

R. F. SNIDER

Department of Chemistry
University of British Columbia
Vancouver, B. C., Canada

Because of the small energy separation of nuclear magnetic (m_I) and rotational (m_J) states (still only 10^{10} sec^{-1} even at magnetic fields of 10^4 gauss) there is little or no phase randomization of the density matrix between these different energy levels during the time between collisions ($\sim 10^{-10}$ sec. at room temperature and one atmosphere pressure). Thus a molecule remembers the phase relation between *m* states from one collision to another. In order to describe the gaseous state appropriately it is then necessary to have a Boltzmann equation which properly accounts for the phase relations between *m* states and is thus an equation for the singlet density matrix which is generally non-diagonal in *m* states. Such a Boltzmann equation was first derived by Waldmann[1] for spin systems and independently by the author[2] for general molecules with degenerate internal states. A brief derivation of this equation following the approach used in reference 2 is given here but use is made of the superoperator[3,4] (operator on quantum mechanical operators)[5] or "operator on Liouville space" notation. Mention is made of work in related areas.

The Quantum Liouville or von Neumann Equation can be written as

$$i\frac{\partial \varrho^{(N)}}{\partial t} = [H^{(N)}, \varrho^{(N)}]_- \equiv L^{(N)}\varrho^{(N)} \tag{1}$$

with normalization $Tr_N\varrho^{(N)} = N!$. $L^{(N)}$ is the superoperator "commutator of $H^{(N)}$ with" and has the property that it is hermitian in the Hilbert–Schmidt space of operators defined as the Hilbert space of operators with inner product

$$\langle A \mid B\rangle_2 \equiv Tr\, A^\dagger B. \tag{2}$$

For a hamiltonian which consists of a sum of one particle and two particle operators, the corresponding Liouville operator becomes

$$L^{(N)} = \sum_{i=1}^{N} L_i^{(1)} + \sum_{i<j} L_{ij}^{(2)} \tag{3}$$

and the first BBGKY equation is (for particle 1)

$$i \frac{\partial \varrho_1^{(1)}}{\partial t} = L_1^{(1)} \varrho_1^{(1)} + Tr_2 L_{12} \varrho_{12}^{(2)} \tag{4}$$

with normalizations $Tr_1 \varrho_1^{(1)} = N, Tr_{12} \varrho_{12}^{(2)} = N(N-1)$. The non-rigorus derivation proceeds by assuming molecular chaos *before* collision-namely at some time $t_0 \ll t$ on a microscopic collision time scale,

$$\varrho_{12}^{(2)}(t_0) = \varrho_1^{(1)}(t_0)\, \varrho_2^{(1)}(t_0) \tag{5}$$

(Boltzmann statistics being assumed throughout). To relate the density matrices at times t and t_0, one assumes only binary collisions are important in which case the second BBGKY equations becomes

$$i \frac{\partial \varrho^{(2)}}{\partial t} = L^{(2)} \varrho^{(2)}. \tag{6}$$

This can be formally integrated to give

$$\begin{aligned} \varrho^{(2)}(t) &= e^{-iL^{(2)}(t-t_0)}\, \varrho^{(1)}(t_0)\, \varrho^{(1)}(t_0) \\ &\simeq e^{-iL^{(2)}(t-t_0)}\, e^{iL_0^{(2)}(t-t_0)} \varrho^{(1)}(t)\, \varrho^{(1)}(t) \end{aligned} \tag{7}$$

where molecular chaos has been used at the time $t_0 \ll t$ and $L_0^{(2)} = L_1^{(1)} + L_2^{(1)}$. Consistent with this approach is the assumption that the macroscopic and microscopic motions can be separated so that $t - t_0$ may be considered long on the microscopic (collision) time scale but negligible on the macroscopic time scale. Thus the limit $t_0 \to -\infty$ of Eq. (7) can be envisaged and indeed this limit has recently been discussed by Jauch, Misra and Gibson[6] for purely scattering problems. Let Ω_L be the corresponding limit superoperator

$$\Omega_L = \lim_{\tau \to \infty} e^{-iL^{(2)}\tau}\, e^{iL_0^{(2)}\tau}. \tag{8}$$

It is possible to show that if the Møller wave operator Ω of scattering theory

$$\Omega = \lim_{\tau \to \infty} e^{-iH^{(2)}\tau}\, e^{iH_0^{(2)}\tau} \tag{9}$$

exists in the strong operator limit, then Ω_L is equivalent to

$$\Omega_L \varrho^{(1)} \varrho^{(1)} = \Omega \varrho^{(1)} \varrho^{(1)} \Omega^\dagger \tag{10}$$

and exists in the Jauch, Misra and Gibson[6] sense. Corresponding to the Lippmann–Schwinger equation

$$\Omega = 1 + \frac{1}{E - H^{(2)} + i\varepsilon} H_{12}^{(2)} \tag{11}$$

for Ω acting on an eigenfunction χ_E of $H_0^{(2)}$ (more correctly in spectral representation of $H_0^{(2)}$) the Moller superoperator Ω_L can be written as

$$\Omega_L = 1 + \frac{1}{\omega - L^{(2)} + i\varepsilon} L_{12}^{(2)} \tag{12}$$

when acting on an eigenoperator A_ω of $L_0^{(2)}$, namely $L_0^{(2)} A_\omega = \omega A_\omega$. The Boltzmann equation now becomes

$$i \frac{\partial \varrho_1^{(1)}}{\partial t} = L_1^{(1)} \varrho_1^{(1)} + Tr_2 L_{12}^{(2)} \Omega_L \varrho_1^{(1)} \varrho_2^{(1)} \tag{13}$$

where

$$L_{12}^{(2)} \Omega_L = L_{12}^{(2)} + L_{12}^{(2)} \frac{1}{\omega - L + i\varepsilon} L_{12}^{(2)} \tag{14}$$

is the binary collision operator. Fano[7] has also obtained this collision (super-) operator in his formal treatment of pressure broadening via correlation functions. This formal expression for the Boltzmann collision operator can be expressed in terms of the ordinary quantum mechanical transition operator $t \equiv H_{12}^{(2)} \Omega$ and the Green's function G for free motion (no $H_{12}^{(2)}$) of the pair of particles

$$\Omega = 1 + Gt, G = \frac{1}{E - H_0^{(2)} + i\varepsilon}, \tag{15}$$

namely,

$$i\hbar \frac{\partial \varrho}{\partial t} = L\varrho + Tr_2[t\varrho\varrho - \varrho\varrho t^+ - Gt\varrho\varrho t^+ + t\varrho\varrho t^+ G^+]. \tag{16}$$

In particular, if ϱ is diagonal in $H^{(1)}$ energy, then one can eliminate G by using

$$G^+ - G = 2\pi i\, \delta(E) \tag{17}$$

and

$$\frac{\partial \varrho}{\partial t} + \frac{iL^{(1)}}{\hbar} \varrho = \frac{2\pi}{\hbar} Tr_2[t\varrho\varrho t^+ \delta(E) + \frac{1}{2\pi i} (t\varrho\varrho - \varrho\varrho t^+)] \tag{18}$$

with the understanding that $\delta(E)$ means that t (and t^+) have non-zero matrix elements only on the $(H_0^{(2)})$ energy shell. The first term on the right hand side is the usual loss term of the Boltzmann equation (it even appears quadratic in t which is like a collision cross-section while the last two terms together is the gain term. In particular, if ϱ is completely diagonal, one can use the optical theorem

$$t^+ - t = 2\pi i\, t^+ \,\delta(E)t \tag{19}$$

to reduce this Boltzmann equation to the Wang Chang–Uhlenbeck[8] equation with degeneracy averaged collision cross-sections.

For practical considerations of NMR, the m states may be treated as energetically degenerate in the collision term so that Eq. (18) is sufficiently general for this purpose. However, at high densities, effects arise due to correlation between J states[9] (and K states in polyatomics) which necessitates the general equation (16). In either case the gas is very close to equilibrium and a linearized form of the equation is sufficient for the description of the relaxation process. This is accomplished by setting[10]

$$\varrho = \varrho_{\text{eq.}} + 1/2[\varrho_{\text{eq.}}, \phi]_+ \tag{20}$$

where $\varrho_{\text{eq.}}$ is the Boltzmann distribution corresponding to the hamiltonian $H^{(1)}$. For perturbations diagonal in $H^{(1)}$ energy, the linear equation for the hermitian perturbation operator is then

$$\frac{\partial \phi}{\partial t} + \frac{i}{\hbar}[H^{(1)}, \phi]_- = -R\phi \tag{21}$$

where R is the positive semi-definite superoperator

$$R(\phi) = \frac{i}{\hbar} Tr_2\, \varrho_{\text{eq.}2}[t(\phi_1 + \phi_2) - (\phi_1 + \phi_2)\,t^+ + 2\pi i t(\phi_1 + \phi_2)\,t^+\,\delta(\text{E})] \tag{22}$$

and ϕ can naturally be thought of as an element of the Hilbert space whose inner product is given by

$$\langle\langle \phi \mid \psi \rangle\rangle \equiv 1/2 Tr \phi^+[\varrho_{\text{eq.}}, \psi]_+. \tag{23}$$

The simplest example in application to NMR is for monatomic gases[11]. Effectively one measures the average value of the nuclear spin operator $\mathbf{I}$ which is affected by the singlet hamiltonian

$$H^{(1)} = \frac{p^2}{2m} - \gamma_I \hbar \mathbf{H} \cdot \mathbf{I}. \tag{24}$$

A one-moment approximation to ϕ is made by setting $\phi = \mathbf{b} \cdot \mathbf{I}$ with $\mathbf{b}$ measuring the deviation of the magnetization from equilibrium. Multiplying Eq. (21) by $\mathbf{I}$ and taking the trace gives

$$\frac{\partial \mathbf{b}}{\partial t} + \gamma_I \mathbf{H} \times \mathbf{b} = -\frac{1}{\mathsf{T}} \cdot \mathbf{b} \tag{25}$$

which is just the Bloch equation of NMR. The relaxation tensor $1/\mathsf{T}$ is given by

$$\frac{1}{\mathsf{T}} = 3 \frac{\langle\langle \mathbf{I} \,|R|\, \mathbf{I} \rangle\rangle}{nI(I+1)} = \frac{\hat{H}\hat{H}}{T_1} + \frac{\mathsf{U} - \hat{H}\hat{H}}{T_2} + \gamma_I \sigma \mathbf{H} \cdot \boldsymbol{\varepsilon} \tag{26}$$

which can be uniquely written in terms of the spin-lattice T_1, spin-spin T_2 relaxation times and the intermolecular chemical shift σ corresponding to the component of $1/\mathsf{T}$ parallel, perpendicular or skew to the field direction $\hat{H}$.

The case for diatomics (and polyatomics) is much more complicated, first of all because the relaxation is dominated by mechanisms coupling the nuclear spin to the rotational angular momentum (direct collisional relaxation of $\mathbf{I}$ is negligible in comparison). This means that both nuclear spin and rotational angular momentum are driven out of equilibrium and moments to describe these correlations must be taken into account. The result of such calculations it to reproduce standard formulae of NMR as given for example by Abragam[12] but with the interpretation that the correlation times usually obtained are now elements of R which in turn can be evaluated in terms of generalized collision cross sections[13].

References

1. L. Waldmann, *Z. Naturforsch.* **12a**, 660 (1959).
2. R. F. Snider, *J. Chem. Phys.* **32**, 1051 (1960).
3. J. A. Crawford, *Nuovo Cim.* **10**, 698 (1958).
4. C. N. Banwell and H. Primas, *Mol. Phys.* **6**, 225 (1963).
5. U. Fano, *Revs. Mod. Phys.* **29**, 74 (1957).
6. J. M. Jauch, B. Misra, and A. G. Gibson, *Helv. Phys. Acta* **41**, 513 (1968).
7. U. Fano, *Phys. Rev.* **131**, 259 (1963).
8. C. S. Wang Chang and G. E. Uhlenbeck, "*Transport phenomena in polyatomic gases*", University of Michigan Report CM-681 (1951).
9. F. M. Chen and R. F. Snider, *J. Chem. Phys.* **50**, 4082 (1969).
10. R. F. Snider, *J. Math. Phys.* **5**, 1580 (1964).
11. F. M. Chen and R. F. Snider, *J. Chem. Phys.* **46**, 3937 (1967).
12. A. Abraham, *The Principles of Nuclear Magnetism,* Oxford University Press, Oxford, England, (1961).
13. F. M. Chen and R. F. Snider, *J. Chem. Phys.* **48**, 3185 (1968).

A one-moment approximation to ϕ_1 is made by setting $\phi_1 = \mathbf{b}\cdot\mathbf{I}$ with $\mathbf{b}$ measuring the deviation of the magnetization from equilibrium. Multiplying Eq. (21) by $\mathbf{I}$ and taking the trace gives

$$\frac{\partial \mathbf{b}}{\partial t} + \gamma_I \mathbf{H} \times \mathbf{b} = -\frac{1}{\mathbf{T}}\cdot\mathbf{b} \tag{25}$$

which is just the Bloch equation of NMR. The relaxation tensor $1/\mathbf{T}$ is given by

$$\frac{1}{\mathbf{T}} = 3\frac{\langle\langle \mathbf{I}|\mathscr{R}|\mathbf{I}\rangle\rangle}{nI(I+1)} = \frac{\hat{H}\hat{H}}{T_1} + \frac{\mathbf{U} - \hat{H}\hat{H}}{T_2} + \gamma_I \mathbf{H}\cdot\boldsymbol{\varepsilon} \tag{26}$$

which can be uniquely written in terms of the spin-lattice T_1, spin-spin T_2 relaxation times and the intermolecular chemical shift ε corresponding to the component of $1/\mathbf{T}$ parallel, perpendicular or skew to the field direction $\hat{H}$.

The case for diatomics (and polyatomics) is much more complicated, first of all because the relaxation is dominated by mechanisms coupling the nuclear spin to the rotational angular momentum (direct collisional relaxation of $\mathbf{I}$ is negligible in comparison). This means that both nuclear spin and rotational angular momentum are driven out of equilibrium and moments to describe these correlations must be taken into account. The result of such calculations is to reproduce standard formulae of NMR as given for example by Abragam[12] but with the interpretation that the correlation times usually obtained are now elements of $\mathscr{R}$ which in turn can be evaluated in terms of generalized collision cross sections[13].

References

1. L. Waldmann, *Z. Naturforsch.* **12a**, 660 (1959).
2. R. F. Snider, *J. Chem. Phys.* **32**, 1051 (1960).
3. J. A. Crawford, *Nuovo Cim.* **10**, 698 (1953).
4. C. N. Banwell and H. Primas, *Mol. Phys.* **6**, 225 (1963).
5. U. Fano, *Revs. Mod. Phys.* **29**, 74 (1957).
6. J. M. Jauch, B. Misra and A. G. Gibson, *Helv. Phys. Acta* **41**, 513 (1968).
7. U. Fano, *Phys. Rev.* **131**, 259 (1963).
8. C. S. Wang Chang and G. E. Uhlenbeck, "*Transport phenomena in polyatomic gases*", University of Michigan Report CM-681 (1951).
9. F. M. Chen and R. F. Snider, *J. Chem. Phys.* **50**, 4082 (1969).
10. R. F. Snider, *J. Math. Phys.* **5**, 1580 (1964).
11. F. M. Chen and R. F. Snider, *J. Chem. Phys.* **46**, 3937 (1967).
12. A. Abragam, *The Principles of Nuclear Magnetism*, Oxford University Press, Oxford, England (1961).
13. P. McGuire and R. F. Snider, *J. Chem. Phys.* **48**, 3185 (1968).

PART C

PROPERTIES AND SOLUTIONS

Contributors

J. V. Sengers
Institute for Molecular Physics
University of Maryland
College Park, Maryland

B. U. Felderhof
Institute for Theoretical Physics
Maliesingel
23 Utrecht
The Netherlands

O. Aono
Department of Physics
Faculty of Science, University of Tokyo
Tokyo, Japan

T. Kihara
Department of Physics
Faculty of Science, University of Tokyo
Tokyo, Japan

S. Harris
College of Engineering
SUNY at Stony Brook
Stony Brook, Long Island, New York

C. R. Willis
Physics Department
Boston University
Boston, Massachusetts

C. Oberman

Plasma Physics Laboratory
Princeton University
Princeton, New Jersey

T. F. Morse

Division of Engineering
Brown University
Providence, Rhode Island

J. Cipolla

Division of Engineering
Brown University
Providence, Rhode Island

PAPER C–1

*Triple collision effects in the transport properties for a gas of hard spheres**

J. V. SENGERS

Institute for Molecular Physics
University of Maryland, College Park, Maryland 20742
and National Bureau of Standards, Washington, D.C. 20234

Abstract

A study is made of the effect of successive correlated binary collisions on the transport properties of gases. In particular, triple collision transport integrals are derived for the thermal conductivity and viscosity of a gas of hard spheres. Preliminary numerical estimates for these collision integrals allow us to make an assessment of the contributions from the various triple collision events.

1 INTRODUCTION

Transport properties of gases are directly related to the transfer of energy and momentum between colliding molecules. In dilute non-ionized gases the transport properties are independent of the density. Roughly speaking the number of molecules available for the transport of thermal energy and momentum is proportional to the density, but the distance over which this transport is effected, the mean free path, is in a dilute gas inversely proportional to the density.

The theoretical description of the transport coefficients of a dilute gas is based on the Boltzmann equation, which describes the rate of change

* The research was supported by the Air Force Systems Command, Arnold Engineering Development Center, Tennessee, under delivery order (40–600) 66–494 at the National Bureau of Standards and under delivery order F40600–69–C–0002 at the University of Maryland.

of the molecular velocity distribution as a result of binary collisions between the molecules. Based on the Boltzmann equation the Chapman–Enskog theory provides a reliable procedure to obtain the transport coefficients of dilute gases[1,2]. This theory is sufficiently well understood that in turn experimental transport coefficients of dilute gases are used to obtain information on the molecular interaction potential.

Experience shows that the transport coefficients do vary with the gas density. From a molecular point of view this density dependence has two origins. First, the effect of triple and higher order collisions on the transport properties can no longer be neglected. Secondly, the difference in position of molecules involved in a collision (spatial inhomogeneity) leads to an additional density dependent contribution, sometimes referred to as collisional transfer.

Prediction of the density dependenc of the transport properties is a task for statistical mechanics. For this purpose the Boltzmann equation has to be generalized to include the two effects mentioned above. Up to the present, only approximate theories have been considered. The earliest attempt is the theory of Enskog for hard spherical molecules[3]. In this theory the spatial correlation between the molecules is approximated by the equilibrium radial distribution function and the velocity correlation is disregarded. Several other approximate theories have been considered to describe the first density correction to the transport coefficients of gases[4]. All these theories attempt to describe the transport coefficients in terms of binary collisions and an explicit evaluation of the multiple collision dynamics is avoided. As a result it is difficult to assess the reliability and predictive power of such descriptions.

It is well known that the equation of state of a gas can be represented by a power series in the density ϱ, the virial expansion:

$$pV = RT[1 + B\varrho + C\varrho^2 + \cdots]. \tag{1.1}$$

The various virial coefficients in this expansion can be related to molecular cluster integrals which are integrals over the configuration of n simultaneously interacting molecules. The characteristic size determining these cluster integrals is the range of the molecular interaction. In analogy to the virial expansion (1.1) a power series expansion for the transport coefficients was proposed by Bogoliubov[5].

To discuss the effect of multiple collisions we can distinguish between two types of multiple collisions:

1) Genuine multiple collisions in which all n molecules are interacting simultaneously. It is evident that such collisions contain again the size of the molecular interaction range as a characteristic length.

2) Multiple collisions that consist of a succession of correlated binary collisions between n particles. After a binary collision the velocities of two molecules will remain correlated until at least the next collision. Thus one can expect that the characteristic length associated with successive correlated binary collisions is no longer the interaction range, but the mean free path. However, in contrast to the interaction range, the mean free path is itself dependent on the density. As a result it is no longer evident that the effect of such multiple collisions can be represented by a power series in the density. Actually[6,7,8], it turns out that these multiple collisions lead to a representation of the transport coefficients which contains terms logarithmic in the density ϱ. In particular, for the thermal conductivity λ and the viscosity η we obtain

$$\lambda = \lambda_0 + \lambda_1\varrho + \lambda_2'\varrho^2 \ln \varrho + \lambda_2\varrho^2 + \cdots$$
$$\eta = \eta_0 + \eta_1\varrho + \eta_2'\varrho^2 \ln \varrho + \eta_2\varrho^2 + \cdots \tag{1.2}$$

Here λ_0 and η_0 are the Chapman–Enskog zero density coefficients which depict the effect of binary collisions only. The first density coefficients λ_1 and η_1 include effects from triple collisions. To predict the transport properties of a moderately dense gas it becomes our task to develop theories and procedures by which the various coefficients λ_i and η_i can be computed for a given molecular interaction potential.

The present paper describes the first results of an effort to obtain quantitative estimates of the effect of multiple collisions, in particular of triple collisions, on the transport properties. Evaluation of the triple collision effects is needed to predict the first density coefficients λ_1 and η_1 in (1.2). Since a study of the effect of triple collisions turns out to be rather complicated, we limit ourselves in this paper to a study of the effects for a very simple molecular model, namely hard spheres.

The transport properties describe the linear response of the gas to the presence of gradients. The dilute gas values λ_0 and η_0 in (1.2) are obtained from the solution of the linearized Boltzmann equation. As a result they can be expressed in terms of *binary collision integrals*. (For hard spheres these results are summarized in Section 3). For repulsive intermolecular potentials the first density corrections are in principle to be determined by solving the Choh–Uhlenbeck integral equation which is a generalization of the linear Boltzmann equation to include the effect of triple collisions. In this paper we develop a procedure to solve this equation for a gas of hard spheres. The first density corrections λ_1 and η_1 are related to well defined *triple collision transport integrals* and preliminary values for these triple collision integrals are presented. The results are discussed in Section 9 and compared with the approximate predictions of the theory of Enskog.

2 THE DISTRIBUTION FUNCTION METHOD

In the kinetic theory of gases the transport coefficients are usually related to the velocity distribution function which specifies the state of the gas[1,2]. This velocity distribution function $f(\mathbf{r}, \mathbf{v}; t)$ is defined such that $f(\mathbf{r}, \mathbf{v}; t)$ $d\mathbf{v}\, d\mathbf{r}$ determines the number of molecules at a time t with position coordinates in the range $d\mathbf{r}$ around $\mathbf{r}$ and with velocities in the range $d\mathbf{v}$ around $\mathbf{v}$. The number density n and the mean velocity $\mathbf{u}$ at $\mathbf{r}$ and t are related to this distribution function by

$$n(\mathbf{r}; t) = \int f(\mathbf{r}, \mathbf{v}; t)\, d\mathbf{v}, \tag{2.1}$$

$$n(\mathbf{r}; t)\, \mathbf{u}(\mathbf{r}; t) = \int \mathbf{v} f(\mathbf{r}, \mathbf{v}; t)\, d\mathbf{v}. \tag{2.2}$$

Denoting the thermal velocity of a molecule by $\mathbf{V} = \mathbf{v} - \mathbf{u}$, the average local kinetic energy is

$$\frac{3}{2} n(\mathbf{r}; t)\, kT(\mathbf{r}; t) = \int \frac{1}{2} m V^2 f(\mathbf{r}, \mathbf{v}; t)\, d\mathbf{v}, \tag{2.3}$$

where m is the mass of the molecules. Since each molecule contributes $\frac{1}{2} m V^2 \mathbf{V}$ to the flow of thermal energy, the total heat flux vector reads

$$\mathbf{Q} = \frac{1}{2} m \int V^2 \mathbf{V} f(\mathbf{r}, \mathbf{v}; t)\, d\mathbf{v}. \tag{2.4}$$

The transport of momentum — or pressure tensor reads

$$\mathsf{P} = m \int \mathbf{V}\mathbf{V} f(\mathbf{r}, \mathbf{v}; t)\, d\mathbf{v}. \tag{2.5}$$

In the equilibrium state the distribution function reduces to the Maxwell–Boltzmann distribution

$$f(V) = n\left(\frac{m}{2\pi kT}\right)^{3/2} \exp\left[-\frac{mV^2}{2kT}\right]. \tag{2.6}$$

In equilibrium the heat flux vector (2.4) vanishes and the pressure tensor becomes diagonal

$$\mathbf{P} = p\mathbf{I}, \tag{2.7}$$

where $\mathbf{I}$ is the unit tensor and p the hydrostatic pressure.

When the gas is slightly out of equilibrium the heat flux vector and the off diagonal elements of the pressure tensor are proportional to the gradients in temperature T and fluid velocity $\mathbf{u}$, respectively:

$$\mathbf{Q} = -\lambda\nabla T = -\lambda\frac{\partial T}{\partial \mathbf{r}}, \tag{2.8}$$

$$\mathbf{P} = p\mathbf{I} - 2\eta \overset{0}{\overline{\overline{\frac{\partial}{\partial \mathbf{r}}\mathbf{u}}}}, \tag{2.9}$$

where $\overset{0}{\overline{\overline{\frac{\partial}{\partial \mathbf{r}}\mathbf{u}}}}$ is the symmetric traceless tensor derived from the dyadic $\frac{\partial}{\partial \mathbf{r}}\mathbf{u}$.[1] These linear relations define the coefficient of thermal conductivity λ and the coefficient of shear viscosity η*. On comparing these definitions with the molecular expressions (2.4) and (2.5) for the fluxes the transport coefficients are related to the distribution function f. For this purpose f is expanded in terms of the gradients of the macroscopic variables and only terms linear in the gradients are retained. In zeroth approximation f reduces to the local Maxwell distribution

$$f_0 = n\left(\frac{m}{2\pi kT}\right)^{3/2} \exp\left[-\frac{m(\mathbf{v}-\mathbf{u})^2}{2kT}\right], \tag{2.10}$$

where the macroscopic variables still depend on $\mathbf{r}$ and t. In first approximation

$$f = f_0(1 + \Phi), \tag{2.11}$$

where Φ is linear in the gradients of T and $\mathbf{u}$. For the time scale under consideration f, and consequently Φ, does not very explicitly with time, but only via the dependence on the macroscopic variables $n(\mathbf{r}; t)$, $\mathbf{u}(\mathbf{r}; t)$ and $T(\mathbf{r}; t)$:

$$\frac{\partial f}{\partial t} = \frac{\partial f}{\partial n}\frac{\partial n}{\partial t} + \frac{\partial f}{\partial \mathbf{u}}\cdot\frac{\partial \mathbf{u}}{\partial t} + \frac{\partial f}{\partial T}\frac{\partial T}{\partial t}. \tag{2.12}$$

* In this paper we do not consider the bulk viscosity, since both its dilute gas value and its first density coefficient vanish.

From the fact that Φ, like f itself, is a scalar function, one concludes that Φ must have the form

$$\Phi(\mathbf{V}) = -A(V)\,\mathbf{V}\cdot\frac{\partial \ln T}{\partial \mathbf{r}} - B(V)\,\mathbf{V}^{\mathrm{o}}\mathbf{V}\colon\frac{\partial}{\partial \mathbf{r}}\mathbf{u}, \tag{2.13}$$

where $A(V)$ and $B(V)$ are scalar functions of the molecular velocity $\mathbf{V}$. To ascertain that n, $\mathbf{u}$ and T in f_0 keep satisfying their definitions (2.1), (2.2) and (2.3) in the presence of gradients, the function Φ is subject to the auxiliary conditions

$$\int f_0 \Phi \, d\mathbf{v} = 0, \tag{2.14}$$

$$\int \mathbf{V} f_0 \Phi \, d\mathbf{v} = 0, \tag{2.15}$$

$$\int V^2 f_0 \Phi \, d\mathbf{v} = 0. \tag{2.16}$$

Specifically, (2.15) implies that

$$\int f_0 V^2 A(V) \, d\mathbf{v} = 0. \tag{2.17}$$

When we substitute this form of the distribution function into the molecular expressions (2.4) and (2.5) for the fluxes and compare the result with the definitions (2.8) and (2.9) of the transport coefficients, we obtain

$$\lambda = \frac{k}{3}\int d\mathbf{V}\, f_0(V)\left(\frac{mV^2}{2kT} - \frac{5}{2}\right) V^2 A(V), \tag{2.18}$$

$$\eta = \frac{m}{10}\int d\mathbf{V}\, f_0(V)\,\mathbf{V}^{\mathrm{o}}\mathbf{V}\colon\mathbf{V}^{\mathrm{o}}\mathbf{V}\, B(V). \tag{2.19}$$

Thus it becomes our task to assess the effect of the triple collisions on the functions $A(V)$ and $B(V)$.

3 TRANSPORT COEFFICIENTS FOR A DILUTE GAS OF HARD SPHERES

The first density correction to the transport coefficients is not only caused by a triple collision contribution to the kinetic fluxes (2.4) and (2.5), but also by an additional collisional transfer contribution from binary collisions. The latter term results from the difference in position of two molecules involved in a binary collision. For a gas of hard spheres the theory of Enskog presents an estimate for both contributions[3]. It turns out that the collisional transfer contribution to the first density contribution as evaluated by Enskog is exact[9]. Therefore, only the effect of triple collisions remains to be evaluated and it is sufficient for our present considerations to limit ourselves to the so called spatially homogeneous case. That is, the distribution-function f is considered as a function of $\mathbf{v}$ ant t and the gradients of the macroscopic variables, but its explicit dependence on the position $\mathbf{r}$ is disregarded.

Before discussing the effect of triple collisions we first summarize how the transport coefficients of a dilute gas are obtained when only binary collisions are taken into account. This will enable us to elucidate how the triple collision transport integrals are generalizations of the well known binary collision transport integrals.

In a dilute gas the time variation of f is described by the Boltzmann equation

$$\frac{\partial f(\mathbf{v}_1; t)}{\partial t} = J(ff), \tag{3.1}$$

where the collision term $J(ff)$ represents the effects of binary collisions. For a gas of hard spherical molecules this collision term $J(ff)$ is given by[1]

$$J(ff) = \int d\mathbf{v}_2 \int d\mathbf{k}\sigma^2(\mathbf{v}_{21} \cdot \mathbf{k}) \left[f(\mathbf{v}_1'; t) f(\mathbf{v}_2'; t) - f(\mathbf{v}_1; t) f(\mathbf{v}_2; t)\right]. \tag{3.2}$$

Here σ is the diameter of the molecules, $\mathbf{v}_{21}$ the relative velocity $\mathbf{v}_2 - \mathbf{v}_1$ and $\mathbf{k}$ the perihelion vector specifying the geometry of a binary collision between

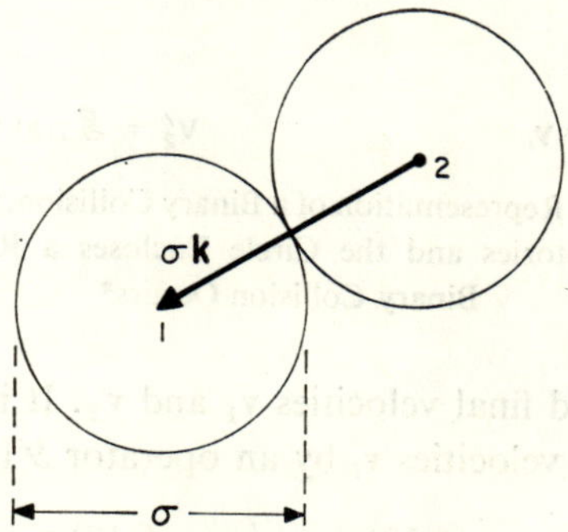

FIG. 1 The Perihelion Vector **k** of a Binary Collision

molecules 1 and 2. For hard spheres this perihelion vector $\mathbf{k}$ is the unit vector in the direction from the center of particle 2 to the center of particle 1 at the time of collision.

The distribution function $f(\mathbf{v}_1; t)$ changes with time, since during a time interval dt certain molecules with velocity $\mathbf{v}_1$ will collide with molecules with velocity $\mathbf{v}_2$. The number of such collisions with perihelion vector $\mathbf{k}$ will be proportional to $f(\mathbf{v}_1; t) f(\mathbf{v}_2; t)$ and the differential cross section

$\sigma^2(\mathbf{v}_{21} \cdot \mathbf{k})\, d\mathbf{k}$. On the other hand, molecules with velocities $\mathbf{v}_1'$ will collide with molecules $\mathbf{v}_2'$ such that after the collision their velocities will become $\mathbf{v}_1$ and $\mathbf{v}_2$. Thus $\mathbf{v}_1'$ and $\mathbf{v}_2'$ are the initial velocities for a binary collision with

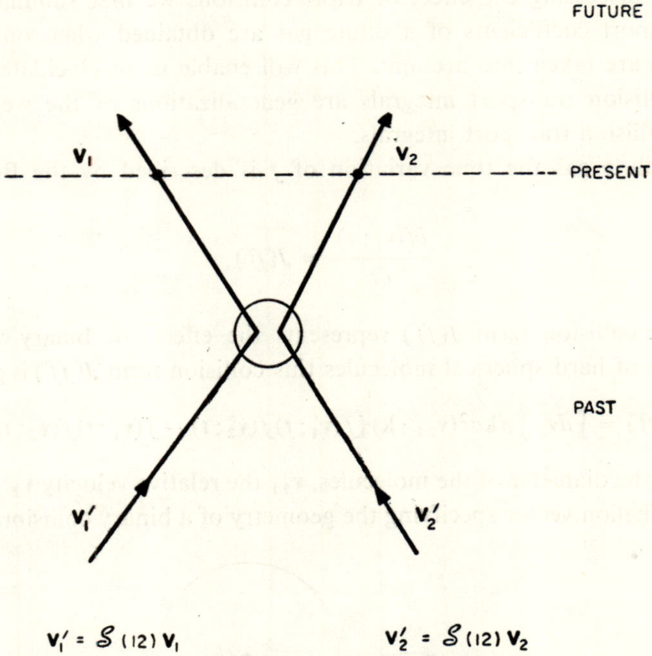

FIG. 2 Schematic Representation of a Binary Collision. The lines represent the Particle Trajectories and the Circle Encloses a Region in which a Binary Collision Occurs*

perihelion vector $\mathbf{k}$ and final velocities $\mathbf{v}_1$ and $\mathbf{v}_2$. It is convenient to relate the velocities $\mathbf{v}_i'$ to the velocities $\mathbf{v}_i$ by an operator $\mathscr{S}(12)$

$$\mathbf{v}_1' = \mathscr{S}(12)\, \mathbf{v}_1; \; \mathbf{v}_2' = \mathscr{S}(12)\, \mathbf{v}_2. \tag{3.3}$$

In general the substitution operator $\mathscr{S}(1 \ldots s)$ associated with s particles is defined as[5]

$$\mathscr{S}(1 \ldots s) = \lim_{t \to \infty} S_{-t}(1 \ldots s) \prod_{i=1}^{s} S_{+t}(i), \tag{3.4}$$

* Using the concept of inverse collisions one usually considers $\mathbf{v}_1', \mathbf{v}_2'$ as the initial velocities for a binary collision with a perihelion vector $-\mathbf{k}$ and final velocities $\mathbf{v}_1, \mathbf{v}_2$. Since a generalization of the concept of inverse collisions to triple collisions is not obvious, we use here the Bogoliubov form[5] of the Boltzmann equation, in which the consideration of inverse encounters is avoided.

where $S_t(1 \ldots s)$ transforms the coordinates and the velocities of s particles into those at a time t later, if the s particles move under their mutual interaction. The effect of the operator $\mathscr{S}(12)$ on $\mathbf{v}_1$ and $\mathbf{v}_2$ is schematically represented by Fig. 2. If the positions and the velocities of the particles are such that they did not collide in the past, the operator $\mathscr{S}(12)$ reduces to one. The particles traverse the trajectories indicated in Fig. 2 from bottom to top. However, since the present is determined by the past, we actually retrace the history of the particles in the past*, that is, we read the diagrams always from top to bottom. The effect of the substitution operators related to triple collisions will be indicated later by similar diagrams. In terms of the substitution operator $\mathscr{S}(12)$ the Boltzmann equation can be written

$$\frac{\partial f(\mathbf{v}_1; t)}{\partial t} = \sigma^2 \int d\mathbf{v}_2 \int d\mathbf{k}(\mathbf{v}_{21} \cdot \mathbf{k})\, T(12) f(\mathbf{v}_1; t) f(\mathbf{v}_2; t), \tag{3.5}$$

with

$$T(12) = \mathscr{S}(12) - 1. \tag{3.6}$$

For a derivation of the transport coefficients it is sufficient to consider the linearized Boltzmann equation. That is, we approximate f by $f_0(1 + \Phi)$ as in (2.11), eliminate the time derivatives of n, $\mathbf{u}$ and T in (2.12) using the Euler equations and retain only terms linear in the gradients[1]. The resulting linearized Boltzmann equation which can be found in many places in the literature, reads

$$I_2(\Phi(\mathbf{V}_1)) = -f_0(V_1)\left[\left(\frac{mV_1^2}{2kT} - \frac{5}{2}\right)\mathbf{V}_1 \cdot \frac{\partial \ln T}{\partial \mathbf{r}_1} + \frac{m}{kT}\mathbf{V}_1^0\mathbf{V}_1 : \frac{\partial}{\partial \mathbf{r}_1}\mathbf{u}\right], \tag{3.7}$$

where I_2 is a linearized binary collision integral operator

$$I_2(\Phi(\mathbf{V}_1)) = -\sigma^2 \int d\mathbf{V}_2 \int d\mathbf{k}(\mathbf{v}_{21} \cdot \mathbf{k}) f_0(V_1) f_0(V_2)\, T(12)\, [\Phi(\mathbf{V}_1) + \Phi(\mathbf{V}_2)]. \tag{3.8}$$

Since the total kinetic energy is invariant under a binary collision, the operator $T(12)$ commutes with $f_0(V_1) f_0(V_2)$.

Using the specific form (2.13) for the perturbation function Φ we obtain linearized integral equations for the functions $A(V_1)$ and $B(V_1)$

$$I_2(\mathbf{V}_1 A^{(0)}(V_1)) = f_0(V_1)\left(\frac{mV_1^2}{2kT} - \frac{5}{2}\right)\mathbf{V}_1,$$

$$I_2(\mathbf{V}_1^0\mathbf{V}_1 B^{(0)}(V_1)) = f_0(V_1)\frac{m}{kT}\mathbf{V}_1^0\mathbf{V}_1, \tag{3.9}$$

* This can be circumvented by changing the direction of time as will be done in a later stage.

where the superscript (0) indicates that we consider only the contribution of zeroth order in the density here.

To solve these integral equations one customarily approximates the solution by a finite sum of Sonine polynomials $S_n^{(m)}$:

$$A^{(0)}(V_1) = -\frac{15}{32}\frac{1}{n\sigma^2}\left(\frac{m}{\pi kT}\right)^{1/2}\sum_{k=1}^{N} a_k^{(0)}(N)\, S_{3/2}^{(k)}(W_1^2),$$

$$B^{(0)}(V_1) = \frac{5}{16}\frac{1}{n\sigma^2}\left(\frac{m}{\pi kT}\right)^{1/2}\frac{m}{kT}\sum_{k=0}^{N-1} b_k^{(0)}(N)\, S_{5/2}^{(k)}(W_1^2), \tag{3.10}$$

where we have introduced a dimensionless velocity

$$\mathbf{W}_i = \left(\frac{m}{2kT}\right)^{1/2}\mathbf{V}_i. \tag{3.11}$$

The Sonine polynominal $S_n^{(m)}(x)$ is defined as the coefficient of t^m in the expansion

$$\sum_{m=0}^{\infty} S_n^{(m)}(x)\, t^m = \frac{1}{(1-t)^{n+1}}\exp\left\{-\frac{xt}{1-t}\right\}. \tag{3.12}$$

They become the eigenfunctions of the linearized Boltzmann operator I_2 when the molecular interaction potential obeys an inverse fourth power law (Maxwell molecules). The number N of Sonine polynomials taken into account in the approximate solution (3.10) is referred to as the order of the Enskog approximation. If we substitute the solution (3.10) into the expressions (2.18) and (2.19) for the transport coefficients, we obtain

$$\lambda_0 = \lambda_0(N) = \frac{75}{64\sigma^2}\left(\frac{k^3T}{m\pi}\right)^{1/2} a_1^{(0)}(N),$$

$$\eta_0 = \eta_0(N) = \frac{5}{16\sigma^2}\left(\frac{mkT}{\pi}\right)^{1/2} b_0^{(0)}(N). \tag{3.13}$$

The values of the coefficients $a_k^{(0)}(N)$ and $b_k^{(0)}(N)$ are determined by a variational method[2]; they should satisfy a set of linear equations

$$\sum_{k=1}^{N} a_k^{(0)}(N)\, a_{kl}^{(0)} = \delta_{l1} \quad (l = 1, \dots, N),$$

$$\sum_{k=0}^{N-1} b_k^{(0)}(N)\, b_{kl}^{(0)} = \delta_{l0} \quad (l = 0, \dots, N-1), \tag{3.14}$$

where δ_{lm} is the Kronecker delta and where $a_{kl}^{(0)}$ and $b_{kl}^{(0)}$ are matrix elements representing inner products associated with the linearized Boltzmann operator:

$$a_{kl}^{(0)} = \frac{1}{2\sqrt{2\pi}} \frac{kT}{mn^2\sigma^2} \int d\mathbf{W}_1 \mathbf{W}_1\, S_{3/2}^{(k)}(W_1^2) \cdot I_2(\mathbf{W}_1 S_{3/2}^{(l)}(W_1^2)),$$

$$b_{kl}^{(0)} = \frac{1}{2\sqrt{2\pi}} \frac{kT}{mn^2\sigma^2} \int d\mathbf{W}_1 \mathbf{W}_1^{\circ} \mathbf{W}_1\, S_{5/2}^{(k)}(W_1^2) : I_2(\mathbf{W}_1^{\circ} \mathbf{W}_1 S_{5/2}^{(l)}(W_1^2)). \quad (3.15)$$

The equations have been normalized such that for hard spheres $a_1^{(0)}(1) = b_0^{(0)}(1) = a_{11}^{(1)} = b_{00}^{(1)} = 1$. The operator I_2 is a symmetric operator, so that

$$a_{kl}^{(0)} = a_{lk}^{(0)};\ b_{kl}^{(0)} = b_{lk}^{(0)}. \quad (3.16)$$

For hard spheres the matrix elements can be readily evaluated. Using the generating function (3.12) for $S_n^{(m)}(W_1^2)$, the elements $a_{kl}^{(0)}$ and $b_{kl}^{(0)}$ are the coefficients of $s^k t^l$ in the expansions[10]

$$\sum_{k=1}^{\infty} \sum_{l=1}^{\infty} a_{kl}^{(0)} s^k t^l =$$

$$= \left\{1 - \frac{1}{2}(s+t)\right\}^{1/2} st(1-st)^{-3} - \frac{1}{4}\left\{1 - \frac{1}{2}(s+t)\right\}^{-1/2} s^2t^2(1-st)^{-2}, \quad (3.17)$$

$$\sum_{k=0}^{\infty} \sum_{l=0}^{\infty} b_{kl}^{(0)} s^k t^l =$$

$$= \left\{1 - \frac{1}{2}(s+t)\right\}^{1/2} \left(1 + \frac{2}{3}st + \frac{1}{3}s^2t^2\right)(1-st)^{-4} -$$

$$\frac{1}{3}\left\{1 - \frac{1}{2}(s+t)\right\}^{-1/2} st(1-st)^{-3} - \frac{1}{48}\left\{1 - \frac{1}{2}(s+t)\right\}^{-3/2} s^2t^2(1-st)^{-2}.$$

For hard spheres the second Enskog approximation amounts to only a few percent. For more realistic molecular interactions the convergence is even faster[2].

4 THE THEORY OF ENSKOG

The transport coefficients λ_0 and η_0 deduced from the Boltzmann equation are independent of the density. A first attempt to generalize the Boltzmann equation to account for the density dependence of the transport coefficients was made by Enskog[3]. Strictly speaking, the number of molecules experiencing a binary collision with perihelion vector $\mathbf{k}$ will be proportional to the pair distribution function $f^{(2)}(\mathbf{r}_1, \mathbf{v}_1, \mathbf{r}_1 - \sigma\mathbf{k}, \mathbf{v}_2; t)$. In equilibrium the pair distribution function $f_0^{(2)}$ is related to the one particle distribution function $f_0^{(1)} = f_0$ by

$$f_0^{(2)}(\mathbf{r}_1, \mathbf{v}_1, \mathbf{r}_1 - \sigma\mathbf{k}, \mathbf{v}_2) = g(\sigma) f_0(v_1) f_0(v_2), \tag{4.1}$$

where $g(\sigma)$ is the equilibrium radial distribution function at a separation σ. This equilibrium radial distribution function is independent of the velocities and depends only on the relative distance $r_{21}(= \sigma$ at a collision). Enskog *assumed* the same relationship between the pair distribution function and f, when the system is not in equilibrium[9]:

$$f^{(2)}(\mathbf{r}_1, \mathbf{v}_1, \mathbf{r}_1 - \sigma\mathbf{k}, \mathbf{v}_2; t) = g(\sigma) f(\mathbf{v}_1; t) f(\mathbf{v}_2; t), \tag{4.2}$$

where $g(\sigma)$ is now the *local* equilibrium radial distribution function depending on the local density and temperature. Thus, again disgarding the spatial dependence of the velocity distribution function f in our present discussion, Enskog generalized the collision term $J(ff)$ in (3.1) to

$$J^E(ff) = \sigma^2 g(\sigma) \int d\mathbf{v}_2 \int d\mathbf{k} (\mathbf{v}_{21} \cdot \mathbf{k})\, T(12) f(\mathbf{v}_1; t) f(\mathbf{v}_2; t) = g(\sigma)\, J(ff). \tag{4.3}$$

If we consider again a perturbation linear in the gradients

$$f = f_0(1 + \Phi), \tag{4.4}$$

the linearized Boltzmann equation (3.7) is to be replaced with

$$g(\sigma)\, I_2(\Phi^E(\mathbf{V}_1)) = -f_0(V_1)\left[\left(\frac{mV_1^2}{2kT} - \frac{5}{2}\right)\mathbf{V}_1 \cdot \frac{\partial \ln T}{\partial \mathbf{r}} + \frac{m}{kT}\mathbf{V}_1^0\mathbf{V}_1 : \frac{\partial}{\partial \mathbf{r}_1}\mathbf{u}\right]. \tag{4.5}$$

This equation is identical to (3.7) if we identify $\Phi^E(\mathbf{V}_1)$ with $g^{-1}(\sigma)\,\Phi(\mathbf{V}_1)$. Consequently, the theory of Enskog predicts that the (kinetic) transport coefficients of a dense gas of hard spheres are related to the dilute gas values by

$$\lambda^E = \frac{\lambda_0}{g(\sigma)}; \; \eta^E = \frac{\eta_0}{g(\sigma)}. \tag{4.6}$$

The value $g(\sigma)$ of the radial distribution function is directly related to the equation of state

$$p = nkT\{1 + b\varrho g(\sigma)\}, \tag{4.7}$$

where

$$b\varrho = \frac{2\pi}{3} n\sigma^3, \tag{4.8}$$

is the covolume of the spheres. The pressure p, as well as $g(\sigma)$ can be expanded into a power series in terms of the density

$$g(\sigma) = 1 + \frac{5}{8} b\varrho + \ldots. \tag{4.9}$$

The coefficient of the first density dependent correction term in $g(\sigma)$ is determined by the so-called excluded volume of a third particle; that is,

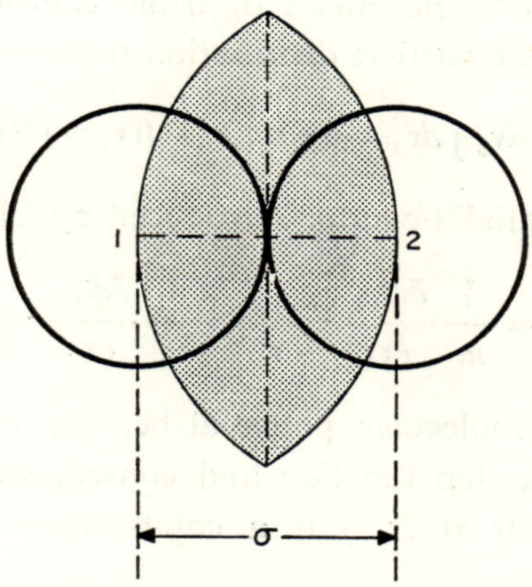

FIG. 3 The Excluded Volume Determining the First Density Correction to $g(\sigma)$

the volume where a third particle would overlap with the two particles 1 and 2 simultaneously.

The power series (4.9) implies a power series for the transport coefficients (4.6):

$$\lambda = \lambda_0 + \lambda_1 b\varrho + \ldots,$$
$$\eta = \eta_0 + \eta_1 b\varrho + \ldots. \tag{4.10}$$

In the sequel we shall refer to λ_1 and η_1 as the first density coefficient of thermal conductivity, resp. viscosity. From (4.6) and (4.9) it is evident that in the Enskog theory λ_1 and η_1 are approximated by

$$\lambda_1^E = -\frac{5}{8}\lambda_0;\ \eta_1^E = -\frac{5}{8}\eta_0, \tag{4.11}$$

independent of the order of the Enskog approximation in λ_0 and η_0.

5 THE CHOH–UHLENBECK INTEGRAL EQUATION

The Enskog theory was based on an ad hoc assumption (4.2). A systematic procedure to take into account successive higher order contributions in the density was proposed by Bogoliubov[5]. For the first density correction the theory was worked out by Choh and Uhlenbeck[12] and further elucidated by Green[13] and Cohen[14]. Limiting ourselves to the first density correction, the Boltzmann equation (3.1) is to be replaced with

$$\frac{\partial f(\mathbf{v}_1;t)}{\partial t} = J(ff) + K(fff), \tag{5.1}$$

where $K(fff)$ represents the effects of triple collisions and is related to the product of *three* one particle distribution functions:

$$K(fff) = \int d\mathbf{v}_2 d\mathbf{v}_3 \int d\mathbf{r}_2 d\mathbf{r}_3 \theta_{12}\tau(12;3) f(\mathbf{v}_1;t) f(\mathbf{v}_2;t) f(\mathbf{v}_3;t). \tag{5.2}$$

Here θ_{12} is an differential operator introduced by Choh and Uhlenbeck[12]

$$\theta_{12} = \frac{1}{m}\frac{\partial\phi_{12}}{\partial\mathbf{r}_1}\cdot\frac{\partial}{\partial\mathbf{v}_1} + \frac{1}{m}\frac{\partial\phi_{12}}{\partial\mathbf{r}_2}\cdot\frac{\partial}{\partial\mathbf{v}_2}, \tag{5.3}$$

where ϕ_{12} is the intermolecular potential between particles 1 and 2. For hard spheres ϕ_{12} is a step function and consequently θ_{12} contains a δ-function. The operator $\tau(12; 3)$ is a combination of substitution operators[13,14]

$$\tau(12;3) = \mathscr{S}(123) - \mathscr{S}(12)\,\mathscr{S}(13) - \mathscr{S}(12)\,\mathscr{S}(23) + \mathscr{S}(12), \tag{5.4}$$

where $\mathscr{S}(123)$, defined by (3.4), is a generalization of the two particle substitution operator $\mathscr{S}(12)$ and includes the interaction between three particles. Apart from neglecting collisions between three particles that actually occur, the Boltzmann equation took into account some binary collisions that actually will not occur, due to the interference of a third particle. Thus in $\tau(12; 3)$ we subtract certain combinations of two particle streaming operators.

To derive the transport coefficients we again linearize the equation by substituting $f = f_0(1 + \Phi)$. As a result the linearized Boltzmann equation (3.7) is replaced with

$$I_2(\Phi(\mathbf{V}_1)) + b\varrho I_3(\Phi(\mathbf{V}_1)) =$$
$$-f_0(V_1)\left[\left(\frac{mV_1^2}{2kT} - \frac{5}{2}\right)\mathbf{V}_1\cdot\frac{\partial \ln T}{\partial \mathbf{r}_1} + \frac{m}{kT}\mathbf{V}_1^0\mathbf{V}_1 : \frac{\partial}{\partial \mathbf{r}_1}\mathbf{u}\right], \tag{5.5}$$

where I_3 is a linearized ternary collision integral operator

$$I_3(\Phi(\mathbf{V}_1)) = -\frac{1}{b\varrho}\int d\mathbf{v}_2\, d\mathbf{v}_3 \int d\mathbf{r}_2\, d\mathbf{r}_3 \theta_{12}\tau(12;3) f_0(V_1) f_0(V_2 f_0(V_3)$$
$$\times [\Phi(\mathbf{V}_1) + \Phi(\mathbf{V}_2) + \Phi(\mathbf{V}_3)]. \tag{5.6}$$

For convenience a factor $(b\varrho)^{-1}$ has been entered in the definition of I_3, so that I_3 has the same dimension as I_2. From the right hand side of (5.5) it is evident that $\Phi(\mathbf{V}_1)$ will have again the form (2.13). To solve the equation we expand the functions $A(V_1)$ and $B(V_1)$ in terms of the density:

$$A(V_1) = A^{(0)}(V_1) + b\varrho A^{(1)}(V_1),$$
$$B(V_1) = B^{(0)}(V_1) + b\varrho B^{(1)}(V_1). \tag{5.7}$$

The corresponding expansion for the transport coefficients was introduced in (4.10). From the expressions (2.18) and (2.19) we see that the first density coefficients λ_1 and η_1 are given by

$$\lambda_1 = \frac{k}{3}\int d\mathbf{V}_1 f_0(V_1)\left(\frac{mV_1^2}{2kT} - \frac{5}{2}\right) V^2 A^{(1)}(V_1),$$
$$\eta_1 = \frac{m}{10}\int d\mathbf{V}_1 f_0(V_1)\, \mathbf{V}_1^0\mathbf{V}_1 : \mathbf{V}_1^0\mathbf{V}_1 B^{(1)}(V_1). \tag{5.8}$$

The integral equations for $A^{(1)}(V_1)$ and $B^{(1)}(V_1)$ are derived by substituting (5.7) into (2.13) and (5.5). Since $A^{(0)}(V_1)$ and $B^{(0)}(V_1)$ satisfy the linearized Boltzmann equation (3.9), we obtain

$$I_2(\mathbf{V}_1 A^{(1)}(V_1)) = -I_3(\mathbf{V}_1 A^{(0)}(V_1)),$$
$$I_2(\mathbf{V}_1^0\mathbf{V}_1 B^{(1)}(V_1)) = -I_3(\mathbf{V}_1^0\mathbf{V}_1 B^{(0)}(V_1)), \tag{5.9}$$

neglecting terms quadratic in the density. In these equations, to which we refer as Choh–Uhlenbeck integral equation, the new ternary collision operator I_3 operates on the functions $A^{(0)}(V_1)$ and $B^{(0)}(V_1)$ that are known from the solution of the linearized Boltzmann equation. The kernel of the Choh-Uhlenbeck integral equation is the same as that of the linearized Boltzmann equation. In principle, one could solve this equation by expanding $A^{(1)}$ and $B^{(1)}$ again in terms of Sonine polynomials[15]. However, we are not interested in $A^{(1)}$ and $B^{(1)}$ as such, but only in the moments (5.8) of these functions, which can be written explicitly in terms of $A^{(0)}$

and $B^{(0)}$*. For this purpose we substitute (3.9) into (5.8)

$$\lambda_1 = \frac{k}{3} \int d\mathbf{V}_1\, \mathbf{V}_1 A^{(1)}(V_1) \cdot I_2(\mathbf{V}_1 A^{(0)}(V_1)),$$

$$\eta_1 = \frac{kT}{10} \int d\mathbf{V}_1\, \mathbf{V}_1^0 \mathbf{V}_1\, B^{(1)}(V_1) : I_2(\mathbf{V}_1^0 \mathbf{V}_1 B^{(0)}(V_1)). \tag{5.10}$$

Since I_2 is a symmetric operator, we can interchange $A^{(0)}$ and $A^{(1)}$, resp. $B^{(0)}$ and $B^{(1)}$, in (5.10). Thus it follows from (5.9) that λ_1 and η_1 are related to $A^{(0)}$ and $\Gamma^{(0)}$ by

$$\boxed{\begin{aligned} \lambda_1 &= -\frac{k}{3} \int d\mathbf{V}_1\, \mathbf{V}_1 A^{(0)}(V_1) \cdot I_3(\mathbf{V}_1 A^{(0)}(V_1)), \\ \eta_1 &= -\frac{kT}{10} \int d\mathbf{V}_1\, \mathbf{V}_1^0 \mathbf{V}_1 B^{(0)}(V_1) : I_3(\mathbf{V}_1^0 \mathbf{V}_1 B^{(0)}(V_1)). \end{aligned}} \tag{5.11}$$

By substituting the Sonine polynomial expansion (3.10) for $A^{(0)}$ and $B^{(0)}$ (with known coefficients) into (5.11) we can evaluate λ_1 and η_1 in successive Enskog approximations. Introducing again dimensionless velocities $\mathbf{W}_i = (m/2kT)^{1/2}\, \mathbf{V}_i$, we obtain in the Nth Enskog approximation

$$\frac{\lambda_1(N)}{\lambda_0(1)} = -\sum_{k=1}^{N} \sum_{l=1}^{N} a_k^{(0)}(N)\, a_l^{(0)}(N)\, a_{kl}^{(1)},$$

$$\frac{\eta_1(N)}{\eta_0(1)} = -\sum_{k=0}^{N-1} \sum_{l=0}^{N-1} b_k^{(0)}(N)\, b_l^{(0)}(N)\, b_{kl}^{(1)}. \tag{5.12}$$

The new matrix elements $a_{kl}^{(1)}$ and $b_{kl}^{(1)}$ are inner products associated with the linearized triple collision integral operator I_3:

$$a_{kl}^{(1)} = \frac{1}{2\sqrt{2\pi}} \frac{kT}{mn^2\sigma^2} \int d\mathbf{W}_1 \mathbf{W}_1 S_{3/2}^{(k)}(W_1^2) \cdot I_3(\mathbf{W}_1 S_{3/2}^{(l)}(W_1^2)),$$

$$b_{kl}^{(1)} = \frac{1}{2\sqrt{2\pi}} \frac{kT}{mn^2\sigma^2} \int d\mathbf{W}_1 \mathbf{W}_1^0 \mathbf{W}_1 S_{5/2}^{(k)}(W_1^2) : I_3(\mathbf{W}_1^0 \mathbf{W}_1 S_{5/2}^{(l)}(W_1^2)). \tag{5.13}$$

These matrix elements are the immediate generalizations of the matrix elements $a_{kl}^{(0)}$ and $b_{kl}^{(0)}$, defined by (3.15), that determine the dilute gas values

* The author is indebted to Dr. M. H. Ernst for this remark.

λ_0 and η_0; they were first introduced in a previous publication[15]. For convenience, we have divided the first density coefficients $\lambda_1(N)$ and $\eta_1(N)$ by the dilute gas values $\lambda_0(1)$ and $\eta_0(1)$, so that the ratios in (5.12) are independent of the temperature.

It is convenient to symmetrize the integrands in (5.13). The operator I_3 can be symmetrized as shown by Green[16]

$$I_3(\Phi(\mathbf{V}_1)) = -\frac{1}{2b\varrho} \int d\mathbf{V}_2 d\mathbf{V}_3 0(123)\, f_0(V_1)\, f_0(V_2)\, f_0(V_3) \sum_{n=1}^{3} \Phi(\mathbf{V}_n), \tag{5.14}$$

where $0(123)$ is a configurational triple collision integral operator

$$0(123) = \int d\mathbf{r}_2\, d\mathbf{r}_3 [\theta_{12}\, \tau(12;3) + \theta_{13}\, \tau(13;2) + \theta_{23}\, \tau(23;1)]. \tag{5.15}$$

The integrands in (5.13) can be further symmetrized by interchanging the velocity variables. Thus we obtain

$$a_{kl}^{(1)} = -\frac{1}{32\pi^6\sigma^5}\left(\frac{m}{kT}\right)^{1/2} \int d\mathbf{W}_1\, d\mathbf{W}_2\, d\mathbf{W}_3 \exp\left[-(W_1^2 + W_2^2 + W_3^2)\right]$$

$$\times \sum_{m=1}^{3} \mathbf{W}_m S_{3/2}^{(k)}(W_m^2) \cdot 0(123) \sum_{n=1}^{3} \mathbf{W}_n S_{3/2}^{(l)}(W_n^2),$$

$$b_{kl}^{(1)} = -\frac{1}{32\pi^6\sigma^5}\left(\frac{m}{kT}\right)^{1/2} \int d\mathbf{W}_1\, d\mathbf{W}_2\, d\mathbf{W}_3 \exp\left[-(W_1^2 + W_2^2 + W_3^2)\right]$$

$$\times \sum_{m=1}^{3} \mathbf{W}_m^0 \mathbf{W}_m S_{5/2}^{(k)}(W_m^2) : 0(123) \sum_{n=1}^{3} \mathbf{W}_n^0 \mathbf{W}_n S_{5/2}^{(l)}(W_n^2), \tag{5.16}$$

where we used the property that $0(123)$ commutes with the kinetic energy and therefore with the product of the Maxwell distributions[15].

There exists a simple relationship between the rigorous expressions (5.11) for λ_1 and η_1 and the approximate expressions (4.11) given by the theory of Enskog. The operator I_3, defined in (5.6), contains an integration over all possible positions $\mathbf{r}_3$ of particle 3. If we limit this integration to those values of $\mathbf{r}_3$ for which both $r_{13} < \sigma$ and $r_{23} < \sigma$, that is, the excluded volume of Fig. 3, the operator I_3 reduces to $\frac{5}{8} I_2$ as shown in a previous paper[15]. Thus the rigorous expressions (5.11) reduce to the values predicted by Enskog, when the configurational integral is restricted to those configurations that determine in equilibrium the first dentity correction to the radial distribution function $g(\sigma)$.

6 CLASSIFICATION OF THE TRIPLE COLLISION EVENTS

The transport coefficients λ_0 and η_0 in the dilute state were determined by the binary collision integrals (3.15). To obtain the first density corrections λ_1 and η_1 we need to evaluate the triple collision integrals (5.16). Apart from the fact that the dynamics of triple collisions is more complicated than that of an individual binary collision, there is another major difference between the two sets of transport collision integrals. In the binary collision integral operator I_2, defined in (3.8), the configurational integration does not extend over all possible positions $\mathbf{r}_2$ of particle 2, but the integration is restricted to a two dimensional surface, specified by the perihelion vector $\mathbf{k}$. For a given $\mathbf{k}$ the operator $T(12)$ in (3.8) can be evaluated at any point along the trajectories of the two particles (provided that the two particles are receding), since one needs only the velocities before and after the binary collision. On the other hand, the triple collision integral operator $0(123)$, defined in (5.15) is still a six dimensional integral and contains the differential operators θ_{ij}; it has not yet been expressed in terms of the velocities of the particles before and after certain collision events. It is possible to convert the six dimensional triple collision integral (5.15) into a five dimensional surface integral which is much more similar to the two dimensional surface integral form of the binary collision integral operator I_2 [15,16]. A derivation of the surface integral form of the triple collision integral operator is presented in Appendix A.

It turns out that the triple collision integral can be decomposed into a sum of six surface integral operators, each of which is associated with a particular triple collision event. To elucidate this result we give first a description of the six triple collision events which are schematically represented in Fig. 4. The positions of the particles (on the five dimensional surface) over which the integration has to be carried out are indicated by the black dots. The direction of the velocities is indicated by the arrows. The lines represent trajectories and are labeled with the corresponding particle number. The circles represent (binary) collisions between the particles whose trajectories are enclosed. The operators in (5.15) transform the present phases of the particles into the phases that the particles would have had in the previous past according to a particular dynamical history. Therefore, we retrace the motion backwards in time and we shall always read the diagrams from top to bottom, just as for the binary collision diagram of Fig. 2. The sequence of successive binary collisions will be ordered as they are encountered in this procedure.

Following Green[13] we distinguish between real and hypothetical collisions. Any binary collision encountered when we retrace the trajectories of the particles taking into account the interaction between all three particles is called a *real* collision. Any binary collision that is encountered only when

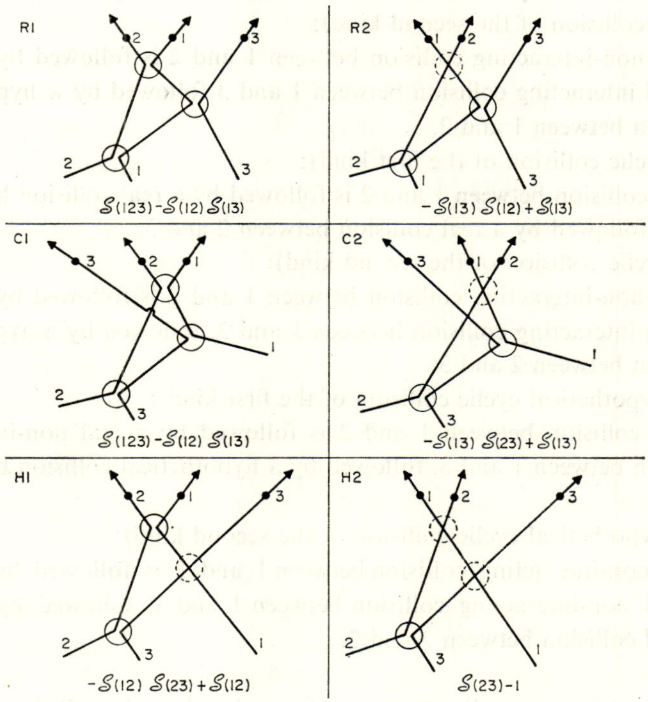

FIG. 4 The Six Triple Collision Events

the effect of some other collision on the trajectories of the particles is disregarded is called a *hypothetical* collision. In addition, we distinguish between interacting and non-interacting collisions. In a *non-interacting* collision the particles are aimed to collide, but we continue the free trajectories of the particles without taking into account the interaction between the particles. In Fig. 4, non-interacting collisions are indicated by a dotted circle. It is evident from the definition that all collisions encountered in the diagrams of Fig. 4 after a non-interacting collision are hypothetical collisions.

With this nomenclature the triple collision events of Fig. 4 can be described as follows (always reading the diagrams from top to bottom!).

R1 (Recollision of the first kind):
A real collision between 1 and 2 is followed by a real collision between 1 and 3, followed by a real collision between 1 and 2 again.
R2 (Recollision of the second kind):
A real non-interacting collision between 1 and 2 is followed by a hypothetical interacting collision between 1 and 3 followed by a hypothetical collision between 1 and 2.
C1 (Cyclic collision of the first kind):
A real collision between 1 and 2 is followed by a real collision between 1 and 3, followed by a real collision between 2 and 3.
C2 (Cyclic collision of the second kind):
A real non-interacting collision between 1 and 2 is followed by a hypothetical interacting collision between 1 and 3, followed by a hypothetical collision between 2 and 3.
H1 (Hypothetical cyclic collision of the first kind):
A real collision between 1 and 2 is followed by a real non-interacting collision between 1 and 3, followed by a hypothetical collision between 2 and 3.
H2 (Hypothetical cyclic collision of the second kind):
A real non-interacting collision between 1 and 2 is followed by a hypothetical non-interacting collision between 1 and 3, followed by a hypothetical collision between 2 and 3.

As shown in Appendix A the configurational triple collision operator $O(123)$, introduced in (5.14), can be replaced with

$$O(123) = 6 \cdot O(12; 13), \tag{6.1}$$

with $O(12;13)$ given by (A.30)

$$O(12; 13) = \sigma^4 \int d\mathbf{k}_1 \, d\mathbf{k}_2 \int d\tau \, \mathbf{V}_{21} \cdot \mathbf{k}_1 \mathbf{V}_{31}(1) \cdot \mathbf{k}_2 \sum_{\mu} T_{\mu}(12; 13). \tag{6.2}$$

The points on the five dimensional surface are specified by the variables $\mathbf{k}_1$, $\mathbf{k}_2$ and τ; $\mathbf{k}_1$ is the perihelion vector of the first collision (between 1 and 2), $\mathbf{k}_2$ the perihelion vector of the second collision (between 1 and 3) and τ the time between the first and the second collision. $\mathbf{V}_{31}(1)$ is the relative velocity of 3 with respect to 1 between the first and the second collision. The summation over μ in (6.2) has to be taken over the diagrams μ = R1, R2, C1, C2, H1, H2. The operator $T_\mu(12;13)$ is only different from zero for phase points whose dynamical history can be represented by the diagram μ. For such phase the definition of $T_\mu(12;13)$ is given in Table I.

TABLE I The triple collision operators $T_\mu(12; 13)$

μ	$T_\mu(12; 13)$
R1	$\mathscr{S}(123) - \mathscr{S}(12)\,\mathscr{S}(13)$
R2	$-\mathscr{S}(13)\,\mathscr{S}(12) + \mathscr{S}(13)$
C1	$\mathscr{S}(123) - \mathscr{S}(12)\,\mathscr{S}(13)$
C2	$-\mathscr{S}(13)\,\mathscr{S}(23) + \mathscr{S}(13)$
H1	$-\mathscr{S}(12)\,\mathscr{S}(23) + \mathscr{S}(12)$
H2	$\mathscr{S}(23) - 1$

These expressions for the operators $T_\mu(12;13)$ in terms of the streaming operators $\mathscr{S}$ are presented in Fig. 4 below each diagram; they are direct generalizations of the operator $T(12)$, defined in (3.6), for the binary collision term. Operating on any function of the velocities, $T_\mu(12;13)$ transforms that function into the difference between its value for the final velocities at the bottom of the diagram and its value for the velocities between the second and the third collision. The effect of the operator has the opposite sign, when the number of non-interacting collisions is odd. Note that in R1 and C1 a fourth collision may occur[17,18]; in that case the trajectories have to be retraced until this fourth collision has also been completed. We consider only those diagrams for which the first collision refers to particles 1 and 2 and the second collision to particles 1 and 3. Any other diagram obtained by permutation of the particle numbers is accounted for by the factor 6 in (6.1). The requirement that the first collision be real means that the particles should not collide, when the trajectories in the diagrams of Fig. 4 are extended into the future[15].

In the previous Section we pointed out that the expressions for λ_1 and η_1 reduce to the predictions of Enskog, if the configurational integral is restricted to the excluded volume of particle 3. To ascertain which triple collision events are taken into account in the Enskog theory, we consider diagram H2 in some detail. In this diagram the first collision (between 1 and 2) is a non-interacting collision. Such a non-interacting collision implies that during a certain time interval particles 1 and 2 move inside each other, when their free trajectories are followed from top to bottom. We call such configurations overlapping configurations. When the time between the first and the second collision is sufficiently small, particle 3 will be aimed to collide with 1, while 1 and 2 are still overlapping. Since the collision between 1 and 3 is also non-interacting, 1 and 3 overlap during a short time interval after the second collision. When in addition the time between the second and the third collision is small, it is possible that 3 will collide with 2, while particles 1 and 2 as well as particles 1 and 3 are still overlapping. Thus, such a collision process leads to the excluded volume configuration (Fig. 3, if we interchange 1 and 3) considered in the Enskog theory. Therefore, as mentioned in a previous paper[15], it can be concluded that the Enskog theory neglects all triple collision events shown in Fig. 4, except for some collisions of the type H2. Among all collisions H2, the theory of Enskog includes only those for which the time between the first and the last collision is so small that, when 3 starts to collide with 2, 1 and 2 as well as 1 and 3 are still overlapping.

The surface integral form (6.1) for $O(123)$ should be substituted into the expressions (5.16) for $a_{kl}^{(1)}$ and $b_{kl}^{(1)}$. In doing so we take the diameter σ of the spheres as the unit of length. Since the velocity was expressed in units $(2kT/m)^{1/2}$, the corresponding dimensionless time τ^* is

$$\tau^* = \frac{1}{\sigma}\left(\frac{2kT}{m}\right)^{1/2} \tau. \tag{6.3}$$

Thus the integrals (5.16) for $a_{kl}^{(1)}$ and $b_{kl}^{(1)}$ become in dimensionless form

$$a_{kl}^{(1)} = -\frac{3\sqrt{2}}{16\pi^6} \int d\mathbf{W}_1\, d\mathbf{W}_2\, d\mathbf{W}_3 \exp\left[-(W_1^2 + W_2^2 + W_3^2)\right]$$
$$\times \sum_{m=1}^{3} \mathbf{W}_m S_{3/2}^{(k)}(W_m^2) \cdot O^*(12;13) \sum_{n=1}^{3} \mathbf{W}_n S_{3/2}^{(l)}(W_n^2),$$
$$b_{kl}^{(1)} = -\frac{3\sqrt{2}}{16\pi^6} \int d\mathbf{W}_1\, d\mathbf{W}_2\, d\mathbf{W}_3 \exp\left[-(W_1^2 + W_2^2 + W_3^2)\right]$$
$$\times \sum_{m=1}^{3} \mathbf{W}_m^0 \mathbf{W}_m S_{5/2}^{(k)}(W_m^2) : O^*(12;13) \sum_{n=1}^{3} \mathbf{W}_n^0 \mathbf{W}_n S_{5/2}^{(l)}(W_n^2), \tag{6.4}$$

with

$$O^*(12;13) = \sum_{\mu} \int d\mathbf{k}_1\, d\mathbf{k}_2 \int d\tau^*\, \mathbf{W}_{21} \cdot \mathbf{k}_1 \mathbf{W}_{31}(1) \cdot \mathbf{k}_2 T_{\mu}(12;13). \tag{6.5}$$

For the evaluation of these triple collision integrals it is more convenient to describe the collisions between the molecules in terms of forward motion, rather than retracing the motion backwards in time. Such a description is obtained if we change the sign of all velocities. The integrands are symmetric in the velocities and therefore invariant under this transformation, provided that we replace t with $-t$ in the definition (3.4) of the operators. For the diagrams in Fig. 4 it means that we change the direction of the velocities as

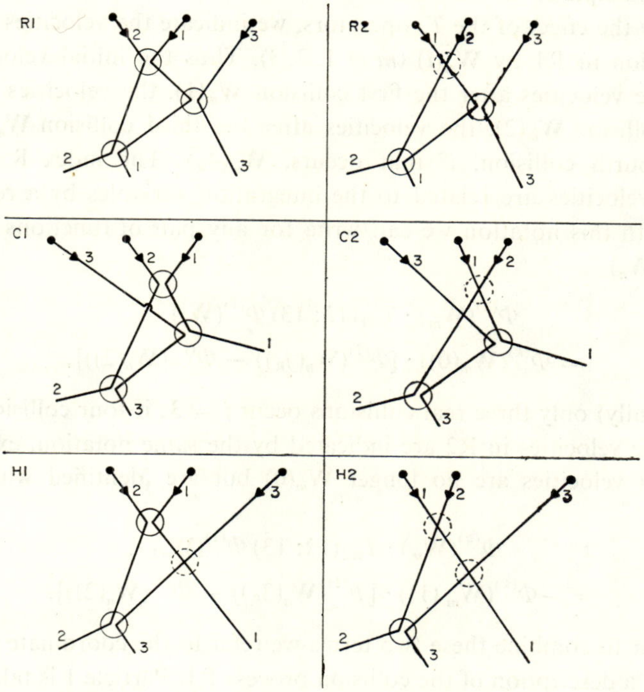

FIG. 5 The Six Triple Collision Events after changing the Direction of the Velocities and Time

shown in Fig. 5, but still read the diagrams from top to bottom. In the sequel of this chapter we shall refer to Fig. 5 in discussing the triple collision events.

7 ANALYSIS OF THE TRIPLE COLLISION TRANSPORT INTEGRALS

In the preceding Section the triple collision integral was decomposed into a sum of integrals associated with the six triple collision events of Fig. 5. However, there exists a close relationship between the diagrams of the first kind (R1, C1 and H1) and those of the second kind (R2, C2, H2). If in R1, C1 or H1 the volocities *after* the first collision are taken as the initial velocities, we obtain the diagrams of the second kind. Minor exceptions to this correspondence will be discussed later. Using this relationship we can combine the contributions from the diagrams of the first kind with those of the second kind[15].

As an example let us consider the recollision events R1 and R2 in Fig. 5. To specify the effect of the T_μ operators, we indicate the velocities after the i-th collision in R1 by $\mathbf{W}_m(i)$ $(m = 1, 2, 3)$. Thus the initial velocities are $\mathbf{W}_m(0)$, the velocities after the first collision $\mathbf{W}_m(1)$, the velocities after the second collision $\mathbf{W}_m(2)$, the velocities after the third collision $\mathbf{W}_m(3_R)$ and after a fourth collision, if that occurs, $\mathbf{W}_m(4_R)$. The index R indicates that the velocities are related to the integration variables by a recollision event. With this notation we can write for any pair of functions $\Phi^{(k)}(\mathbf{W}_m)$ and $\Phi^{(l)}(\mathbf{W}_n)$

$$\Phi^{(k)}(\mathbf{W}_m) \cdot T_{R1}(12; 13)\, \Phi^{(l)}(\mathbf{W}_n)$$
$$= \Phi^{(k)}(\mathbf{W}_m(0)) \cdot [\Phi^{(l)}(\mathbf{W}_n(j_R)) - \Phi^{(l)}(\mathbf{W}_n(2))]. \tag{7.1}$$

If (as usually) only three real collisions occur $j = 3$; if four collisions occur $j = 4$. The velocities in R2 are indicated by the same notation, except that the initial velocities are no longer $\mathbf{W}_m(0)$ but are identified with $\mathbf{W}_m(1)$. Thus

$$\Phi^{(k)}(\mathbf{W}_m) \cdot T_{R2}(12; 13)\, \Phi^{(l)}(\mathbf{W}_n)$$
$$= -\Phi^{(k)}(\mathbf{W}_m(1)) \cdot [\Phi^{(l)}(\mathbf{W}_n(3_R)) - \Phi^{(l)}(\mathbf{W}_n(2))]. \tag{7.2}$$

In order to combine these two terms we refer to the coordinate system of Fig. 6 for a description of the collision process R1. Particle 1 is taken at rest at the origin after the first collision. The Z-axis is taken in the direction of $\mathbf{W}_{21}(1) = \mathbf{W}_2(1) - \mathbf{W}_1(1)$. The X-axis is chosen such that the perihelion vector $\mathbf{k}_1$ of the first collision is in the XZ plane as indicated in the figure; θ_1 is the angle between $-\mathbf{k}_1$ and the Z-axis. The perihelion vector $\mathbf{k}_2$ of the second collision lies in general outside the XZ plane. In this coordinate system particle 2 came originally from A. At the time of the first collision particle 2 is at B and particle 3 at C with relative position vectors*

$$\mathbf{r}_{21}(1) = -\mathbf{k}_1; \mathbf{r}_{31}(1) = -\mathbf{k}_2 - \tau^*\mathbf{W}_{31}(1);$$
$$\mathbf{r}_{32}(1) = \mathbf{k}_1 - \mathbf{k}_2 - \tau^*\mathbf{W}_{31}(1). \tag{7.3}$$

In order for a recollision to occur, particle 3 should collide with 1 at D after a time τ^* such that particle 1 will collide again with 2.

In the collision event R2 the first collision is a non-interacting collision. Let the perihelion vector of this non-interacting collision be $\mathbf{k}'_1$, and the time

* In this paper $\mathbf{W}_{ij} = \mathbf{W}_i - \mathbf{W}_j$ and $\mathbf{r}_{ij} = \mathbf{r} - \mathbf{r}_j$.

between the first and the second collision $\tau^{*\prime}$. This implies that in Fig. 6 particle 2 came originally from A′. The relative position vectors at the time

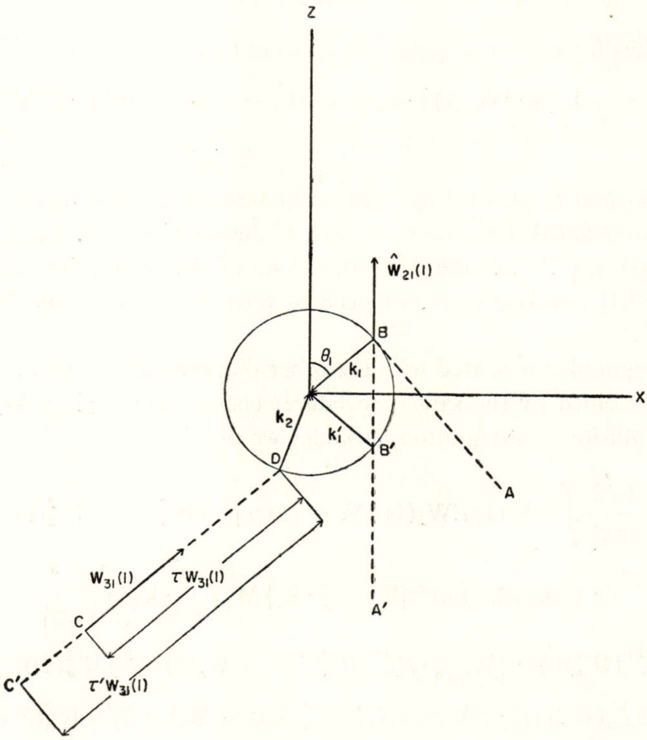

FIG. 6 Coordinate System for the Triple Collision Events. The Figure Represents the Situation Immediately after the First Collision

that the first collision begins are now indicated by $\mathbf{r}_{ij}(0)$. Thus

$$\mathbf{r}_{21}(0) = -\mathbf{k}_1'; \quad \mathbf{r}_{31}(0) = -\mathbf{k}_2 - \tau^{*\prime}\mathbf{W}_{31}(1);$$

$$\mathbf{r}_{32}(0) = \mathbf{k}_1' - \mathbf{k}_2 - \tau^{*\prime}\mathbf{W}_{31}(1). \tag{7.4}$$

We require that particle 3 collides with 1 at D such that 1 and 2 will collide again. The same event as that described earlier is obtained if $\mathbf{k}_1'$ and $\tau^{*\prime}$ are chosen such that*

$$\mathbf{k}_1' = \mathbf{k}_1 - 2\hat{\mathbf{W}}_{21}(1) \cdot \mathbf{k}_1; \quad \tau^{*\prime} = \tau^* + \tau_0^*, \tag{7.5}$$

---

* In this paper a superscript indicates a unit vector: $\hat{\mathbf{a}} = \mathbf{a}/a$.

with

$$\tau_0^* = \frac{2|\hat{\mathbf{W}}_{21}(1)\cdot\mathbf{k}_1|}{W_{21}(1)}. \tag{7.6}$$

This implies for the initial positions $\mathbf{r}_{ij}(0)$ in (7.4)

$$\mathbf{r}_{21}(0) = -\mathbf{k}_1 + 2\hat{\mathbf{W}}_{21}(1)\cdot\mathbf{k}_1;\quad \mathbf{r}_{31}(0) = -\mathbf{k}_2 - (\tau^* + \tau_0^*)\,\mathbf{W}_{31}(1);$$
$$\mathbf{r}_{32}(0) = \mathbf{r}_{31}(0) - \mathbf{r}_{21}(0). \tag{7.7}$$

We take $\mathbf{k}_1$ and τ^*, defined by (7.5) as the new integration variables for the integrals associated with the diagrams of the second kind. Then the velocities $\mathbf{W}_m(i)$ in (7.2) are the same functions of the integration variables as those in (7.1) and one does not need to retrace the diagrams R1 and R2 separately.

The integrands associated with the other diagrams of the first and second can be combined by the same procedure. Using the velocities $\mathbf{W}_m(1)$ after the first collision as integration variables we obtain

$$a_{kl}^{(1)} = -\frac{3\sqrt{2}}{16\pi^6}\int d\mathbf{W}_1(1)\,d\mathbf{W}_2(1)\,d\mathbf{W}_3(1)\exp\left[-(W_1^2(1) + W_2^2(1) + W_3^2(1)\right]$$
$$\cdot\int d\mathbf{k}_1\,d\mathbf{k}_2\int d\tau^*\,|\mathbf{W}_{21}(1)\cdot\mathbf{k}_1|\,\mathbf{W}_{31}(1)\cdot\mathbf{k}_2\sum_{m=1}^{3}\sum_{n=1}^{3} \tag{7.8}$$
$$\cdot\,[\mathbf{W}_m(0)\,S_{3/2}^{(k)}(W_m^2(0))\cdot\{\mathbf{W}_n(j_R)\,S_{3/2}^{(l)}(W_n^2(j_R)) - \mathbf{W}_n(2)\,S_{3/2}^{(l)}(W_n^2(2))\} \tag{R1}$$
$$-\mathbf{W}_m(1)\,S_{3/2}^{(k)}(W_m^2(1))\cdot\{\mathbf{W}_n(3_R)\,S_{3/2}^{(l)}(W_n^2(3_R)) - \mathbf{W}_n(2)\,S_{3/2}^{(l)}(W_n^2(2))\} \tag{R2}$$
$$+\mathbf{W}_m(0)\,S_{3/2}^{(k)}(W_m^2(0))\cdot\{\mathbf{W}_n(j_C)\,S_{3/2}^{(l)}(W_n^2(j_C)) - \mathbf{W}_n(2)\,S_{3/2}^{(l)}(W_n^2(2))\} \tag{C1}$$
$$-\mathbf{W}_m(1)\,S_{3/2}^{(k)}(W_m^2(1))\cdot\{\mathbf{W}_n(3_C)\,S_{3/2}^{(l)}(W_n^2(3_C)) - \mathbf{W}_n(2)\,S_{3/2}^{(l)}(W_n^2(2))\} \tag{C2}$$
$$-\mathbf{W}_m(0)\,S_{3/2}^{(k)}(W_m^2(0))\cdot\{\mathbf{W}_n(3_H)\,S_{3/2}^{(l)}(W_n^2(3_H)) - \mathbf{W}_n(1)\,S_{3/2}^{(l)}(W_n^2(1))\} \tag{H1}$$
$$+\mathbf{W}_m(1)\,S_{3/2}^{(k)}(W_m^2(1))\cdot\{\mathbf{W}_n(3_H)\,S_{3/2}^{(l)}(W_n^2(3_H)) - \mathbf{W}_n(1)\,S_{3/2}^{(l)}(W_n^2(1))\}], \tag{H2}$$

and a similar expression for $b_{kl}^{(1)}$.

To evaluate the various terms in the integrand we only have to retrace three triple collision events to which we refer as recollisions, cyclic collisions and hypothetical cyclic collisions. Nevertheless, we specify in (7.8) whether a term is associated with a diagram $\mu 1$ of the first kind or $\mu 2$ of the second kind. The reason is that there are some differences in the region of integration over the variables. For instance, in C2 and H2 it is possible that 3 starts to collide with 1 while 1 and 2 are still overlapping. This phenomenon occurs when $0 < \tau^{*\prime} < \tau_0^*$. For such a phase point there exists no diagram of the first kind as a counter part. Thus it follows from (7.5) that negative values of τ^* such that $-\tau_0^* < \tau^* < 0$ are to be included in the contribution from C2 and H2, but not in any of the other terms. Apart from the phase points leading to overlapping configurations, the integration region associated with a diagram of the second kind is very similar to that associated with the corresponding diagram of the first kind. The remaining differences arise from the effect of various possible fourth collisions between the three particles, the effect of which is small.

The integration region will be specified by formulating the conditions that the integration variables should satisfy. Some of these conditions should be imposed on all integrals, namely

$$\mathbf{W}_{21}(1) \cdot \mathbf{k}_1 < 0, \tag{7.9}$$

$$\mathbf{W}_{31}(1) \cdot \mathbf{k}_2 > 0. \tag{7.10}$$

Condition (7.9) guarantees that the first collision and (7.10) that the second collision will take place. Furthermore as noted above

$$\tau^* > -\tau_0^* \quad (\mu = \text{C2, H2}), \tag{7.11}$$

and

$$\tau^* > 0 \quad (\mu = \text{R1, R2, C1, H1}). \tag{7.12}$$

(A negative value of τ^* is not allowed for R2, since it implies that the third collision would precede the second collision).

The requirement that a third collision should take place leads to a second class of conditions. They differ, of course, whether we consider a recollision, a cyclic collision or a hypothetical cyclic collision and are most conveniently expressed in terms of the positions and velocities at (just after) the time of the second collision. The initial positions for a diagram of the first kind were given by (7.3) and those for a diagram of the second kind by (7.7). The positions at the time of the second collision are the same for all diagrams:

$$\mathbf{r}_{21}(2) = -\mathbf{k}_1 + \tau^*\mathbf{W}_{21}(1); \quad \mathbf{r}_{31}(2) = -\mathbf{k}_2; \quad \mathbf{r}_{32}(2) = \mathbf{k}_1 - \mathbf{k}_2 - \tau^*\mathbf{W}_{21}(1). \tag{7.13}$$

The basic equations describing a collision between two hard spheres are presented in Appendix B. Applying the inequalities (B.3) and (B.4) we see that for *recollisions* (μ = R1, R2)

$$\{\mathbf{W}_{21}(2) \times \mathbf{r}_{21}(2)\}^2 - W_{21}^2(2) < 0, \tag{7.14}$$

$$\mathbf{W}_{21}(2) \cdot \mathbf{r}_{21}(2) < 0, \tag{7.15}$$

for *cyclic* collisions (μ = C1, C2)

$$\{\mathbf{W}_{32}(2) \times \mathbf{r}_{32}(2)\}^2 - W_{32}^2(2) < 0, \tag{7.16}$$

$$\mathbf{W}_{32}(2) \cdot \mathbf{r}_{32}(2) < 0, \tag{7.17}$$

and for *hypothetical cyclic* collisions (μ = H1, H2)

$$\{\mathbf{W}_{32}(1) \times \mathbf{r}_{32}(2)\}^2 - W_{32}^2(1) < 0, \tag{7.18}$$

$$\mathbf{W}_{32}(1) \cdot \mathbf{r}_{32}(2) < 0. \tag{7.19}$$

For recollisions condition (7.15) combined with (7.10) can be reduced to

$$\mathbf{k}_1 \cdot \mathbf{k}_2 < 0, \tag{7.20}$$

$$\mathbf{W}_{31}(1) \cdot \mathbf{k}_2 > \frac{\mathbf{W}_{21}(1) \cdot \mathbf{k}_1}{\mathbf{k}_1 \cdot \mathbf{k}_2}, \tag{7.21}$$

as pointed out by Weinstock[19].

The initial phases should be such that particles 2 and 3 are not overlapping. Thus

$$r_{32}^2(1) > 1 \quad (\mu = \text{R1, C1, H1}), \tag{7.22}$$

$$r_{32}^2(0) > 1 \quad (\mu = \text{R2, C2, H2}). \tag{7.23}$$

For the cyclic and hypothetical cyclic collisions, particles 2 and 3 should also not overlap at the time of the second collision, since the third collision should not precede the second one.

$$r_{32}^2(2) > 1. \tag{7.24}$$

The triple collision events were defined in Section 6 together with a distinction between real and hypothetical collisions. This distinction leads to some minor additional conditions:

i) In order that the first collision be real, the initial phases in Fig. 5 should be such that the particles did not experience a collision in their past history.

ii) In order that in R1, C1 and H1 the second collision be real, particle 3 should not collide with 2 before colliding with 1.

iii) In order that in R1 and C1 the third collision be real, no other collision should interfere along the path of particle 2.

Although these conditions together guarantee the proper triple collision events, some of them are redundant. Therefore, not all conditions need to be imposed explicitly. For instance, when in R1 the phases of the particles are such that the third collision (between 1 and 2) will occur it is no longer possible for 3 to collide with 2 before the second collision[27,28]. On the other hand, in C1 this possibility has to be excluded explicitly.

We made an effort to reduce the number of conditions to be imposed. The conditions retained are presented in Table II. The conditions (μ.1) through (μ.5) determine the basic phenomenon. The remaining conditions exclude possible fourth collisions that would interfere with the triple collision sequence desired. It turns out that one will never encounter more than *four* successive collisions, independently of the interacting or non-interacting nature of the collisions[27,28].

We note that the integration regions associated with R1 and R2 are identical, except for the exclusion of a possible fourth collision in the previous past [(R1.6) and (R2.6)], the probability of which is very small[18,20]. The related integrands in (7.8) only differ in the possibility of a fourth collision succeeding the third one. Except for possible negative values of τ^*, the integrations associated with C1 and C2, resp. H1 and H2, are also very similar.

The position vectors $\mathbf{r}_{ij}(n)$ are related to the integration variables by (7.3), (7.7) and (7.13). The velocities $\mathbf{W}_{ij}(0)$ and $\mathbf{W}_{ij}(2)$ are related to the integration variables $\mathbf{k}_1$, $\mathbf{k}_2$ and $\mathbf{W}_i(1)$ by

$$\begin{aligned} \mathbf{W}_{21}(0) &= \mathbf{W}_{21}(1) - 2\mathbf{W}_{21}(1)\cdot\mathbf{k}_1\mathbf{k}_1, \\ \mathbf{W}_{31}(0) &= \mathbf{W}_{31}(1) - \mathbf{W}_{21}(1)\cdot\mathbf{k}_1\mathbf{k}_1, \\ \mathbf{W}_{32}(0) &= \mathbf{W}_{32}(1) + \mathbf{W}_{21}(1)\cdot\mathbf{k}_1\mathbf{k}_1, \end{aligned} \tag{7.25}$$

and

$$\begin{aligned} \mathbf{W}_{21}(2) &= \mathbf{W}_{21}(1) - \mathbf{W}_{31}(1)\cdot\mathbf{k}_2\mathbf{k}_2, \\ \mathbf{W}_{31}(2) &= \mathbf{W}_{31}(1) - 2\mathbf{W}_{21}(1)\cdot\mathbf{k}_2\mathbf{k}_2, \\ \mathbf{W}_{32}(2) &= \mathbf{W}_{32}(1) - \mathbf{W}_{31}(1)\cdot\mathbf{k}_2\mathbf{k}_2. \end{aligned} \tag{7.26}$$

The integrals (7.8) together with the conditions of Table II specify the triple collision transport integrals to be evaluated.

TABLE II Conditions specifying the integration region for the triple collision transport integrals

Recollisions of the first kind R1

$$\tau^* > 0 \tag{R1.1}$$

$$\mathbf{W}_{21}(1)\cdot\mathbf{k}_1 < 0 \tag{R1.2}$$

$$\mathbf{k}_1\cdot\mathbf{k}_2 < 0 \tag{R1.3}$$

$$\mathbf{W}_{31}(1)\cdot\mathbf{k}_2 > \frac{\mathbf{W}_{21}(1)\cdot\mathbf{k}_1}{\mathbf{k}_1\cdot\mathbf{k}_2} \tag{R1.4}$$

$$\{\mathbf{W}_{21}(2)\times\mathbf{r}_{21}(2)\}^2 - W_{21}^2(2) < 0 \tag{R1.5}$$

$$\mathbf{W}_{32}(0)\cdot\mathbf{r}_{32}(1) < 0 \quad \text{or} \quad \{\mathbf{W}_{32}(0)\times\mathbf{r}_{32}(1)\}^2 - W_{32}^2(0) > 0 \tag{R1.6}$$

TABLE II (continued)

Recollisions of the second kind R2

$$\tau^* > 0 \quad \text{(R2.1)}$$

$$\mathbf{W}_{21}(1) \cdot \mathbf{k}_1 < 0 \quad \text{(R2.2)}$$

$$\mathbf{k}_1 \cdot \mathbf{k}_2 < 0 \quad \text{(R2.3)}$$

$$\mathbf{W}_{31}(1) \cdot \mathbf{k}_2 > \frac{\mathbf{W}_{21}(1) \cdot \mathbf{k}_1}{\mathbf{k}_1 \cdot \mathbf{k}_2} \quad \text{(R2.4)}$$

$$\{\mathbf{W}_{21}(2) \times \mathbf{r}_{21}(2)\}^2 - W_{21}^2(2) < 0 \quad \text{(R2.5)}$$

$$\mathbf{W}_{32}(1) \cdot \mathbf{r}_{32}(0) < 0 \quad \text{or} \quad \{\mathbf{W}_{32}(1) \times \mathbf{r}_{32}(0)\}^2 - W_{32}^2(1) > 0 \quad \text{(R2.6)}$$

Cyclic collisions of the first kind C1

$$\tau^* > 0 \quad \text{(C1.1)}$$

$$\mathbf{W}_{21}(1) \cdot \mathbf{k}_1 < 0 \quad \text{(C1.2)}$$

$$\mathbf{W}_{31}(1) \cdot \mathbf{k}_2 > 0 \quad \text{(C1.3)}$$

$$\mathbf{W}_{32}(2) \cdot \mathbf{r}_{32}(2) < 0 \quad \text{(C1.4)}$$

$$\{\mathbf{W}_{32}(2) \times \mathbf{r}_{32}(2)\}^2 - W_{32}^2(2) < 0 \quad \text{(C1.5)}$$

$$r_{32}^2(1) > 1 \quad \text{(C1.6)}$$

$$\mathbf{W}_{31}(0) \cdot \mathbf{r}_{31}(1) < 0 \quad \text{or} \quad \{\mathbf{W}_{31}(0) \times \mathbf{r}_{31}(1)\}^2 - W_{31}^2(0) > 0 \quad \text{(C1.7)}$$

3 should not collide with 2 before colliding with 1 (C1.8)

2 should not recollide with 1 before colliding with 3 (C1.9)

Cyclic collisions of the second kind C2

$$\tau^* > -\tau_0^* \quad \text{(C2.1)}$$

$$\mathbf{W}_{21}(1) \cdot \mathbf{k}_1 < 0 \quad \text{(C2.2)}$$

$$\mathbf{W}_{31}(1) \cdot \mathbf{k}_2 > 0 \quad \text{(C2.3)}$$

$$\mathbf{W}_{32}(2) \cdot \mathbf{r}_{32}(2) < 0 \quad \text{(C2.4)}$$

$$\{\mathbf{W}_{32}(2) \times \mathbf{r}_{32}(2)\}^2 - W_{32}^2(2) < 0 \quad \text{(C2.5)}$$

$$r_{32}^2(0) > 1 \quad \text{(C2.6)}$$

$$r_{32}^2(2) > 1 \quad \text{(C2.7)}$$

$$\mathbf{W}_{32}(1) \cdot \mathbf{r}_{32}(0) < 0 \quad \text{or} \quad \{\mathbf{W}_{32}(1) \times \mathbf{r}_{32}(0)\}^2 - W_{32}^2(1) > 0 \quad \text{(C2.8)}$$

Hypothetical cyclic collisions of the first kind H1

$$\tau^* > 0 \quad \text{(H1.1)}$$

$$\mathbf{W}_{21}(1) \cdot \mathbf{k}_1 < 0 \quad \text{(H1.2)}$$

$$\mathbf{W}_{31}(1) \cdot \mathbf{k}_2 > 0 \quad \text{(H1.3)}$$

$$\mathbf{W}_{32}(1) \cdot \mathbf{r}_{32}(2) < 0 \quad \text{(H1.4)}$$

$$\{\mathbf{W}_{32}(1) \times \mathbf{r}_{32}(2)\}^2 - W_{32}^2(1) < 0 \quad \text{(H1.5)}$$

TABLE II (continued)

$$r_{32}^2(1) > 1 \tag{H1.6}$$

$$r_{32}^2(2) > 1 \tag{H1.7}$$

$$\mathbf{W}_{31}(0) \cdot \mathbf{r}_{31}(1) < 0 \quad \text{or} \quad \{\mathbf{W}_{31}(0) \times \mathbf{r}_{31}(1)\}^2 - W_{31}^2(0) > 0 \tag{H1.8}$$

Hypothetical cyclic collisions of the second kind H2

$$\tau^* > -\tau_0^* \tag{H2.1}$$

$$\mathbf{W}_{21}(1) \cdot \mathbf{k}_1 < 0 \tag{H2.2}$$

$$\mathbf{W}_{31}(1) \cdot \mathbf{k}_2 > 0 \tag{H2.3}$$

$$\mathbf{W}_{32}(1) \cdot \mathbf{r}_{32}(2) < 0 \tag{H2.4}$$

$$\{\mathbf{W}_{32}(1) \times \mathbf{r}_{32}(2)\}^2 - W_{32}^2(1) < 0 \tag{H2.5}$$

$$r_{32}^2(0) > 1 \tag{H2.6}$$

$$r_{32}^2(2) > 1 \tag{H2.7}$$

8 EVALUATION OF THE TRIPLE COLLISION TRANSPORT INTEGRALS IN THE FIRST ENSKOG APPROXIMATION

The triple collision integrals (7.8) which contain still fourteen integrations, can be reduced to seven dimensional integrals. First, the operators $T_\mu(12;13)$ commute with the total momentum, and the conditions of Table II, specifying the integration region, are independent of the total momentum. Thus we can integrate over the total momentum or velocity and express the resulting integrals in terms of the relative velocities $\mathbf{W}_{21}(1)$ and $\mathbf{W}_{31}(1)$. Secondly, the integrand is a scalar, invariant under rotations of the coordinate system. Thus we can integrate over all possible directions of $\mathbf{W}_{21}(1)$ and evaluate the resulting integrand with $\mathbf{W}_{21}(1)$ taken in the Z-direction as in Fig. 6. Furthermore, we can integrate over all directions of the component of $\mathbf{k}_1$ in the XY plane and take $\mathbf{k}_1$ in the XZ plane as in Fig. 6. Finally, for hard spheres the angle of deflection, caused by a binary collision, is independent of the magnitude of the relative velocity. This enables us to integrate over all values $W_{21}(1) = (|\mathbf{W}_{21}(1)|)$ and to evaluate the resulting integrand for collision events for which $\hat{\mathbf{W}}_{21}(1)$ is a unit vector.

For the matrix elements $a_{11}^{(1)}$ and $b_{00}^{(1)}$, determining the first Enskog approximations $\lambda_1(1)$ and $\eta_1(1)$ in (5.12), this reduction was carried out in a previous paper[15]. Denoting the angle between $-\mathbf{k}_1$ and the Z-axis by θ_1

(see Fig. 6) we obtain

$$\frac{\lambda_1(1)}{\lambda_0(1)} = -a_{11}^{(1)} = \frac{2}{3}\frac{\eta_1(1)}{\eta_0(1)} + \frac{385}{\pi^2}\frac{3^7}{2^{14}}\int d\mathbf{W}_{31}(1)\int_0^{\pi/2} d\theta_1 \sin\theta_1 \int d\mathbf{k}_2$$

$$\times \int d\tau \cos\theta_1\, \mathbf{W}_{31}(1)\cdot\mathbf{k}_2 E^{-13/2}\sum_\mu \mathbf{K}\cdot T_\mu(12;13)\,\mathbf{K},$$

$$\frac{\eta_1(1)}{\eta_0(1)} = -b_{00}^{(1)} = \frac{35}{\pi^2}\frac{3^6}{2^{10}}\int d\mathbf{W}_{31}(1)\int_0^{\pi/2} d\theta_1 \sin\theta_1 \int d\mathbf{k}_2 \int d\tau \cos\theta_1$$

$$\times\, \mathbf{W}_{31}(1)\cdot\mathbf{k}_2 E^{-11/2}\sum_\mu \mathsf{L} : T_\mu(12;13)\,\mathsf{L}. \tag{8.1}$$

For convenience, we have introduced the abbreviations

$$E = W_{21}^2 + W_{31}^2 - \mathbf{W}_{21}\cdot\mathbf{W}_{31}, \tag{8.2}$$

$$\mathbf{K} = \mathbf{W}_{21}\left(W_{21}^2 - \frac{2}{3}E\right) + \mathbf{W}_{31}\left(W_{31}^2 - \frac{2}{3}E\right), \tag{8.3}$$

$$\mathsf{L} = \mathbf{W}_{21}\mathbf{W}_{21} + \mathbf{W}_{31}\mathbf{W}_{31} - \frac{1}{2}(\mathbf{W}_{21}\mathbf{W}_{31} + \mathbf{W}_{31}\mathbf{W}_{21}). \tag{8.4}$$

The operators $T_\mu(12; 13)$ commute with E which represents the kinetic energy. In (8.1) τ is the scaled dimensionless time $\tau^* W_{21}(1)$. Just as in (7.8) the effect of the $T_\mu(12; 13)$ operators can be written explicitly as

$$\begin{aligned}
\sum_\mu \mathbf{K}\cdot T_\mu(12;13)\,\mathbf{K} = {} & \mathbf{K}(0)\cdot\{\mathbf{K}(j_R) - \mathbf{K}(2)\} && \text{(R1)}\\
& -\mathbf{K}(1)\cdot\{\mathbf{K}(3_R) - \mathbf{K}(2)\} && \text{(R2)}\\
& +\mathbf{K}(0)\cdot\{\mathbf{K}(j_C) - \mathbf{K}(2)\} && \text{(C1)}\\
& -\mathbf{K}(1)\cdot\{\mathbf{K}(3_C) - \mathbf{K}(2)\} && \text{(C2)}\\
& -\mathbf{K}(0)\cdot\{\mathbf{K}(3_H) - \mathbf{K}(1)\} && \text{(H1)}\\
& +\mathbf{K}(1)\cdot\{\mathbf{K}(3_H) - \mathbf{K}(1)\}, && \text{(H2)}
\end{aligned} \tag{8.5}$$

$$\begin{aligned}
\sum_\mu \mathsf{L} : T_\mu(12;13)\,\mathsf{L} = {} & \mathsf{L}(0) : \{\mathsf{L}(j_R) - \mathsf{L}(2)\} && \text{(R1)}\\
& -\mathsf{L}(1) : \{\mathsf{L}(3_R) - \mathsf{L}(2)\} && \text{(R2)}\\
& +\mathsf{L}(0) : \{\mathsf{L}(j_C) - \mathsf{L}(2)\} && \text{(C1)}\\
& -\mathsf{L}(1) : \{\mathsf{L}(3_C) - \mathsf{L}(2)\} && \text{(C2)}\\
& -\mathsf{L}(0) : \{\mathsf{L}(3_H) - \mathsf{L}(1)\} && \text{(H1)}\\
& +\mathsf{L}(1) : \{\mathsf{L}(3_H) - \mathsf{L}(1)\}, && \text{(H2)}
\end{aligned} \tag{8.6}$$

where again each term has to be integrated subject to the conditions of Table II. The argument n in $\mathbf{K}(n)$ and $\mathbf{L}(n)$ indicates for which velocities $\mathbf{W}_{21}(n)$, $\mathbf{W}_{31}(n)$ these quantities should be evaluated. The kinetic energy E is invariant during the collision process and thus independent of n.

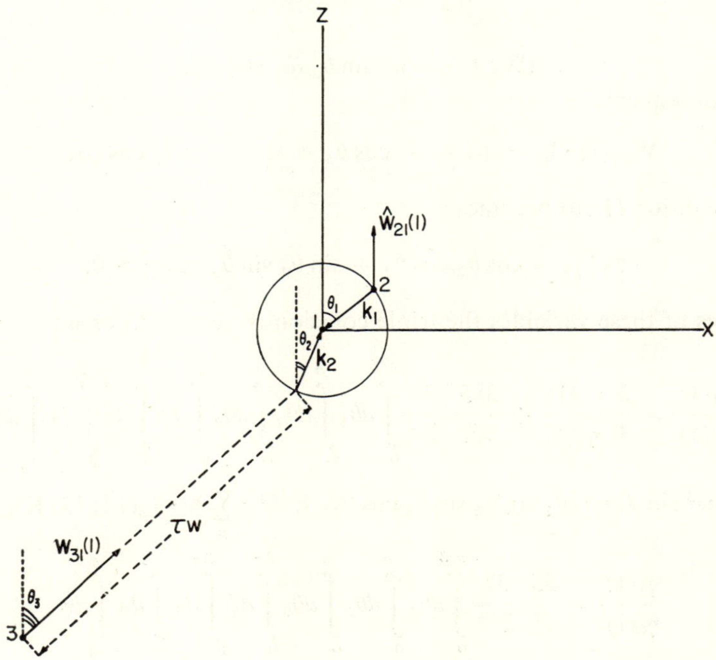

FIG. 7 Coordinate System for the Evaluation of the Triple Collision Integrals. The Figure Represents the Situation Immediately after the First Collision

We specify the perihelion vector $\mathbf{k}_2$ by its polar angle θ_2 with the Z-axis (see Fig. 7) and by its azimuthal angle ϕ with the XZ plane as initial plane. Hence

$$k_{2x} = \sin\theta_2 \cos\phi, \quad k_{2y} = \sin\theta_2 \sin\phi, \quad k_{2z} = \cos\theta_2, \tag{8.7}$$

and

$$d\mathbf{k}_2 = \sin\theta_2 \, d\theta_2 \, d\phi. \tag{8.8}$$

The velocity $\mathbf{W}_{31}(1)$ is specified by its magnitude w, its polar angle θ_3 with the Z-axis and the aximuthal angle ψ in the XY plane, measured with respect to the plane through $\mathbf{k}_2$ and the Z-axis as initial plane. Thus

$$\begin{aligned} W_{31}(1)_x &= w \sin\theta_3 \cos(\phi + \psi), \\ W_{31}(1)_y &= w \sin\theta_3 \sin(\phi + \psi), \\ W_{31}(1)_z &= w \cos\theta_3, \end{aligned} \tag{8.9}$$

and

$$d\mathbf{W}_{31}(1) = w^2 \sin\theta_3 \, dw \, d\theta_3 \, d\psi. \tag{8.10}$$

As a consequence

$$\mathbf{W}_{31}(1) \cdot \mathbf{k}_2 = w(\cos\theta_2 \cos\theta_3 + \sin\theta_2 \sin\theta_3 \cos\psi), \tag{8.11}$$

and condition (7.10) becomes

$$\cos\theta_{23} \equiv \cos\theta_2 \cos\theta_3 + \sin\theta_2 \sin\theta_3 \cos\psi > 0. \tag{8.12}$$

In terms of these variables the triple collision integrals (8.1) read

$$\frac{\lambda_1(1)}{\lambda_0(1)} = \frac{2}{3}\frac{\eta_1(1)}{\eta_0(1)} + \frac{385}{\pi^2}\frac{3^7}{2^{14}} \int_0^{\pi/2} d\theta_1 \int_0^{\pi} d\theta_2 \int_0^{\pi} d\theta_3 \int_0^{2\pi} d\phi \int_0^{2\pi} d\psi \int_0^{\infty} dw \int_{-\tau_0}^{\infty} d\tau$$

$$\times w^3 \sin\theta_1 \cos\theta_1 \sin\theta_2 \sin\theta_3 \cos\theta_{23} \, E^{-13/2} \sum_\mu \mathbf{K} \cdot T_\mu(12;13)\,\mathbf{K}, \tag{8.13}$$

$$\frac{\eta_1(1)}{\eta_0(1)} = \frac{35}{\pi^2}\frac{3^6}{2^{10}} \int_0^{\pi/2} d\theta_1 \int_0^{\pi} d\theta_2 \int_0^{\pi} d\theta_3 \int_0^{2\pi} d\phi \int_0^{2\pi} d\psi \int_0^{\infty} dw \int_{-\tau_0}^{\infty} d\tau$$

$$\times w^3 \sin\theta_1 \cos\theta_1 \sin\theta_2 \sin\theta_3 \cos\theta_{23} \, E^{-11/2} \sum_\mu \mathsf{L} : T_\mu(12;13)\,\mathsf{L},$$

with

$$E = 1 + w^2 - w\cos\theta_3. \tag{8.14}$$

Of course, in addition the integration variables are still subject to the conditions of Table II.

A further analytical reduction of the number of integrations without making any approximations seems rather complicated. Therefore, to carry out the remaining seven integrations we take resort to a numerical procedure.

The integration over the velocity w extends to infinity. In addition, the time τ between collisions can also become infinite, when two of the molecules are moving in parallel directions. In order to apply a numerical integration procedure, we transform the integral into an integral over a finite integration region. For large values of w the integrand decreases as w^{-4}; this suggests the transformation

$$u = \frac{1}{1 + w^3}, \tag{8.15}$$

where $0 < u < 1$. To reduce the integration over τ to a finite interval we proceed as follows. For a given $\mathbf{k}_1$, $\mathbf{k}_2$ and $\mathbf{W}_{31}(1)$ (θ_1, θ_2, θ_3, ϕ, ψ and w in (8.13)) we solve the inequalities (μ.4) and (μ.5) in Table II for the variable τ (together with the requirement $\tau > -\tau_0$):

$$\begin{aligned} &T_{\mathrm{R,Min}} < \tau < T_{\mathrm{R,Max}} \quad (\mu = \mathrm{R1, R2}), \\ &T_{\mathrm{C,Min}} < \tau < T_{\mathrm{C,Max}} \quad (\mu = \mathrm{C1, C2}), \\ &T_{\mathrm{H,Min}} < \tau < T_{\mathrm{H,Max}} \quad (\mu = \mathrm{H1, H2}). \end{aligned} \tag{8.16}$$

These inequalities determine the range of τ values for which a third collision can occur when $\mathbf{k}_1$, $\mathbf{k}_2$ and $\mathbf{W}_{31}(1)$ are specified. It turns out that $T_{\mathrm{R,Min}} = 0$ independent of the other variables. As a new integration variable we take $\tilde{\tau}$ such that

$$\tau = T_{\nu,\mathrm{Min}} + \tilde{\tau}\, \Delta T_\nu \quad (\nu = \mathrm{R, C, H}), \tag{8.17}$$

where

$$\Delta T_\nu = T_{\nu,\mathrm{Max}} - T_{\nu,\mathrm{Min}}. \tag{8.18}$$

Hence

$$\frac{\lambda_1(1)}{\lambda_0(1)} = \frac{2}{3}\frac{\eta_1(1)}{\eta_0(1)} + \frac{385}{\pi^2}\frac{3^6}{2^{14}} \int_0^{\pi/2} d\theta_1 \int_0^{\pi} d\theta_2 \int_0^{\pi} d\theta_3 \int_0^{2\pi} d\phi \int_0^{2\pi} d\psi \int_0^1 du \int_0^1 d\tilde{\tau}$$

$$\times \frac{\sin\theta_1 \cos\theta_1 \sin\theta_2 \sin\theta_3\, w \cos\theta_{23}}{u^2 E^{13/2}} \sum_\mu \Delta T_\mu\, \mathbf{K} \cdot T_\mu(12;13)\, \mathbf{K}, \tag{8.19}$$

$$\frac{\eta_1(1)}{\eta_0(1)} = \frac{35}{\pi^2} \cdot \frac{3^5}{2^{10}} \int_0^{\pi/2} d\theta_1 \int_0^{\pi} d\theta_2 \int_0^{\pi} d\theta_3 \int_0^{2\pi} d\phi \int_0^{2\pi} d\psi \int_0^1 du \int_0^1 d\tilde{\tau}$$

$$\times \frac{\sin\theta_1 \cos\theta_1 \sin\theta_2 \sin\theta_3\, w \cos\theta_{23}}{u^2 E^{11/2}} \sum_\mu \Delta T_\mu\, \mathbf{L} : T_\mu(12;13)\, \mathbf{L},$$

where $\Delta T_{\mathrm{R1}} = \Delta T_{\mathrm{R2}} = \Delta T_{\mathrm{R}}$, etc.

As mentioned earlier the time intervals ΔT_μ become infinite for certain points in the integration region. For recollisions this happens when $\theta_2 \to 0$,

where

$$\Delta T_{\mathrm{R}} \propto \frac{1}{\mathbf{W}_{21}(2) \times \mathbf{W}_{21}(1)} \propto \frac{1}{\sin \theta_2};$$

for hypothetical cyclic collisions when $\theta_3 \to 0$, where

$$\Delta T_{\mathrm{H}} \propto \frac{1}{\mathbf{W}_{31}(1) \times \mathbf{W}_{21}(1)} \propto \frac{1}{\sin \theta_3};$$

and for cyclic collisions when ζ, the angle between $\mathbf{W}_{32}(2)$ and $\mathbf{W}_{21}(1)$, approaches zero

$$\Delta T_{\mathrm{C}} \propto \frac{1}{\mathbf{W}_{32}(2) \times \mathbf{W}_{21}(1)} \propto \frac{1}{\sin \zeta}.$$

The behavior of the integrand near these singularities can be characterized by

$$\sin \theta_2 \sin \theta_3 \left(\frac{a_{\mathrm{R}}}{\sin \theta_2} + \frac{a_{\mathrm{C}}}{\sin \zeta} + \frac{a_{\mathrm{H}}}{\sin \theta_3} \right).$$

The coordinate system introduced was designed to remove any singular behavior of the integrand for $\theta_2 = 0$ and $\theta_3 = 0$. However, although the integral is convergent the integrand is singular near $\zeta = 0$ and therefore, not yet suitable for numerical integration. Since

$$\lim_{\zeta \to 0} \sin \zeta = \lim_{\theta_2 \to \pi/2} \lim_{\phi \to 0} \left\{ \left(\frac{\pi}{2} - \theta_2 \right)^2 \cos^2 \theta_3 + \psi^2 \sin^2 \theta_3 \right\}^{1/2},$$

the singularity can be removed by the transformation

$$\begin{aligned} \varrho \cos \chi &= \pi - 2\theta_2, \\ \varrho \sin \chi &= \tfrac{1}{2} \psi. \end{aligned} \tag{8.20}$$

ϱ and χ are the polar coordinates associated with the variables $\pi - 2\theta_2$ and $\psi/2$.

$$\int_0^{\pi} d\theta_2 \int_0^{2\pi} d\psi = \int_0^{\pi} d\chi \int_0^{\varrho_{\mathrm{Max}}} \varrho \, d\varrho, \tag{8.21}$$

with

$$\varrho_{\mathrm{Max}} = \mathrm{Min} \left\{ \frac{\pi}{\sin \chi}, \frac{\pi}{|\cos \chi|} \right\}. \tag{8.22}$$

Thus the limiting behavior of the integrand around $\zeta = 0$ or $\varrho = 0$ becomes

$$\frac{d\theta_2\, d\psi}{\sin \zeta} = \frac{d\chi\, d\varrho}{\{\frac{1}{4}\cos^2 \chi \cos^2 \theta_3 + 4 \sin^2 \chi \sin^2 \theta_3\}^{1/2}},$$

which is well behaved for any value of χ. Introducing ϱ such that

$$\varrho = \tilde{\varrho}\, \varrho_{\mathrm{Max}}, \tag{8.23}$$

the final form of the triple collision integrals to be considered for numerical evaluation becomes

$$\frac{\lambda_1(1)}{\lambda_0(1)} - \frac{2}{3}\frac{\eta_1(1)}{\eta_0(1)} = \frac{385}{\pi^2}\frac{3^6}{2^{14}} \int_0^{\pi/2} d\theta_1 \int_0^{\pi} d\theta_3 \int_0^{2\pi} d\phi \int_0^{\pi} d\chi \int_0^{1} d\tilde{\varrho} \int_0^{1} du \int_0^{1} d\tilde{\tau}$$

$$\times \frac{\tilde{\varrho}\, \varrho_{\mathrm{Max}} \sin \theta_1 \cos \theta_1 \sin \theta_2 \sin \theta_3\, w \cos \theta_{23}}{u^2 E^{13/2}} \sum_{\mu} \Delta T_\mu\, \mathbf{K} \cdot T_\mu(12; 13)\, \mathbf{K},$$

$$\frac{\eta_1(1)}{\eta_0(1)} = \frac{35}{\pi^2}\frac{3^5}{2^{10}} \int_0^{\pi/2} d\theta_1 \int_0^{\pi} d\theta_3 \int_0^{2\pi} d\phi \int_0^{\pi} d\chi \int_0^{1} d\tilde{\varrho} \int_0^{1} du \int_0^{1} d\tilde{\tau} \tag{8.24}$$

$$\times \frac{\tilde{\varrho}\, \varrho_{\mathrm{Max}} \sin \theta_1 \cos \theta_1 \sin \theta_2 \sin \theta_3\, w \cos \theta_{23}}{u^2 E^{11/2}} \sum \Delta T_\mu\, \mathsf{L} : T_\mu(12; 13)\, \mathsf{L}.$$

The seven dimensional integration region is specified by

$$\begin{aligned} &0 < \theta_1 < \pi/2, \\ &0 < \theta_3 < \pi, \\ &0 < \phi < 2\pi, \\ &0 < \chi < \pi, \\ &0 < \tilde{\varrho} < 1, \\ &0 < u < 1, \\ &0 < \tilde{\tau} < 1. \end{aligned} \tag{8.25}$$

Condition (μ.2) in Table II is taken care of by $0 < \theta_1 < \pi/2$. Conditions (μ.4) and (μ.5) are satisfied by $0 < \tilde{\tau} < 1$ (together with $\tau > 0$ for R and $\tau > -\tau_0$ for C and H). The remaining conditions will be imposed explicitly. When one of these conditions is violated the corresponding term in the integrand will be taken to be zero. We remark that the triple collision transport integrals (7.8), together with the conditions of Table II, represent a complete solution of the Choh-Uhlenbeck integral equation for λ_1 and η_1, in which all dynamical effects are taken into account. Since we did not have any reliable *a priori* estimate of the effects of the various triple collision events, we retained the complete collision integrals (8.24) together with all conditions of Table II in the numerical integration. From the results obtained we can suggest *a posteriori* approximations that would not change the values significantly as will be discussed in Section 9.

A computer program was devised that yields the value of the integrand for any point of the integration region (8.25). The following checks were included in the computer program:

1) For each triple collision event we verified conservation of kinetic energy.

2) For each triple collision event we verified conservation of angular momentum.

3) For each triple collision event subject to the conditions of Table II we verified that all other (redundant) conditions mentioned in Section 7 were satisfied as well.

To obtain an estimate for the integrals we need to select a sampling procedure. At first we tried a method of Haselgrove for numerical integration[21]. However, we did not obtain a satisfactory convergence for the estimates. As an alternative we decided upon a modified Monte Carlo quadrature suggested by Haber[22]*. Consider the integral

$$I = \int_0^1 \dots \int_0^1 f(\mathbf{x})\, d\mathbf{x}, \tag{8.26}$$

over the n dimensional vector $\mathbf{x}$. The n dimensional integration region is to be pictured as divided into $N = k^n$ subcubes. In the i-th subcube two independent random points are chosen $\mathbf{x}_i$ and $\mathbf{y}_i$. In addition we consider the points $\mathbf{x}_i'$ and $\mathbf{y}_i'$ which are, respectively, the points symmetrically opposite

* The author is indebted to Dr. S. Haber for his advice and to Mr. F. P. Karriker for his assistance in carrying out the computations.

to $\mathbf{x}_i$ and $\mathbf{y}_i$ in the subcube. A first estimate of the integral I is

$$J_1(k) = \frac{1}{2N} \sum_{i=1}^{N} \{f(\mathbf{x}_i) + f(\mathbf{y}_i)\}, \tag{8.27}$$

with an estimated standard deviation

$$\sigma_1 = \frac{1}{2N} [\sum_{i=1}^{N} \{f(\mathbf{x}_i) - f(\mathbf{y}_i)\}^2]^{1/2}. \tag{8.28}$$

A second better estimate is obtained by

$$J_2(k) = \frac{1}{2N} \sum_{i=1}^{N} \left\{ \frac{f(\mathbf{x}_i) + f(\mathbf{x}_i')}{2} + \frac{f(\mathbf{y}_i) + f(\mathbf{y}_i')}{2} \right\}, \tag{8.29}$$

with an estimated standard deviation

$$\sigma_2 = \frac{1}{2N} \sum_{i=1}^{N} \left[\left\{ \frac{f(\mathbf{x}_i) + f(\mathbf{x}_i')}{2} - \frac{f(\mathbf{y}_i) + f(\mathbf{y}_i')}{2} \right\}^2 \right]^{1/2}. \tag{8.30}$$

To test the reliability of this procedure for our triple collision transport integrals, we first evaluate the seven dimensional integrals associated with diagram H2 with the additional conditions $r_{21}^2(3) < 1$ and $r_{31}^2(3) < 1$ at the time of the third collision. As pointed out in Section 6 under these conditions we should regain the values predicted by Enskog. Thus for these integrals we can compare the estimates with the exact answer. The results obtained for $k = 2$, 3, 4, 5 and 6 are presented in Table III together with the corresponding standard deviations. It turns out that the estimates $k = 2$ and $k = 3$ are not yet reliable which is not surprising. However, the estimates for $k > 4$ are all within each others standard deviations and do agree with the exact result.

Subsequently, we applied this procedure to the complete triple collision integrals (8.24). In practice, we substract from H2 the contributions contained in the Enskog theory and compute directly the difference between the complete triple collision integral and the value predicted by Enskog. Since we expect possible cancellations between the contributions of different diagrams (confirmed by the results) we combine for each phase point all six terms of the integrand to obtain a maximum precision for the total integral. From the test calculation we conclude that we should disregard the estimates for $k = 2$ and $k = 3$. In the preliminary computations reported here, we obtained four successive estimates $J_1(4)$, $J_2(4)$, $J_1(5)$, $J_2(5)$, indicated respectively by $i = 1, 2, 3, 4$ in Tables IV and V. The results obtained for the total triple collision integrals are presented in Table IV. In addition, we made an estimate of the contributions of the six individual triple collision events; these results are reported in Table V. We see that the four different estimates agree with each other, as was the case for the test calculation in Table III. Nevertheless, to obtain reliable numerical values a much more detailed sampling of the integration region is required. Alternate procedures to calculate these collision integrals are presently being explored. But our preliminary calculations allows us to make some interesting conclusions regarding the relative contribution of various triple collision events.

TABLE III Test of the numerical integration procedure

	$\frac{\lambda_1(1)}{\lambda_0(1)} - \frac{2}{3}\frac{\eta_1(1)}{\eta_0(1)}$ Enskog	$\frac{\eta_1(1)}{\eta_0(1)}$ Enskog
$J_1(2)$	-0.32 ± 0.29	-0.53 ± 0.35
$J_2(2)$	-0.13 ± 0.15	-0.49 ± 0.21
$J_1(3)$	-0.16 ± 0.02	-0.54 ± 0.05
$J_2(3)$	-0.17 ± 0.04	-0.55 ± 0.06
$J_1(4)$	-0.203 ± 0.017	-0.634 ± 0.025
$J_2(4)$	-0.211 ± 0.012	-0.634 ± 0.019
$J_1(5)$	-0.215 ± 0.008	-0.637 ± 0.011
$J_2(5)$	-0.213 ± 0.007	-0.634 ± 0.009
$J_1(6)$	-0.213 ± 0.007	-0.629 ± 0.009
exact	-0.208	-0.625

TABLE IV Estimates for the triple collision transport integrals

i	$\frac{\lambda_1(1)}{\lambda_0(1)} - \frac{2}{3}\frac{\eta_1(1)}{\eta_0(1)}$	$\frac{\eta_1(1)}{\eta_0(1)}$
1	-0.191 ± 0.010	-0.612 ± 0.012
2	-0.199 ± 0.010	-0.614 ± 0.010
3	-0.190 ± 0.004	-0.598 ± 0.005
4	-0.192 ± 0.004	-0.601 ± 0.004

TABLE V Estimates for the contributions of the separate triple collision events

μ	i	$\dfrac{\lambda_1(1)}{\lambda_0(1)} - \dfrac{2}{3}\dfrac{\eta_1(1)}{\eta_0(1)}$	$\dfrac{\eta_1(1)}{\eta_0(1)}$
R1	1	-0.048 ± 0.007	-0.042 ± 0.006
R1	2	-0.050 ± 0.005	-0.039 ± 0.004
R1	3	-0.051 ± 0.003	-0.039 ± 0.003
R1	4	-0.049 ± 0.002	-0.040 ± 0.002
R2	1	$+0.044 \pm 0.006$	$+0.035 \pm 0.007$
R2	2	$+0.045 \pm 0.005$	$+0.032 \pm 0.004$
R2	3	$+0.047 \pm 0.003$	$+0.033 \pm 0.003$
R2	4	$+0.046 \pm 0.002$	$+0.032 \pm 0.002$
C1	1	-0.023 ± 0.003	-0.036 ± 0.004
C1	2	-0.028 ± 0.006	-0.039 ± 0.005
C1	3	-0.035 ± 0.006	-0.037 ± 0.002
C1	4	-0.032 ± 0.004	-0.037 ± 0.002
C2	1	$+0.057 \pm 0.004$	$+0.147 \pm 0.006$
C2	2	$+0.064 \pm 0.007$	$+0.152 \pm 0.006$
C2	3	$+0.069 \pm 0.006$	$+0.151 \pm 0.003$
C2	4	$+0.069 \pm 0.004$	$+0.152 \pm 0.003$
H1	1	$+0.135 \pm 0.016$	$+0.197 \pm 0.018$
H1	2	$+0.141 \pm 0.012$	$+0.206 \pm 0.013$
H1	3	$+0.150 \pm 0.008$	$+0.219 \pm 0.009$
H1	4	$+0.145 \pm 0.006$	$+0.211 \pm 0.007$
H2	1	-0.355 ± 0.017	-0.913 ± 0.021
H2	2	-0.371 ± 0.014	-0.926 ± 0.016
H2	3	-0.370 ± 0.008	-0.924 ± 0.010
H2	4	-0.370 ± 0.006	-0.920 ± 0.008

9 DISCUSSION OF THE RESULTS

In the preceding Section we obtained preliminary numerical estimates for the triple collision transport integrals. Retaining the best estimates we conclude*

$$\frac{\lambda_1(1)}{\lambda_0(1)} = -0.593 \pm 0.014, \qquad \frac{\eta_1(1)}{\eta_0(1)} = -0.601 \pm 0.008, \tag{9.1}$$

where, for the error estimate, we have replaced the standard deviation by a 95 per cent confidence interval.

* In addition, there is a collisional transfer contribution $+6/5$ to λ_1/λ_0 and $+4/5$ to η_1/η_0.

The triple collision integrals could be decomposed into integrals associated with the six triple collision events of Fig. 5. The best estimates for the contributions of the separate events are

μ	$\frac{\lambda_1(1)}{\lambda_0(1)}$	$\frac{\eta_1(1)}{\eta_0(1)}$
R1	-0.076 ± 0.007	-0.040 ± 0.004
R2	$+0.067 \pm 0.007$	$+0.033 \pm 0.004$
C1	-0.057 ± 0.011	-0.037 ± 0.004
C2	$+0.172 \pm 0.012$	$+0.152 \pm 0.006$
H1	$+0.286 \pm 0.021$	$+0.211 \pm 0.015$
H2	-0.983 ± 0.023	-0.920 ± 0.015

(9.2)

The theory of Enskog neglects the effects of all diagrams except for a piece of diagram H2. To facilitate a comparison with the Enskog prediction $-5/8$ (4.11) we rewrite (9.1) as

$$\frac{\lambda_1(1)}{\lambda_0(1)} = -0.625[1 - (5 \pm 2.2)\%]; \quad \frac{\eta_1(1)}{\eta_0(1)} = -0.625[1 - (4 \pm 1.3)\%]. \tag{9.3}$$

As mentioned in the previous Section a further study of these collision integrals is in progress. We do not yet consider the preliminary results presented here as definite. Nevertheless, we feel confident enough to draw the following conclusions.

As a first conclusion, we see that despite the approximations involved, the theory of Enskog presents a surprisingly good estimate for the first density coefficient, namely within 5%. From (9.2) it appears that the complete contribution from diagram H2 would have overestimated the triple collision integrals by as much as 50%. This additional contribution is cancelled against contributions from the remaining diagrams.

Secondly, the difference between the results (9.3) obtained from the Choh–Uhlenbeck integral equation and those from the Enskog theory are larger than the error estimates. Thus we confirm the theoretical expectation that the theory of Enskog is not strictly correct. Recently, Alder and Wainwright obtained an estimate for the first density coefficient of self diffusion from molecular dynamics calculations[23]. The derivations from the theory of Enskog found here for the thermal conductivity and viscosity have the same sign as the deviation found by Alder and Wainwright for the self diffusion. We have extended the method of this report to include the self diffusion; preliminary calculations carried out so far are in substantial agreement with the molecular dynamics results.

As pointed out in Section 7 the integrals associated with R1 and R2 could not be combined into one single integral due to the effect of a possible fourth collision preceding or succeeding the three correlated binary collisions. In a quantitative analysis Cohen, Murphy and Foch concluded that in a succession of three collisions pertaining to our diagrams R1 and C1, the probability of a fourth collision is only of the order of one or two parts in a thousand[24]. Our computer results confirm that the probability of a fourth collision is rather small both for real and hypothetical collisions. Thus to a good approximation, certainly within the present precision, we can combine these two integrals and forget the fourth collisions. Similarly, we expect that in first approximation we could combine C1 and C2, resp. H1 and H2, provided we exclude the overlapping configurations which should be treated separately. However, we have not made an explicit quantitative investigation of the contributions to C2 and H2 when the overlapping configurations are excluded.

The results (9.2) reveal an appreciable difference between the contributions R1, R2 and C1 on one hand and the contributions C2, H1 and H2 on the other hand. We note that in C2, H1 and H2 a binary collision can occur, while two other particles are overlapping (see Fig. 5). In C2 1 and 3 are allowed to collide while 1 and 2 are still inside each other. In H1 particle 2 and 3 are allowed to collide while 1 and 3 are overlapping. In H2 particles 1 and 3 are allowed to collide while 1 and 2 are overlapping, and 2 and 3 are allowed to collide while 1 and 2 and/or 1 and 3 are overlapping. Collisions with such overlapping configurations are not included in R1, R2 and C1. Our computations indicate that the probability for triple collision events with overlapping configurations is rather large and they determine the major contributions to the triple collision integrals. On the other hand, the contributions from successive correlated binary collisions that are spatially separated are very small. In (9.2) this is illustrated by the results for R1 and R2 which do not include overlapping configurations and whose contributions almost cancel each other.

These results suggest *a posteriori* that the Enskog values could be considered as a valid *first* approximation to the triple collision integrals; this first approximation is determined by those events in H2, where in the third collision 2 and 3 are colliding, while both 2 and 3 are overlapping with particle 1. As a *second* approximation those triple collision events suggest themselves which contain collisions between two particles while *one* of them is overlapping with the third particle. This second approximation contains contributions from C2, H1 and H2. As a *third* approximation one could then consider the effect of three successive binary collisions for which all pairs are spatially separated. The various effects of possible fourth collisions could be treated as an even higher order correction. A precise formulation of this expansion was recently completed by Gillespie and Sengers[29].

We should remark that we have evaluated λ_1 and η_1 only in the first Enskog approximation, that is, we have taken only the first Sonine polynomial contribution. As discussed in Section 3, the first Enskog approximation is a good one in the case of the dilute gas values λ_0 and η_0, obtained from the linearized Boltzmann equation. We have made some studies of the effect of the higher order Enskog approximation of the solution of the Choh–Uhlenbeck equation for a two dimensional model of hard disks[25]. The preliminary results lead us to expect that this expansion is suitable to obtain the first density coefficients λ_1 and η_1. Since the Enskog values $-5/8$ are independent of the Enskog approximation, we only need the expansion into Sonine polynomials for the term determining the deviation from the Enskog theory. We expect that the second Enskog approximation would not change the values (9.1) by more than a few tenths of a percent. To justify this suggestion we hope to obtain better estimates for the triple collision integrals, including the effect of the second Enskog approximation, in the future.

Acknowledgement

The author wants to thank Professor M. S. Green who suggested the subject of this research. He is indebted to Dr. S. Haber for this advice in selecting a numerical integration method and to Mr. F. P. Karriker for his assistance in carrying out the computations. During this research the author had the benefit of many stimulating discussions with Professor J. R. Dorfman.

APPENDIX A SURFACE INTEGRAL FORM OF THE TRIPLE COLLISION INTEGRAL

The fact that the six dimensional configurational triple collision integral can be converted into a five dimensional surface integral was pointed out by Green[16].

The streaming operators $\mathscr{S}(1 \dots s)$ are mathematically defined as[5]

$$\mathscr{S}(1..s) = \lim_{\tau\to\infty} \mathscr{S}_\tau(1..s) = \lim_{\tau\to\infty} e^{-\tau\mathscr{H}_s} \cdot e^{+\tau\mathscr{H}_s(0)}, \tag{A.1}$$

where $\mathscr{H}_s(1 \dots s)$ is a "Hamilton" operator

$$\mathscr{H}_s(1..s) = \sum_{i=1}^{s} \mathbf{v}_i \cdot \frac{\partial}{\partial \mathbf{r}_i} - \sum_{i<j}^{s} \theta_{ij}, \tag{A.2}$$

and $\mathscr{H}_s^{(0)}$ that part of $\mathscr{H}_s$ which is associated with the free streaming of s particles

$$\mathscr{H}_s^{(0)}(1..s) = \sum_{i=j}^{s} \mathbf{v}_i \cdot \frac{\partial}{\partial \mathbf{r}_i}. \tag{A.3}$$

We assume* that for large values of τ the streaming operators become independent of τ

$$\lim_{\tau\to\infty} \frac{d}{d\tau} \mathscr{S}_\tau(1..s) = \mathscr{H}_s(1..s)\, \mathscr{S}(1..s) - \mathscr{S}(1..s)\, \mathscr{H}_s^{(0)}(1..s) = 0, \tag{A.4}$$

so that[26]

$$\mathscr{H}_s^{(0)}(1..s)\, \mathscr{S}(1..s) - \mathscr{S}(1..s)\, \mathscr{H}_s^{(0)}(1..s) = \sum_{i<j}^{s} \theta_{ij}(1..s). \tag{A.5}$$

In particular

$$\mathscr{H}_3^{(0)}(123)\, \mathscr{S}(123) - \mathscr{S}(123)\, \mathscr{H}_3^{(0)}(123) = (\theta_{12} + \theta_{13} + \theta_{23})\, \mathscr{S}(123), \tag{A.6}$$

and

$$\mathscr{H}_3^{(0)}(123)\, \mathscr{S}(\alpha) - \mathscr{S}(\alpha)\, \mathscr{H}_3^{(0)}(123) = \theta_\alpha \mathscr{S}(\alpha), \tag{A.7}$$

where α is either 12, 13 or 23. Using these relations the operator $0(123)$ given by (5.15) can be rewritten into the form[16]

$$\begin{aligned} 0(123 = \int d\mathbf{r}_{21}\, d\mathbf{r}_{31} [&\mathscr{H}_3^{(0)} U(123) - U(123)\, \mathscr{H}_3^{(0)} \\ &- \sum_{\alpha\neq\beta}\sum \{\mathscr{H}_3^{(0)} U(\alpha) - U(\alpha)\, \mathscr{H}_3^{(0)}\}\, U(\beta)]. \end{aligned} \tag{A.8}$$

For convenience we have introduced the Ursell operators

$$\begin{aligned} U(\alpha) &= \mathscr{S}(\alpha) - 1, \\ U(123) &= \mathscr{S}(123) - \mathscr{S}(12) - \mathscr{S}(13) - \mathscr{S}(23) + 2. \end{aligned} \tag{A.9}$$

---

* In this assumption we neglect the collisions which are taking place exactly at the time $-\tau$. Therefore, we make an error, referred to as the "initial" error term by Dorfman and Cohen.[6] This initial error term does not contribute to the first density coefficients.

The summations over α and β are to be taken over the pairs 12, 13 and 23. The integrand depends only on the relative positions and the integration can be carried out over any two relative position vectors, for which we have chosen $\mathbf{r}_{21}$ and $\mathbf{r}_{31}$. Since in (5.14) the operator $O(123)$ operates on a function of the velocities alone, the term $U(123)\,\mathscr{H}_3^{(0)}$ vanishes in (A.8).

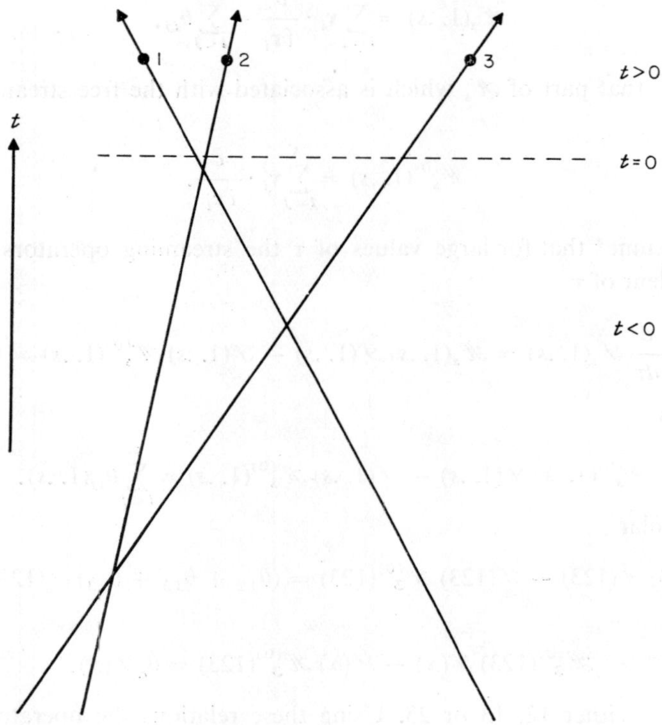

FIG. A1 Diagram Representing the Free Trajectories of Particles 1, 2 and 3

In order to define an appropriate transformation for the integration variables, we consider for fixed $\mathbf{V}_1$, $\mathbf{V}_2$, $\mathbf{V}_3$ the positions of particles 1, 2 and 3 which are related to each other by free particle motion. This is schematically illustrated in Fig. A1, where the time t associated with the free motion of the particles is taken to increase when the diagram is read from bottom to top. However, since the $\mathscr{S}$ operators transform the present phases of the particles to phases in the previous history of the motion, we shall retrace the trajectories backwards in time and always read the diagrams from top to bottom as in the main text.

We remark that the three free particle trajectories have at least one "intersection", by which we mean that along the free trajectories the relative distance of at least one pair of particles becomes equal to σ for some value of t. Otherwise, the particles are not dynamically correlated and the Ursell operators in (A.8) would vanish for all positions along these trajectories. Suppose that when the trajectories in Fig. A1 are retraced from top to bottom, the distance r_{21} is the first distance which becomes equal to σ. Then, the positions of the particles along the free trajectories can be characterized by

$$\mathbf{r}_{21} = \mathbf{r}_{21}(t) = \exp(t\mathscr{H}_3^{(0)})\,\hat{\mathbf{r}}_{21}(0) = \hat{\mathbf{r}}_{21}(0) + t\mathbf{V}_{21},$$
$$\mathbf{r}_{31} = \mathbf{r}_{31}(t) = \exp(t\mathscr{H}_3^{(0)})\,\mathbf{r}_{31}(0) = \mathbf{r}_{31}(0) + t\mathbf{V}_{31}, \tag{A.11}$$

where the notation $\hat{\mathbf{r}}_{21}(0)$ indicates that the distance $r_{21}(t)$ is equal to σ for $t = 0$. We take $\hat{\mathbf{r}}_{21}(0)$, $\mathbf{r}_{31}(0)$ and t as the new integration variables. In this new coordinate system $d\hat{\mathbf{r}}_{21}(0)$ is a two dimensional surface element. The Jacobian of this transformation is

$$d\mathbf{r}_{21}\,d\mathbf{r}_{31} = \mathbf{V}_{21}\cdot\hat{\mathbf{r}}_{21}(0)\,d\hat{\mathbf{r}}_{21}(0)\,d\mathbf{r}_{31}(0)\,dt. \tag{A.11}$$

From the expression (5.15) for $0(123)$ we see that the integrand will give contributions in the three regions for which, respectively, $r_{21} < \sigma$, $r_{31} < \sigma$ and $r_{32} < \sigma$. Therefore, the upper limit for the integration over t can be taken at any value $t > 0$ corresponding to phases of *receding* particles. The lower limit has to be taken at a sufficiently negative value of t, so that the free trajectories will no longer intersect when retraced further backwards in time. As a consequence, all Ursell operators vanish at the lower limit of the t-integration.

In this new coordinate system

$$\mathscr{H}_3^{(0)} = \frac{\partial}{\partial t}, \tag{A.12}$$

since for any function $g(\mathbf{r}_{21}, \mathbf{r}_{31}, \mathbf{V}_1, \mathbf{V}_2, \mathbf{V}_3)$:

$$\mathscr{H}_3^{(0)} g(\mathbf{r}_{21}, \mathbf{r}_{31}, \mathbf{V}_1, \mathbf{V}_2, \mathbf{V}_3) = \mathscr{H}_3^{(0)} e^{\tau\mathscr{H}_3(0)}\, g(\hat{\mathbf{r}}_{21}(0), \mathbf{r}_{31}(0), \mathbf{V}_1, \mathbf{V}_2, \mathbf{V}_3)$$
$$= \frac{\partial}{\partial t} e^{t\mathscr{H}_3(0)}\, g(\hat{\mathbf{r}}_{21}(0), \mathbf{r}_{31}(0), \mathbf{V}_1, \mathbf{V}_2, \mathbf{V}_3) = \frac{\partial}{\partial t} g(\mathbf{r}_{21}, \mathbf{r}_{31}, \mathbf{V}_1, \mathbf{V}_2, \mathbf{V}_3).$$

Therefore, as pointed out by Green[16], Eq. (A.8) transforms to

$$0(123) = \int d\hat{\mathbf{r}}_{21}(0)\,d\mathbf{r}_{31}(0)\,\mathbf{V}_{21}\cdot\hat{\mathbf{r}}_{21}(0)\int dt$$
$$\times\left[\frac{\partial}{\partial t}U(123) - \sum_{\alpha\neq\beta}\sum\left\{\frac{\partial}{\partial t}U(\alpha) - U(\alpha)\frac{\partial}{\partial t}\right\}U(\beta)\right]. \tag{A.13}$$

In the first term of (A.13) the t-integration can be carried out immediately. The second term can be simplified by partial integration.

$$0(123) = \int d\hat{\mathbf{r}}_{21}(0)\, d\mathbf{r}_{31}(0)\, \mathbf{V}_{21} \cdot \hat{\mathbf{r}}_{21}(0) \left[U(123) - \sum_{\alpha \neq \beta} \sum \int dt \left\{ \frac{\partial}{\partial t} U(\alpha) \right\} U(\beta) \right]. \tag{A.14}$$

To elucidate the result for the second time integral we shall consider as an example

$$0' = \int dt \left[\frac{\partial}{\partial t} U(13) \right] U(23) + \int dt \left[\frac{\partial}{\partial t} U(23) \right] U(13). \tag{A.15}$$

We consider the three possible successions of two binary collisions in Figs. A2, A3 and A4 separately. We first evaluate (A.15) for phase points for which one encounters along the free trajectories the two collisions indicated in Fig. A2*. In the figure a circle indicates a collision between the particles whose trajectories are enclosed. In the case under consideration both particles 1 and 3 and particles 2 and 3 are aimed to collide. We have indicated the free trajectories by dotted lines when they have to be distinguished from the actual trajectories, if we would take into account the interaction between 1 and 3. Following the particles along their free trajectories, particles 1 and 3 are inside each other ($r_{31}(t) < \sigma$) during the time interval $t_2 < t < t_1$ and particles 2 and 3 are inside each other during the time interval $t_4 < t < t_3$.

For the evaluation of the first integral in (A.15) we have to study the variation of $U(13)$ as a function of t along the free trajectories. For $t > t_1$, $U(13)$ has a certain value which is constant, since the operators ultimately operate on functions of the velocities alone. For $t_2 < t < t_1$ the operator $\mathscr{S}(13)$ creates infinite velocities; this contribution vanishes as a result of the normalization of the distribution function and $U(13)$ can be taken as -1. Finally, for $t < t_2$, particles 1 and 3 are no longer aimed to collide and $U(13) = 0$. Thus

$$\begin{aligned} t > t_1 &: U(13) = U(13) \\ t_2 < t < t_1 &: U(13) = -1 \\ t < t_2 &: U(13) = 0. \end{aligned} \tag{A.16}$$

---

* In diagram A2 (when read from top to bottom) we disregard the possibility that the collision between 2 and 3 may cause a deflected path of 3 to collide with the original path of 1. The contribution of such an event is taken into account by diagram A3 (upon interchanging 1 and 2).

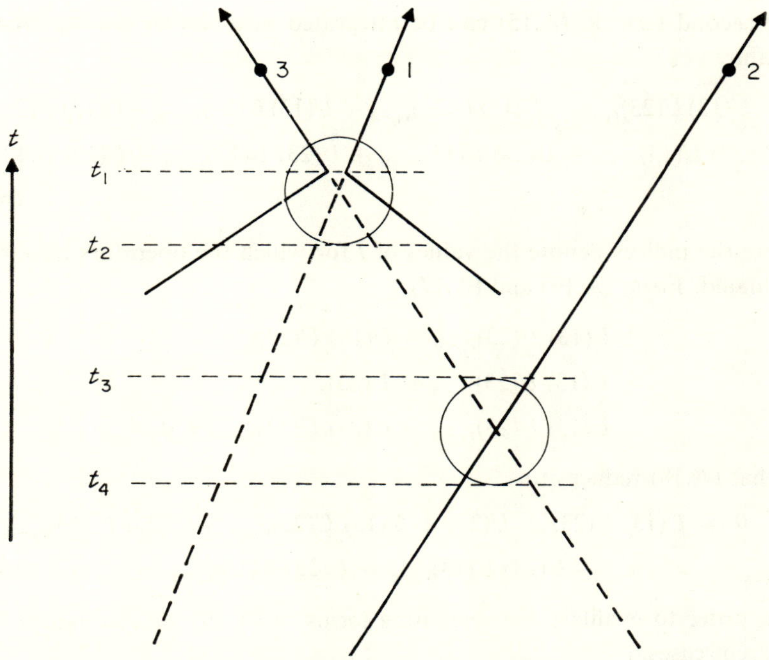

FIG. A2 Diagram of Successive Collisions Leading to a Contribution $U(13)\,U(23)$ to the Surface Integral

Similarly

$$\begin{aligned} t > t_3 &: U(23) = U(23) \\ t_4 < t < t_3 &: U(23) = -1 \\ t < t_4 &: U(23) = 0. \end{aligned} \tag{A.17}$$

It follows from (A.16) that the first integrand in (A.15) only contributes for $t = t_1$ and $t = t_2$:

$$\int dt \left[\frac{\partial}{\partial t} U(13) \right] U(23)$$

$$= \lim_{\delta \to +0} \int_{t_1-\delta}^{t_1+\delta} dt \left[\frac{\partial}{\partial t} U(13) \right] U(23) + \lim_{\delta \to +0} \int_{t_2-\delta}^{t_2+\delta} dt \left[\frac{\partial}{\partial t} U(13) \right] U(23).$$

For hard spheres we can take δ arbitrarily small, so that $U(23)$ does not vary during the interval δ. Thus

$$\int dt \left[\frac{\partial}{\partial t} U(13) \right] U(23) = U(13)\, U(23) \Big|_{t_1-\delta}^{t_1+\delta} + U(13)\, U(23) \Big|_{t_2-\delta}^{t_2+\delta}.$$

The second term in (A.15) can be integrated in a similar way so that we obtain

$$0' = U(13)\,U(23)_{t_1+\delta} - U(13)\,U(23)_{t_1-\delta} + U(13)\,U(23)_{t_2+\delta} - U(13)\,U(23)_{t_2-\delta}$$
$$+ U(23)\,U(13)_{t_3+\delta} - U(23)\,U(13)_{t_3-\delta} + U(23)\,U(13)_{t_4+\delta} - U(23)\,U(13)_{t_4-\delta} \tag{A.18}$$

where the indices denote the values of t for which the operators have to be evaluated. From (A.16) and (A.17)

$$U(13)\,U(23)_{t_1+\delta} = U(13)\,U(23),$$
$$U(13)\,U(23)_{t_1-\delta} = U(23),$$
$$U(13)\,U(23)_{t_2-\delta} = U(23)\,U(13)_{t_4+\delta} = 0,$$

so that (A.18) reduces to

$$0' = U(13)\,U(23) + U(23) + U(13)\,U(23)_{t_2+\delta} + U(23)\,U(13)_{t_3+\delta}$$
$$- U(23)\,U(13)_{t_3-\delta} + U(23)\,U(13)_{t_4+\delta}. \tag{A.19}$$

In order to evaluate the remaining terms in (A.19) we distinguish three different cases.

Case I $\quad t_4 < t_3 < t_2 < t_1$

In this case the two collisions are spatially separated. From (A.16) and (A.17) we conclude

$$U(13)\,U(23)_{t_2+\delta} = -U(23),$$
$$U(23)\,U(13)_{t_3+\delta} = U(23)\,U(13)_{t_3-\delta} = U(23)\,U(13)_{t_4-\delta} = 0,$$

so that (A.19) becomes

$$0' = U(13)\,U(23). \tag{A.20}$$

Case II $\quad t_4 < t_2 < t_3 < t_1$

In this case the collision between 2 and 3 starts while 1 and 3 are still overlapping. In this case

$$U(13)\,U(23)_{t_2+\delta} = +1; \quad U(23)\,U(13)_{t_3+\delta} = -U(23);$$
$$U(23)\,U(13)_{t_3-\delta} = +1; \quad U(23)\,U(13)_{t_4+\delta} = 0,$$

so that (A.19) reduces to (A.20).

Case III $\quad t_2 < t_4 < t_3 < t_1$

In this case the collision between 2 and 3 not only begins while 1 and 3 are overlapping, but 2 and 3 have also separated before 1 and 3 leave their overlapping configurations. We have

$$U(13)\,U(23)_{t_2+\delta} = 0; \qquad U(23)\,U(13)_{t_3+\delta} = -U(23);$$
$$U(23)\,U(13)_{t_3-\delta} = +1; \quad U(23)\,U(13)_{t_4+\delta} = +1,$$

so that (A.19) reduces again to (A.20).

Thus we conclude that for phase points represented by Fig. A2 the integral (A.15) leads to the result (A.20) regardless whether or not the collisions are

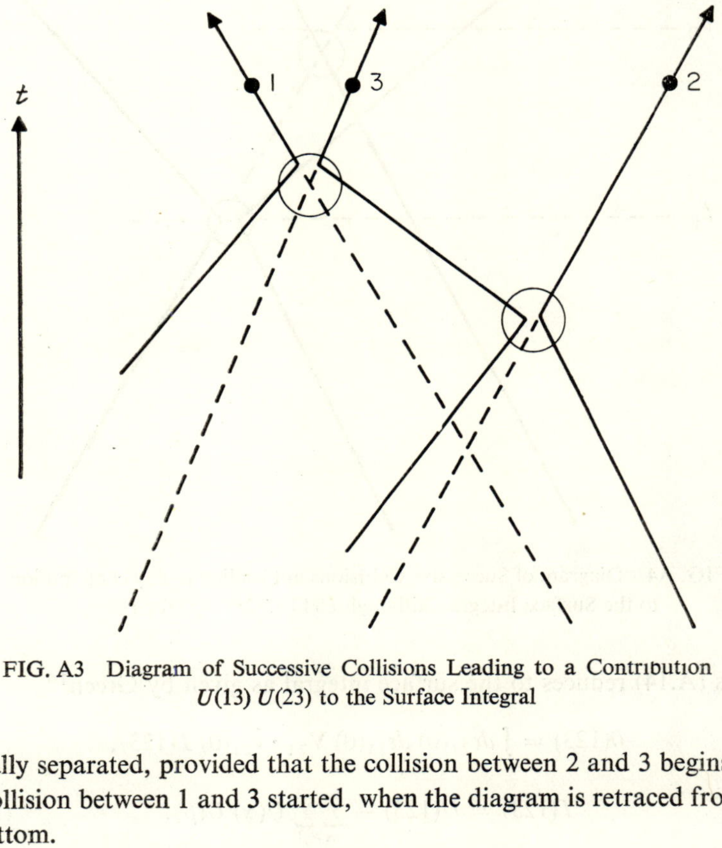

FIG. A3 Diagram of Successive Collisions Leading to a Contribution $U(13)\,U(23)$ to the Surface Integral

spatially separated, provided that the collision between 2 and 3 begins after the collision between 1 and 3 started, when the diagram is retraced from top to bottom.

Using the same method one can prove that also

$$0' = U(13)\,U(23), \tag{A.21}$$

for phase points represented by the diagram of Fig. A3. When the integration is carried out for phase points represented by Fig. A4, one finds[16]

$$0' = 0. \tag{A.22}$$

Thus events of this kind do not contribute to the surface integral, although the operator $U(13)\,U(23)$ itself would not vanish for $t > t_0$.

The contribution of any diagram containing two other successive binary collisions is obtained by permutation of the particle numbers. Thus for hard

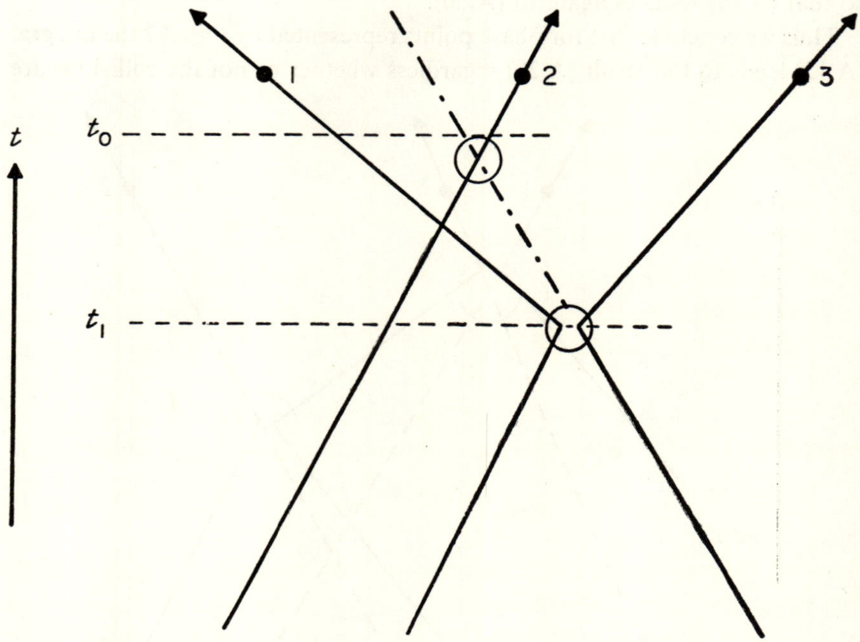

FIG. A4 Diagram of Successive Collisions not leading to any contribution to the Surface Integral, although $U(13)U(23) \neq 0$ for $t > t_0$

spheres (A.14) reduces to the surface integral as given by Green[16]

$$O(123) = \int d\hat{\mathbf{r}}_{21}(0)\, d\mathbf{r}_{31}(0)\, \mathbf{V}_{21} \cdot \hat{\mathbf{r}}_{21}(0)\, T(123), \tag{A.23}$$

with

$$T(123) = U(123) - \sum_{\alpha>\beta}\sum U(\alpha)\, U(\beta). \tag{A.24}$$

The notation $\alpha > \beta$ indicates that the product $U(\alpha)\,U(\beta)$ should only be evaluated for successive binary collisions involving pairs α and β for which the α-collision occurs prior to the β-collision when the diagrams are retraced backwards in time. For example, if $\alpha = 13$ and $\beta = 23$, $U(13)\,U(23)$ should only be evaluated for the collision sequences shown in Fig. A2 and A3 (regardless, of course, of the occurrence of other collisions).

To evaluate the triple collision operator $T(123)$ in more detail we introduce the following notation. Each binary collision in a collision sequence representing the dynamical history of a phase point will be indicated by listing the pairs whose interaction has to be taken into account successively for that particular collision to occur. For example, in Fig. A2 the first collision is denoted by (13) and the second collision by (23). In Fig. A3 the first collision is again denoted by (13), but the second collision is now denoted by (13) (23), since the collision between 2 and 3 occurs after the interaction between 1 and 3 has been taken into account. With this notation (A.24) can be written more specifically as

$$T(123) = U(123) - \sum_{(\alpha)>(\alpha)}\sum_{(\beta)} \mathscr{S}(\alpha)\, U(\beta) + \sum_{(\alpha)>(\beta)}\sum U(\beta) \quad (\alpha \neq \beta), \tag{A.25}$$

where the summation in the second term is to be taken over all pairs of collisions (α) and $(\alpha)\,(\beta)$ for which (α) occurs prior to $(\alpha)\,(\beta)$, and the summation in the third term over all pairs of collision (α) and (β) such that (α) occurs prior to (β).

The triple collision operator $T(123)$ can be decomposed into a sum of operators, each of which is associated with a particular triple collision event:

$$T(123) = \sum_{\mu}\sum_{\text{perm}} T_\mu(\alpha_1; \alpha_2). \tag{A.26}$$

Here we have indicated the possible pairs of particles by α_i $(i = 1, 2, 3)$. The operators $T_\mu(\alpha_1; \alpha_2)$ are defined in Section 6 for the specific example $\alpha_1 = 12$ and $\alpha_2 = 13$; α_1, α_2 refer to the pairs of particles involved in the first, resp. second collision. The first summation in (A.26) is to be taken over the six diagrams μ = R1, R2, C1, C2, H1, H2 and the second summation over the six permutations of the particles 1, 2 and 3.

The validity of the addition theorem (A.26) for an arbitrary phase point can be demonstrated as follows. Let (α_1) be the first real collision in the diagram representing the dynamical history. If there is a second real collision, we assume that to be $(\alpha_1)\,(\alpha_2)$. As a consequence, $(\alpha_1) > (\alpha_1)\,(\alpha_2)$. We also have to consider hypothetical collisions of the type (α) and $(\alpha)\,(\beta)$. If there is only one hypothetical collision of the type (α), we suppose it is (α_2). If there are two hypothetical collisions of this type, we assume $(\alpha_2) > (\alpha_3)$. Then from (A.25) $T(123)$ can be written as

$$T(123) = [\mathscr{S}(123) - \mathscr{S}(\alpha_1) - \mathscr{S}(\alpha_1)\, U(\alpha_2)] - \mathscr{S}(\alpha_1)\, U(\alpha_3)_{(\alpha_1)>(\alpha_1)\,(\alpha_3)}$$
$$- \sum_{\substack{(\alpha)>(\alpha)\\ \alpha\neq\alpha_1}}\sum_{(\beta)} \mathscr{S}(\alpha)\, U(\beta) + U(\alpha_3), \quad (\alpha \neq \beta), \tag{A.27}$$

in which the conditions concerning the time order of the events are indicated where necessary. The sum of the first three terms between the brackets is only different from zero if there are three or more real collisions, thus if the conditions for either R1($\alpha_1;\alpha_2$) or C1($\alpha_1;\alpha_2$) are satisfied. In that case

$$\mathscr{S}(123) - \mathscr{S}(\alpha_1) - \mathscr{S}(\alpha_1)\,U(\alpha_2) = \mathscr{S}(123) - \mathscr{S}(\alpha_1)\mathscr{S}(\alpha_2),$$

which is just the contribution $T_{\mathrm{R1}}(\alpha_1;\alpha_2)$ or $T_{\mathrm{C1}}(\alpha_1;\alpha_2)$. The next term $\mathscr{S}(\alpha_1)\,U(\alpha_3)$ is different from zero, if there is a collision $(\alpha_1)\,(\alpha_3)$ such that $(\alpha_1) > (\alpha_1)\,(\alpha_3)$. This implies also that $(\alpha_1) > (\alpha_1)\,(\alpha_2) > (\alpha_1)\,(\alpha_3)$, since otherwise $(\alpha_1)\,(\alpha_3)$ would be a real collision contrary to our assumption. Thus the conditions for diagram H1($\alpha_1;\alpha_2$) are satisfied with contribution $T_{\mathrm{H1}}(\alpha_1;\alpha_2) = -\mathscr{S}(\alpha_1)\,U(\alpha_3)$. Similarly, one easily verifies that the terms $-\mathscr{S}(\alpha)\,U(\beta)$, with $(\alpha) > (\alpha)\,(\beta)$ and $\alpha \neq \alpha_1$, are only different from zero if the conditions for diagram R2($\alpha_1;\alpha$), when $\beta = \alpha_1$, or for diagram C2($\alpha_1;\alpha$), when $\beta \neq \alpha_1$, are satisfied, leading to the contribution $T_{\mathrm{R2}}(\alpha_1;\alpha) = -\mathscr{S}(\alpha)\,U(\alpha_1)$ or $T_{\mathrm{C2}}(\alpha_1;\alpha) = -\mathscr{S}(\alpha)\,U(\beta)$. Finally, the last term $U(\alpha_3)$ does not vanish if there is a collision (α_3). As a consequence of our assumption this implies $(\alpha_2) > (\alpha_3)$, so that the conditions for diagram H2($\alpha_1;\alpha_2$) are satisfied with the contribution $T_{\mathrm{H2}}(\alpha_1;\alpha_2) = U(\alpha_3)$. This concludes the proof that $T(123)$ can be represented as a sum of operators associated with the basic triple collision diagrams, as expressed by (A.26).

The result (A.26) for $T(123)$ has to be substituted into expression (A.23) for $0(123)$. Since the integrals in (5.16) for λ_1 and η_1 are symmetric in all three particles, it is sufficient to evaluate (A.26) only for one permutation of the particles and to multiply the result by 6. Thus, we replace $0(123)$ in (5.16) by

$$0(123) = 6\,0(12;13) = 6\sum_{\mu}\int d\hat{\mathbf{r}}_{21}(0)\,d\mathbf{r}_{31}(0)\,\mathbf{V}_{21}\cdot\hat{\mathbf{r}}_{21}(0)\,T_{\mu}(12;13). \quad \text{(A.28)}$$

In (A.28), $\hat{\mathbf{r}}_{21}(0)$ and $\mathbf{r}_{31}(0)$ represent the relative position vectors at the time of the first collision in the diagrams of Fig. 4. Let $\mathbf{k}_1$ be the perihelion vector of the first collision, $\mathbf{k}_2$ the perihelion vector of the second collision, and τ the time between the first and the second collision. To specify the collision cylinder associated with the second collision, we have to make a distinction between the diagrams of the first kind in which this collision is real and the diagrams of the second kind in which this collision is hypothetical. Therefore, we define

$$\begin{aligned} \mathbf{V}_{31}(1) &= \mathbf{V}_{31} - (\mathbf{V}_{21}\cdot\mathbf{k}_1)\,\mathbf{k}_1, \quad &(\mu = \mathrm{R1, C1, H1}),\\ \mathbf{V}_{31}(1) &= \mathbf{V}_{31}, &(\mu = \mathrm{R2, C2, H2}). \end{aligned} \quad \text{(A.29)}$$

Then $\mathbf{V}_{31}(1)$ is the relevant relative velocity to be considered for the second collision. In terms of these new variables, (A.28) transforms to

$$O(12; 13) = \sum_{\mu} \sigma^4 \int d\mathbf{k}_1 \int d\mathbf{k}_2 \int d\tau \mathbf{V}_{21} \cdot \mathbf{k}_1 \mathbf{V}_{31}(1) \cdot \mathbf{k}_2 T_{\mu}(12; 13). \quad \text{(A.30)}$$

It should be kept in mind that the integral in (A.23), and consequently in (A.30), is to be carried out over phase points corresponding to receding particles. This implies that in the evaluation of (A.30) one has to verify for any given $\mathbf{k}_1, \mathbf{k}_2$, and τ, that the particles will not suffer any collision in the future.

APPENDIX B DYNAMICS OF A COLLISION BETWEEN TWO HARD SPHERES

Consider two hard spheres 1 and 2 with unit diameter. The relative position of 2 with respect to 1 is $\mathbf{r}_{21} = \mathbf{r}_2 - \mathbf{r}_1$ and the relative velocity

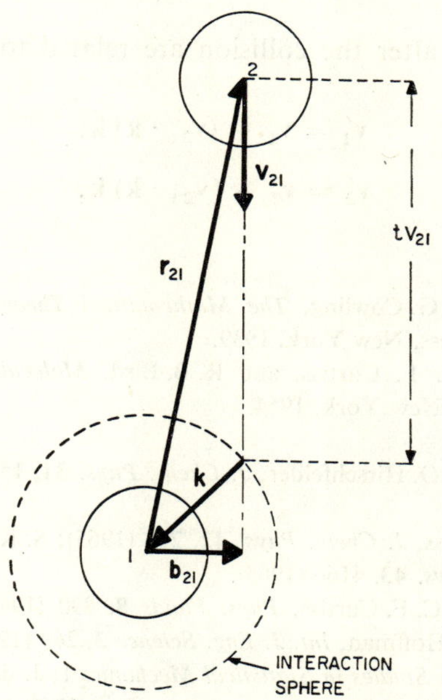

FIG. B1 Geometry of a Collision between two Hard Spheres

$\mathbf{v}_{21} = \mathbf{v}_2 - \mathbf{v}_1$ at the initial time $t = 0$. The "impact vector" $\mathbf{b}_{21}$ (see Fig. B1) of the collision between 1 and 2 is related to the initial position

$\mathbf{r}_{21}$ and initial velocity $\mathbf{v}_{21}$ by

$$\mathbf{b}_{21} = \frac{(\mathbf{v}_{21} \times \mathbf{r}_{21}) \times \mathbf{v}_{21}}{v_{21}^2}, \tag{B.1}$$

and the perihelion vector $\mathbf{k}$ of the collision

$$\mathbf{k} = -\mathbf{b}_{21} + \hat{\mathbf{v}}_{21}\sqrt{1 - b_{21}^2}. \tag{B.2}$$

The two particles will experience a collision, if they approach each other

$$\mathbf{v}_{21} \cdot \mathbf{r}_{21} < 0, \tag{B.3}$$

and if the impact parameter b_{21} is smaller than the diameter ($\sigma = 1$)

$$(\mathbf{v}_{21} \times \mathbf{r}_{21})^2 - v_{21}^2 < 0. \tag{B.4}$$

The time t at which the collision will take place is

$$t = -\frac{\hat{\mathbf{v}}_{21} \cdot \mathbf{r}_{21} + \sqrt{1 - b_{21}^2}}{v_{21}}. \tag{B.5}$$

The velocities $\mathbf{v}_1'$, $\mathbf{v}_2'$ after the collision are related to the initial velocities $\mathbf{v}_1, \mathbf{v}_2$ by

$$\mathbf{v}_1' = \mathbf{v}_1 + (\mathbf{v}_{21} \cdot \mathbf{k})\,\mathbf{k},$$

$$\mathbf{v}_2' = \mathbf{v}_2 - (\mathbf{v}_{21} \cdot \mathbf{k})\,\mathbf{k}. \tag{B.6}$$

References

1. S. Chapman and T. G. Cowling, *The Mathematical Theory of Nonuniform Gases*, Cambridge Univ. Press, New York, 1939.
2. J. O. Hirschfelder, C. F. Curtiss, and R. B. Bird, *Molecular Theory of Gases and Liquids*, John Wiley, New York, 1954.
3. Ref. 1, Chapter 16.
4. D. E. Stogryn and J. O. Hirschfelder, *J. Chem. Phys.* **31**, 1531, 1545 (1959); **33**, 942 (1960).
   S. K. Kim and J. Ross, *J. Chem. Phys.* **42**, 263 (1965); S. K. Kim, G. P. Flynn and J. Ross, *J. Chem. Phys.* **43**, 4166 (1965).
   D. K. Hoffman and C. F. Curtiss, *Phys. Fluids* **8**, 890 (1965); C. F. Curtiss, M. B. McElroy and D. K. Hoffman, *Int. J. Eng. Science* **3**, 269 (1965).
5. N. N. Bogoliubov, in *Studies in Statistical Mechanics* I; J. de Boer and G. E. Uhlenbeck, eds., North Holland Publ. Comp. Amsterdam, 1962, p. 5.
6. J. R. Dorfman and E. G. D. Cohen, *J. Math. Phys.* **8**, 282 (1967).
7. J. Weinstock, *Phys. Rev.* **140**, A460 (1965).
   R. Goldman and E. A. Frieman, *J. Math. Phys.* **8**, 1410 (1967).
8. J. V. Sengers, *Phys. Rev. Letters* **15**, 515 (1965).
   K. Kawasaki and I. Oppenheim, *Phys. Rev.* **139**, A1763 (1965).

9. J. V. Sengers and E. G. D. Cohen, *Physica* **27**, 230 (1961).
10. D. Burnett, *Proc. London Math. Soc.* **39**, 385 (1935).
11. Ref. 1, Section 10.21.
12. S. T. Choh and G. E. Uhlenbeck, *The kinetic theory of dense gases*, Univ. Michigan, Ann Arbor, Michigan, Navy Theoretical Physics Contract No. Nonr 1224 (15), (1958).
13. M. S. Green, *J. Chem. Phys.* **25**, 836 (1956); M. S. Green and R. A. Piccirelli. *Phys. Rev.* **132**, 1388 (1963).
14. E. G. D. Cohen, *Physica* **28**, 1025, 1045 (1962).
15. J. V. Sengers, *Phys. Fluids* **9**, 1333 (1966).
16. M. S. Green, *Phys. Rev.* **136**, A905 (1964).
17. G. Sandri, R. D. Sullivan and P. Norem, *Phys. Rev. Letters* **13**, 743 (1964).
18. E. G. D. Cohen, in *Boulder Lectures in Theoretical Physics* **8A**, Univ. Colorado Press, 1966, p. 145.
19. J. Weinstock, *Phys. Rev.* **132**, 470 (1963).
20. G. Sandri and A. H. Kritz, *Phys. Rev.* **150**, 92 (1966).
21. C. B. Haselgrove, *Math. Computation* **15**, 323 (1961).
22. S. Haber, *Math. Computation* **20**, 361 (1966).
23. B. J. Alder and T. E. Wainwright, *Phys. Rev. Letters* **18**, 988 (1967).
24. Ref. 18, Appendix I.
25. This work was carried out in collaboration with Dr. H. T. Wood. A preliminary account can be found in a technical report AEDC–TR–69–68, prepared for the Arnold Engineering Development Center, Tennessee (1969).
26. R. A. Piccirelli, *J. Math. Phys.* **1**, 922 (1966).
27. J. V. Sengers, D. T. Gillespie and W. R. Hoegy, *Phys. Letters* **32A**, 387 (1970).
28. W. R. Hoegy and J. V. Sengers, *Phys. Rev.* **A2** (1970), in press.
29. D. T, Gillespie and J. V. Sengers, *Proc. 5th Symp. Thermophysical Properties*, C. F. Bonilla, ed., ASME, New York, 1970, p. 42.

9. J. V. Sengers and E. G. D. Cohen, *Physica* **27**, 230 (1961).
10. D. Burnett, *Proc. London Math. Soc.* **39**, 385 (1935).
11. Ref. 1, Section 10.
12. S. T. Choh and G. E. Uhlenbeck, *The kinetic theory of dense gases*, Univ. Michigan, Ann Arbor, Michigan, Navy Theoretical Physics Contract No. Nonr 1224(15) (1958).
13. M. S. Green, *J. Chem. Phys.* **25**, 836 (1956); M. S. Green and R. A. Piccirelli, *Phys. Rev.* **132**, 1388 (1963).
14. E. G. D. Cohen, *Physica* **28**, 1025, 1045 (1962).
15. J. V. Sengers, *Phys. Fluids* **9**, 1333 (1966).
16. M. S. Green, *Phys. Rev.* **136**, A905 (1964).
17. G. Sandri, R. D. Sullivan and P. Norem, *Phys. Rev. Letters* **13**, 743 (1964).
18. E. G. D. Cohen, in *Boulder Lectures in Theoretical Physics* 8A, Univ. Colorado Press, 1966, p. 145.
19. J. Weinstock, *Phys. Rev.* **132**, 470 (1963).
20. G. Sandri and A. H. Kritz, *Phys. Rev.* **150**, 92 (1966).
21. C. B. Haselgrove, *Math. Computation* **15**, 323 (1961).
22. S. Haber, *Math. Computation* **20**, 361 (1966).
23. B. J. Alder and T. E. Wainwright, *Phys. Rev. Letters* **18**, 988 (1967).
24. Ref. 15, Appendix I.
25. This work was carried out in collaboration with Dr. H. T. Wood. A preliminary account can be found in a technical report AEDC-TR-69-68, prepared for the Arnold Engineering Development Center, Tennessee (1969).
26. R. A. Piccirelli, *J. Math. Phys.* **7**, 922 (1966).
27. J. V. Sengers, D. T. Gillespie and W. R. Hoegy, *Phys. Letters* **32A**, 387 (1970).
28. W. R. Hoegy and J. V. Sengers, *Phys. Rev.* **A 2** (1970), in press.
29. D. T. Gillespie and J. V. Sengers, *Proc. 5th Symp. Thermophysical Properties*, C. F. Bonilla, ed., ASME, New York, 1970, p. 42.

PAPER C–2

Constants of the linearized motion of Vlasov-plasmas

B. U. FELDERHOF*
Baker Laboratory, Cornell University
Ithaca, New York 14850

CONSTANTS OF THE LINEARIZED MOTION OF VLASOV-PLASMAS

In recent years considerable attention has been given to the problem of finding constants of the motion of the linearized Vlasov equation, possibly coupled to Maxwell's equations[1–5]. The main motivation of this search for constants was the hope that they might be useful in the study of plasma stability. In the one-dimensional electrostatic case it has in fact been shown by Völk[4] how a particular constant may be constructed which leads to a necessary and sufficient stability criterion, namely the Penrose criterion. Here we shall be concerned only with finding the constants and shall not deal with their possible application.

Previous work on the problem was either restricted to relatively simple one-dimensional electrostatic situations or in the more general three-dimensional case, including coupling between plasma and radiation fields, required rather elaborate calculations. We have found a straightforward formal scheme which directly leads to a large classe of constants of the motion. The method works quite generally and may be applied to both relativistic and nonrelativistic classical plasmas, and also to quantum-plasmas. In order to demonstrate the method we shall in this brief report assume the simplest of worlds, namely a classical nonrelativistic gas of electrons interacting via the Coulomb force and moving in the presence

* On leave of absence from the Institute for Theoretical Physics, Rijksuniversiteit, Utrecht, the Netherlands.

of a static charge density. We shall quote some results for the more interesting case in which the plasma is coupled to the radiation fields.

From a formal point of view it turns out to be useful to stress the Hamiltonian nature of the one-electron equations of motion and to employ canonical variables. Therefore we write the nonlinear Vlasov equation for the one-electron distribution function $f(\mathbf{r}, \mathbf{p}, t)$ in the Liouville form

$$\frac{\partial f}{\partial t} = \frac{\partial H}{\partial \mathbf{r}} \cdot \frac{\partial f}{\partial \mathbf{p}} - \frac{\partial H}{\partial \mathbf{p}} \cdot \frac{\partial f}{\partial \mathbf{r}} = [H, f], \tag{1}$$

where the notation [,] indicates Poisson brackets as defined by the first line. The time-dependent one-electron Hamiltonian H is given by

$$H(\mathbf{r}, \mathbf{p}, t) = \frac{p^2}{2m} + e\phi(\mathbf{r}, t) \tag{2}$$

where m is the electron mass, e the electron charge, and where the electrostatic potential $\phi(\mathbf{r}, t)$ is given by the Coulomb integral

$$\phi(\mathbf{r}, t) = \int \frac{\varrho(\mathbf{r}', t)}{|\mathbf{r} - \mathbf{r}'|} d\mathbf{r}' \tag{3}$$

with the charge density $\varrho(\mathbf{r}, t)$ determined self-consistently by the electron distribution,

$$\varrho(\mathbf{r}, t) = e \int f(\mathbf{r}, \mathbf{p}, t)\, d\mathbf{p} + \hat{\varrho}(\mathbf{r}), \tag{4}$$

where $\hat{\varrho}(\mathbf{r})$ is a static charge density caused by the ions and any additional charges which may be present. Of course, the general solution of the nonlinear set of Eqs (1) to (4) is not known, but in simple situations one may find particular time-independent solutions $f_0(\mathbf{r}, \mathbf{p})$.

Suppose one has found such a solution $f_0(\mathbf{r}, \mathbf{p})$, which therefore satisfies

$$[H_0, f_0] = 0 \tag{5}$$

with

$$H_0(\mathbf{r}, \mathbf{p}) = \frac{p^2}{2m} + e\,\phi_0(\mathbf{r}), \tag{6}$$

where $\phi_0(\mathbf{r})$ is determined self-consistently from $f_0(\mathbf{r}, \mathbf{p})$ via (3) and (4). For solutions close to such a time-independent solution one may write

$$f = f_0 + f_1 \tag{7}$$

and linearize the Vlasov equation

$$\frac{\partial f_1}{\partial t} = [H_0, f_1] + [H_1, f_0] \tag{8}$$

where the perturbed Hamiltonian

$$H_1 = e\,\phi_1(\mathbf{r}, t) \tag{9}$$

is determined via the Coulomb integral (3) from the first order charge density

$$\varrho_1(\mathbf{r}, t) = e \int f_1(\mathbf{r}, \mathbf{p}, t)\, d\mathbf{p}. \tag{10}$$

The question we want to answer is: Can one find constants of the motion for the linearized equation? We shall be particularly interested in constants bilinear in f_1, since the electrostatic energy is bilinear in f_1 and one may therefore expect such constants to have some physical meaning.

Formally one may write Eq. (8) in a Schrödinger form

$$i\frac{\partial}{\partial t}|f_1\rangle = S|f_1\rangle \tag{11}$$

where S is a linear streaming operator. If we introduce the scalar product

$$\langle g \mid f\rangle = \int g^*(\mathbf{r}, \mathbf{p})\, f(\mathbf{r}, \mathbf{p})\, d\mathbf{r}\, d\mathbf{p} \tag{12}$$

the general bilinear expression is of the form

$$\langle f_1 | O | f_1\rangle \tag{13}$$

where O is a linear operator. We now ask for operators O for which the expression (13) is independent of time. Direct differentiation and use of (11) does not lead to a simple condition on O, since it may readily be shown that the streaming operator S is non-hermitian in the scalar product (12). The condition may, however, be simplified by studying the streaming operator in more detail.

From (8) it follows that one may write

$$S = L_0 + CU \tag{14}$$

where L_0 is the unperturbed Liouville operator

$$L_0|f_1\rangle = i\,|\,[H_0, f_1]\rangle, \tag{15}$$

the operator U serves to define the perturbed Hamiltonian H_1 in terms of f_1,

$$U|f_1\rangle = |H_1\rangle = \left| e\int \frac{f_1(\mathbf{r}', \mathbf{p}', t)}{|\mathbf{r} - \mathbf{r}'|}\, d\mathbf{r}'\, d\mathbf{p}' \right\rangle, \tag{16}$$

and finally the operator C takes the Poisson brackets with the unperturbed distribution f_0,

$$C|g\rangle = -i\,|[f_0, g]\rangle. \tag{17}$$

It is easily checked that the three operators L_0, C and U are hermitian in the scalar product (12). Hence the hermitian conjugate of S is given by

$$S^\dagger = L_0 + UC. \tag{18}$$

Moreover, one finds the important relation

$$CS^\dagger = SC. \tag{19}$$

This is obvious for the last terms in S and $S^\dagger$ and one has

$$CL_0 = L_0 C \tag{20}$$

on account of the Jacobi identity and Eq. (5), i.e., the fact that f_0 is a time-independent self-consistent solution of the nonlinear Vlasov equation.

The relation (19) shows that the operator C^{-1} may be used as a weighting factor as follows. Writing the general operator O in the form

$$O = C^{-1}T \tag{21}$$

one finds that the "expectation value" of T defined by

$$\langle T\rangle = \tfrac{1}{2}\langle f_1|C^{-1}T|f_1\rangle \tag{22}$$

is independent of time if and only if

$$ST = TS. \tag{23}$$

As in quantum mechanics such commuting operators may be found on the basis of invariance considerations. In the present case the invariance properties of the streaming operator S are intimately related to the symmetries of the equilibrium solution $f_0(\mathbf{r}, \mathbf{p})$. A first obvious constant is the expectation value of S itself. This corresponds to time translation invariance. Obviously S commutes with itself. If the time-independent solution is invariant under spatial translations in one or more directions one finds that S commutes with the corresponding component of the translation operator

$$\mathbf{P} = -i\,\partial/\partial\mathbf{r}. \tag{24}$$

Similarly if f_0 is invariant under spatial rotations about some axis one finds that S commutes with the corresponding component of the rotation operator

$$\mathbf{J} = -i\mathbf{r} \times \frac{\partial}{\partial \mathbf{r}} - i\mathbf{p} \times \frac{\partial}{\partial \mathbf{p}}. \tag{25}$$

Once one has found a number of commuting operators, a whole class of constants may be found since any function of operators which commute with S again commutes.

The expectation values (22) of the operators S, $\mathbf{P}$ and $\mathbf{J}$ may readily be evaluated explicitly but we shall here quote results for the rather more interesting case in which the plasma is coupled to the radiation field. The above simpler case is of course included as the case where the radiation fields vanish identically. We define a function $\tilde{f}_1(\mathbf{r}, \mathbf{p}, t)$ by the operation

$$|\tilde{f}_1\rangle = iC^{-1}\,|f_1\rangle. \tag{26}$$

This in general involves solving a linear first order partial differential equation. The explicit form of the constants mentioned above is

$$\begin{aligned}
\langle S\rangle &= \frac{1}{2}\int \tilde{f}_1[H_0, f_1]\, d\mathbf{r}\, d\mathbf{p} + \frac{1}{8\pi}\int [\mathbf{E}_1^2 + \mathbf{B}_1^2]\, d\mathbf{r} \\
\langle \mathbf{P}\rangle &= -\frac{1}{2}\int \tilde{f}_1 \frac{\partial f_1}{\partial \mathbf{r}}\, d\mathbf{r}\, d\mathbf{p} + \frac{1}{4\pi c}\int [\mathbf{E}_1 \times \mathbf{B}_1 - 4\pi\varrho_1 \mathbf{A}_1]\, d\mathbf{r} \\
\langle \mathbf{J}\rangle &= -\frac{1}{2}\int \tilde{f}_1\left(\mathbf{r} \times \frac{\partial}{\partial \mathbf{r}} + \mathbf{p} \times \frac{\partial}{\partial \mathbf{p}}\right) f_1\, d\mathbf{r}\, d\mathbf{p} \\
&\quad + \frac{1}{4\pi c}\int [\mathbf{r} \times (\mathbf{E}_1 \times \mathbf{B}_1) - 4\pi\varrho_1 \mathbf{r} \times \mathbf{A}_1]\, d\mathbf{r},
\end{aligned} \tag{27}$$

where $\mathbf{A}_1(\mathbf{r}, t)$ is the first order vector potential in the Coulomb gauge, $\operatorname{div} \mathbf{A}_1 = 0$. The first constant is related to the plasma free energy, the second constant to the plasma linear momentum, and the third constant to the plasma angular momentum. The first two constants were obtained earlier along a different route by Oberman and Kruskal[5]. Related constants were obtained by Buneman[6] by a completely different method, namely by describing the flow of particles in μ-space from the Lagrangian rather than the Eulerian point of view. A more detailed account of the present work is being published elsewhere[7].

Acknowledgements

This work was supported by the National Science Foundation and by the Advanced Research Projects Agency through the Materials Science Center at Cornell University.

References

1. T. K. Fowler, *J. Math. Phys.* **4**, 550 (1963).
2. J. Dawson and C. Oberman, *Phys. Fluids* **7**, 773 (1964).
3. K. M. Case, *Phys. Fluids* **8**, 96 (1965).
4. H. J. Völk, *Phys. Fluids* **10**, 1697 (1967).
5. C. Oberman and M. Kruskal, *J. Math. Phys.* **6**, 327 (1965).
6. O. Buneman, in *Relativistic Plasmas*, edited by O. Buneman and W. B. Pardo, (W. A. Benjamin, Inc., New York, 1968), p. 41.
7. B. U. Felderhof, *Phys. Fluids.* **12,** 1684 (1969).

PAPER C–3

A review of the unified theory of relaxations in plasmas

TARO KIHARA

Department of Physics, Faculty of Science University of Tokyo, Japan

OSAMU AONO*

Statistical Physics Laboratory, Department of Physics State University of New York, Buffalo, New York

Abstract

A unification of the collision theory and the wave theory leads to a more exact theory, in which no cut-off precedures of diverging integrals are needed, and which gives exact Coulomb logarithms in kinetic coefficients. The kinetic equation has several expressions contingent on the method of unification. The coulomb logarithm depends on external fields and non-uniformity in space and time. The quantum effect is fully taken into account. Typical examples are given.

1 INTRODUCTION

In a hot plasma, the potential energy between the constituent particles is, on the average, much smaller than their kinetic energy. In this respect, the plasma is similar to the rarefied gas. However, if a relaxation in plasmas is treated within the framework of the molecular theory of gases, which is based on binary collisions between the constituent particles, then the relaxation rate diverges logarithmically, a difficulty which is known as the divergence of the Coulomb logarithm. This divergence arises from the fact that collective interactions due to the long range nature of Coulomb potential are not included in the collision theory.

* Present address: University of Tokyo, Tokyo, Japan.

Owing to the collective interaction, the electric field around a charged particle in the plasma is screened so that the relaxation rate becomes finite. Since the number of particles which screen the field of the charge is very large, it should be possible to ascribe the screening to the macroscopic dielectric properties of the plasma. In fact, the collective interaction may be treated as an interaction between a particle and fluctuating fields in the plasma as a dielectric medium; such theory is to be called the wave theory. In the vicinity of the particle of interest, however, the fluctuation is much disturbed, an effect which is not included in the wave theory. Therefore, the relaxation rate based on the wave theory also has an logarithmic divergence.

If we employ cut-off procedures, the rate of relaxation on the kinetic coefficient becomes finite but still contains an unknown factor in the Coulomb logarithm $\ln (cl/r_0)$. Here c is the unknown numerical factor of order unity; l is the screening length given by the wave theory; and r_0 is the close collision radius in the collision theory. Therefore, the cutoff theories apply only when $\ln (l/r_0) \gg 1$, or they are of logarithmic accuracy.

A theory which can give the exact value of the constant c was developed by the present authors[1~5]. They unified the collision theory and the wave theory into such a theory, on the basis of the complementarity of the two theories, and named it the unified theory. Since the unification method is not unique, the kinetic equation in the unified theory has many expressions; some of them were obtained by Baldwin[6], Frieman and Book[7], Guernsey[8], Perkins[9], and Bernstein and Ahearne[10]. The fact that their kinetic equations belong to the unified theory was proved by Gould and DeWitt[11], and more completely by the present authors[4,12]. Examples having the exact Coulomb logarithms were obtained exclusively by the present authors and their collaborators.

In the present paper, we review the unified theory, which is exact when the argument of the Coulomb logarithm is large; the logarithm itself may not necessarily be large. In the next section, the background knowledge is briefly reviewed. From a formula in Section 3 which shows the principle of the unified theory, all the kinetic equations that were obtained up to the present are derived systematically in Section 4; the relation between them are made clear. In Section 5, the structure of the kinetic equation is investigated. Effects of external fields and quantum effects are referred in Section 6. The remaining sections treat typical solutions of the kinetic equation.

2 THEORIES OF LOGARITHMIC ACCURACY

For later use, three approximate theories are briefly reviewed in this section.

The kinetic equation for distribution function f may be expressed in the form

$$\frac{df}{dt} = I. \tag{2.1}$$

Here f is a function of the time, the position and the momentum; d/dt denotes the differentiation with respect to time along the free orbit of the particle in an external and a self-consistent fields; and **I** is due to the interaction.

The simplest approximation for the interaction term is $I = 0$: that is, $df/dt = 0$, which was proposed first by Jeans[13], and later by Vlasov[14]. This simple equation is useful for phenomena such as plasma oscillations. We do not, however, discuss the details, because we are interested in the interaction term **I**.

Hereafter we consider only a uniform plasma in the absence of any external fields. However, non-uniform plasmas in external fields can be treated with a straigthforward generalization. Explicit expressions will be given for an electron plasma, but they apply to multi-component plasmas by making trivial changes in meaning of notations.

Collision theory

If we assume that the interactions are binary, the interaction term **I** reduces to Boltzmann's collision term I_0:

$$I_0 = \iint \{f(\mathbf{p}')\,f(\mathbf{q}') - f(\mathbf{p})f(\mathbf{q})\}\, g\, d\mathbf{b}\, d\mathbf{q}. \tag{2.2}$$

Here **p** and **q** are the momenta of the colliding particles; the prime denotes the momentum after the collision; $\mathbf{g} = (\mathbf{p} - \mathbf{q})/m$ is the relative velocity, m being the mass; and **b** is a vector whose direction lies in the plane of orbit and is perpendicular to **g**. The collision term I_0 can formally be written as

$$I_0 = \int_{b>0} B(b)\, db, \tag{2.3}$$

after integrations with respect to **q** and the direction of **b**. When the impact parameter is much larger than the close collision radius r_0 which is defined by the impact parameter corresponding to the right-angle deflection, the in tegrand becomes

$$B(b) = A/b, \quad \text{for} \quad b \gg r_0, \tag{2.4}$$

$$A = 2\pi\, e^4 \int \frac{\partial}{\partial \mathbf{p}} \cdot \frac{1}{g}\left(1 - \frac{\mathbf{gg}}{g^2}\right) \cdot \left(\frac{\partial}{\partial \mathbf{p}} - \frac{\partial}{\partial \mathbf{q}}\right) f(\mathbf{p})\, f(\mathbf{q})\, d\mathbf{q}, \tag{2.5}$$

where e being the electronic charge. Therefore, the interaction term I_0 diverges logarithmically, when we let the upper limit of b tend to infinity, as pointed out first by Gruner[15].

This divergence is caused by the fact that collective interactions are not included in the collision theory, and has been avoided in the past by cutting off the range of integration at some large value of the impact parameter. In the earliest treatment, this was taken as the interparticle separation[16], but more recently as a screening length l such as the Debye length in the thermal equilibrium case[17–19]. Another method to avoid the divergence was the use of the Debye potential.[20–22] This method is not more exact than, but equivalent to, the cut-off procedure, because the potential between two moving particles is not of Debye-type.

Thus we have a finite result containing a "Coulomb logarithm" whose argument is the ratio of the cut-off length to a close collision radius r_0. However, the cut-off result is only of logarithmic accuracy, i.e. valid only for large values of the logarithm itself and not in general for large values of its argument. Since we are interested in a more exact theory, we need to know about the collective interaction.

Wave theory

The collective interaction can be treated on the wave theory, which was proposed by Balescu[23], Lenard[24], and others[25–28]. The interaction term I^0 of the kinetic equation in the wave theory is given by an integral in the wave number space

$$I^0 = 2e^4 \iint \frac{\mathbf{k}}{k^2} \cdot \frac{\partial}{\partial \mathbf{p}} \frac{\delta(\mathbf{k}\cdot\mathbf{g})}{|\varepsilon(\mathbf{k}, \mathbf{k}\cdot\mathbf{p}/m)|^2} \frac{\mathbf{k}}{k^2} \cdot \left(\frac{\partial}{\partial \mathbf{p}} - \frac{\partial}{\partial \mathbf{q}}\right) f(\mathbf{p})\, f(\mathbf{q})\, d\mathbf{k}\, d\mathbf{q}. \tag{2.6}$$

Here ε is the dielectric permittivity of the plasma

$$\varepsilon(\mathbf{k}, \omega) = 1 + \lim_{\nu \to 0_+} \frac{1}{k^2} \int \frac{4\pi e^2 \mathbf{k} \cdot \partial f(\mathbf{p})/\partial \mathbf{p}}{\omega - \mathbf{k}\cdot\mathbf{p}/m + i\nu}\, d\mathbf{p}, \tag{2.7}$$

which is a function of the wave number $\mathbf{k}$ and the frequency ω. The fact that ε is not identically equal to unity shows the screening effect due to the collective interaction; an average of the integral on the right side of (2.7) is the inverse square of the screening length l.

Integrating (2.6) with respect to $\mathbf{q}$ and the angular component of $\mathbf{k}$, we can write

$$I^0 = \int_{k>0} K(k)\, dk. \tag{2.8}$$

Since $\varepsilon \to 1$ as $k \to \infty$, the integrand reduces to

$$K(k) = A/k, \quad \text{for} \quad k \gg l^{-1}. \tag{2.9}$$

where A is the same as has been given in (2.5). Therefore, the interaction term I^0 in the wave theory also diverges logarithmically; this divergence is due to an improper treatment of close collisions. To avoid the divergence within the framework of the wave theory, we have to cut off the range of integration near the inverse of the close collision radius. Thus the result is again of logarithmic accuracy; in fact, one divergence has merely been replaced by another. However, the wave theory is complementary to the collision theory. Recognizing this complementarity, we can unify the two theories into a more accurate theory.

Impulse approximation

If we neglect orbital curvatures of colliding particles in the collision theory, we have an interaction term I_0^0 in the impulse approximation

$$I_0^0 = \int A/b\, db, \tag{2.10}$$

which diverges at small b as well as large b owing to the neglect of both screening effects due to the collective interaction and the curvature effect at close collisions. An alternative expression for I_0^0 can be obtained from (2.6) by putting $\varepsilon = 1$:

$$I_0^0 = \int A/k\, dk, \tag{2.11}$$

because the orbital curvature has not been included in the wave theory.

Since the integration should be limited in the region[18] $r_0 \ll (b \text{ or } k^{-1}) \ll l$, we have the same result of logarithmic accuracy as before. In this sense, the impulse approximation is equivalent to the collision and the wave theories.

3 UNIFICATION

By reducing the Liouville equation, we have a hierarchy of equations for distribution functions[29]; the first two members of the hierarchy are

$$\frac{df(x_1)}{dt} = \int [\phi_{12}, f(x_1, x_2)]\, dx_2, \tag{3.1}$$

$$\frac{df(x_1, x_2)}{dt} = \int [\phi_{13} + \phi_{23}, f(x_1, x_2, x_3)]\, dx_3 + [\phi_{12}, f(x_1, x_2)]. \tag{3.2}$$

Here d/dt denotes the differentiation with respect to time along the particle orbits; $x_i (i = 1, 2, 3)$ is a six dimensional vector denoting the position $\mathbf{r}_i$ and the momentum $\mathbf{p}_i$ of the i-th particle; ϕ_{ij} is the potential; and the notation [,] means the Poisson brackets. The right side of (3.1) is to be called the interaction term **I**. The second term on the right side of (3.2) shows the direct interaction between the particles 1 and 2, while the first term presents the screening effect due to the surrounding particles.

To truncate the hierarchy, we consider as follows. The direct interaction between two particles becomes infinitely large as the particle separation $r = |\mathbf{r}_2 - \mathbf{r}_1|$ decreases. Therefore, there exists such a length l as the screening effects is negligible if

$$r \ll l. \tag{3.3}$$

We call this length l the screening length, which will be the Debye length in thermal equilibrium cases. In the region (3.3), the equation (3.2) reduces to

$$\frac{df(x_1, x_2)}{dt} = [\phi_{12}, f(x_1, x_2)]. \tag{3.4}$$

Substituting the solution $f_0(x_1, x_2)$ of this equation into (3.1), we have the kinetic equation in the collision theory: that is, the interaction term reduces to Boltzmann's collision term I_0 [29]

$$I_0 = \int [\phi_{12}, f_0(x_1, x_2)]\, dx_2. \tag{3.5}$$

When the distance r becomes large, the correlation between the two particles vanishes. Therefore, if

$$r \gg r_0, \tag{3.6}$$

the distribution functions reduce to

$$f(x_1, x_2) = f(x_1)\, f(x_2) \tag{3.7}$$

$$f(x_1, x_2, x_3) = f(x_1)\, f(x_2, x_3) + f(x_2)\, f(x_1, x_3) - f(x_1)\, f(x_2)\, f(x_3), \tag{3.8}$$

where r_0 is the close collision radius which is a distance where the potential is comparable to the kinetic energy of relative motion. This fact may rather be regarded as a boundary condition. In the region (3.6), we have

$$\frac{df(x_1, x_2)}{dt} = \int [\phi_{13} + \phi_{23}, f(x_1) f(x_2, x_3) + f(x_2) f(x_1, x_3)] \, dx_3 + [\phi_{12}, f(x_1) f(x_2)], \tag{3.9}$$

where the last term in (3.8) has vanished because of uniformity of the system. Using the solution $f^0(x_1, x_2)$ of this equation, we have the interaction term I^0 in the wave theory[23,24]

$$I^0 = \int [\phi_{12}, f^0(x_1, x_2)] \, dx_2 = \int [\phi_{12}, \int F^0(\mathbf{k}) \, e^{i\mathbf{k}\cdot\mathbf{r}} \, d\mathbf{k}] \, dx_2, \tag{3.10}$$

where $F^0(\mathbf{k})$ is the Fourier transform of $f^0(x_1, x_2)$

$$F^0(\mathbf{k}) = (2\pi)^{-3} \int f^0(x_1, x_2) \, e^{-i\mathbf{k}\cdot\mathbf{r}} \, d\mathbf{r}, \quad \mathbf{r} = \mathbf{r}_2 - \mathbf{r}_1. \tag{3.11}$$

It should be emphasized that the conditions (3.3) and (3.6) are absolutely important to (3.4) and (3.9); neglect of the conditions leads to the divergences of I_0 and I^0.

We assume that

$$r_0 \ll l, \tag{3.12}$$

an assumption which is the most important in the present paper. Then, the two regions (3.3) and (3.6) cover the whole region of the particle separation. An average of the ratio r_0/l is called the plasma parameter.

On defining **a** distance a such as

$$r_0 \ll a \ll l, \tag{3.13}$$

we see that the contributions from the regions $r < a$ and $r > a$ to the interaction term **I** are given by (3.5) and (3.10), respectively: that is

$$I = \int_{r<a} [\phi, f_0] \, dx + \int_{r>a} [\phi, f^0] \, dx, \tag{3.14}$$

where we have omitted unnessary suficies and arguments. Each term depends on **a**, but the sum does not, because $f_0 = f^0$ in the region $r_0 \ll r \ll l$. The first term includes the close collision, and the second does the collective interaction. Therefore, the interaction term (3.14) is convergent, and the accuracy is better than the logarithmic accuracy. Thus, we have unified the two theories into a more exact theory, namely the unified theory.

Bernstein and Ahearne[10] have obtained a formula similar to (3.14). They divided the two-particle distribution function into two parts in the

Fourier-transformed space: the one part of large wave numbers corresponding to the collision theory, and the other to the wave theory. However, their final result has an error, which can be corrected as follows. The second term on the right side of (3.14) is written as

$$\int K(k) \frac{\sin ka}{ka} dk, \tag{3.15}$$

where $K(k)$ has been defined in (2.8), and use has been made of the formula

$$\int \frac{\mathbf{r}}{r^3} e^{i\mathbf{k}\cdot\mathbf{r}} d\mathbf{r} = -\frac{4\pi i \mathbf{k}}{k^2} \int \frac{d}{dr} \frac{\sin kr}{kr} dr. \tag{3.16}$$

If we take a wave number k_0 in the region $a^{-1} \gg k_0 \gg l^{-1}$, the equation (2.9) holds for $k > k_0$, and $\sin ka/ka = 1$ for $k < k_0$. Therefore, we have

$$I = \int_{r>a} [\phi, f_0]\, dx + \int_0^{k_0} K(k)\, dk + \int_{k_0}^{\infty} \frac{A}{k} \frac{\sin ka}{ka} dk. \tag{3.17}$$

The last term is calculated to be $A(1 - \ln \gamma k_0 a)$ for $k_0 a \ll 1$, where $\ln \gamma = 0.577$ being Euler's constant, but they missed the term $\ln \gamma k_0 a$.

A simple example[30] has been solved on the basis of the formula (3.14). However, it is very difficult, in general, to calculate the two-particle distribution f_0 in terms of the particle separation. We need, therefore, further consideration to obtain a soluble expression.

4 EXPLICIT EXPRESSION

For the purpose of obtaining convenient expression for the unified interaction term (3.14), we introduce a two-particle distribution function $f_0^0(x_1, x_2)$ defined by the solution of the equation

$$\frac{df(x_1, x_2)}{dt} = [\phi_{12}, f(x_1) f(x_2)], \tag{4.1}$$

which does not include close collisions nor collective interactions. Substitution of f_0^0 into (3.1) gives a kinetic equation on the impulse approximation with the interaction term I_0^0

$$I_0^0 = \int [\phi, f_0^0]\, dx. \tag{4.2}$$

From (3.4) and (3.9), we see that

$$f_0^0 = \begin{cases} f^0 & \text{for} \quad r \ll l \\ f_0 & \text{for} \quad r \gg r_0. \end{cases} \tag{4.3}$$

By use of this function f_0^0, we can rewrite (3.14) as

$$I = \int [\phi, f_0 + f^0 - f_0^0]\, dx, \tag{4.4}$$

where we may disregard the conditions (3.3) and (3.6).

Now the region of integration is extended to the infinite space, we may formally write

$$I = \int B(b)\, db + \int K(k)\, dk - I_0^0. \tag{4.5}$$

This scheme was first suggested by Hubbard[31] and confirmed by several authors[7,32]. Since all the three terms on the right side are individually divergent, and since there are no functional relations between **b** and **k**, care must be taken to insure that the divergences are correctly cancelled; an exact procedure of the cancellation was found first by the present authors. The variety of the expressions arises from the various methods of canceling the divergences.

Dual expression for I_0^0

From (4.1) and (4.2), we have easily

$$I_0^0 = \iint \left[\phi, \int \frac{e^2}{2\pi i}\, \delta(\mathbf{k}\cdot\mathbf{g})\, \frac{\mathbf{k}}{k^2} \cdot \left(\frac{\partial}{\partial \mathbf{p}} - \frac{\partial}{\partial \mathbf{q}} \right) f(\mathbf{p})\, f(\mathbf{q})\, e^{i\mathbf{k}\cdot\mathbf{r}}\, d\mathbf{k} \right] d\mathbf{q}\, d\mathbf{r}. \tag{4.6}$$

The integration with respect to **r** can be performed in the cylindrical coordinate system with the axis in the direction of the relative velocity **g**; the distance b from the axis is the impact parameter. By use of the formula

$$\delta(\mathbf{k}\cdot\mathbf{g}) \int \frac{\mathbf{r}}{r^3}\, e^{i\mathbf{k}\cdot\mathbf{r}}\, d\mathbf{r} = \delta(\mathbf{k}\cdot\mathbf{g})\, \frac{4\pi i \mathbf{k}}{k} \int J_1(kb)\, db, \tag{4.7}$$

the expression (4.6) reduces to

$$I_0^0 = A \iint J_1(kb)\, db\, dk, \tag{4.8}$$

where A has been given in (2.5), and J_1 is the Bessel function. Using $\int_0^\infty J_1(x)\, dx = 1$, we recover (2.10) and (2.11).

Substituting (4.8) into (4.5), we end up with

$$I = \int_0^{b_0} B(b)\, db + \int_{b_0}^{\infty} \left\{ B(b) - A \int_0^{\infty} J_1(kb)\, dk \right\} db$$

$$+ \int_0^{\infty} \left\{ K(k) - A \int_0^{b_0} J_1(kb)\, db \right\} dk, \tag{4.9}$$

where b_0 is an arbitrary parameter; the right side is independent of b_0. This procedure of canceling the divergences in (4.5) was found by Aono[1], and two examples of the exact Coulomb logarithms were obtained.

Symmetric interaction term

The expression (4.8) can further be transformed into

$$I_0^0 = \int db\{1 - \exp(-\tfrac{1}{2}b^2/a^2)\}\,A/b + \int dk\{1 - \exp(-\tfrac{1}{2}k^2a^2)\}\,A/k, \quad (4.10)$$

by use of the identity

$$1 - \exp(-\tfrac{1}{2}k^2a^2) = k\int_0^\infty J_1(kb)\exp(-\tfrac{1}{2}b^2/a^2)\,db. \quad (4.11)$$

If the constant **a** is in the region

$$r_0 \ll a \ll l, \quad (4.12)$$

the functions in the brackets of (4.10) are appreciable, only when $b \gg r_0$ (or $k \gg l^{-1}$), where (2.4) [or (2.9)] holds. Therefore, we have

$$I = \int_0^\infty B(b)\exp(-\tfrac{1}{2}b^2/a^2)\,db + \int_0^\infty K(k)\,exp(-\tfrac{1}{2}k^2a^2)\,dk, \quad (4.13)$$

which was obtained by Kihara[2,3]. By use of this symmetric expression, several examples have been solved[2,3,33,34].

Impact parameter in the wave theory

Since the wave number **k** in the wave theory is perpendicular to **g**, we may use the formula (4.7) to integrate (3.10), the distance b corresponding to the impact parameter. Then, we have

$$I^0 = \iint K(k)\,kJ_1(kb)\,dk\,db. \quad (4.14)$$

Therefore, the unified interaction term is transformed into an integral with respect to the impact parameter

$$I = \int_0^\infty \{B(b) - A/b + \int_0^\infty K(k)\,kJ_1(kb)\,dk\}\,db. \quad (4.15)$$

This transformation corresponds to the treatment of Frieman and Book[7].

Modification of collision theory

We define two-particle functions $f_\varkappa$ and $f_\varkappa^0$ by the solutions of the equations corresponding to (3.4) and (4.1)

$$\frac{df_\varkappa}{dt} = [\phi_\varkappa, f_\varkappa(x_1, x_2)], \tag{4.16}$$

$$\frac{df_\varkappa^0}{dt} = [\phi_\varkappa, f(x_1)\, f(x_2)]. \tag{4.17}$$

Here $\phi_\varkappa$ is a potential which satisfies the conditions

$$\phi_\varkappa = \phi \quad \text{for} \quad r \ll \varkappa^{-1},$$
$$|\phi_\varkappa| \ll |\phi| \quad \text{for} \quad r \gg \varkappa^{-1}, \tag{4.18}$$

where the length $\varkappa^{-1}$ is much larger than r_0

$$\varkappa^{-1} \gg r_0. \tag{4.19}$$

Since the difference between $[\phi, f_0]$ and $[\phi_\varkappa, f_\varkappa]$ is appreciable only when $r \gtrsim \varkappa^{-1}$ ($\gg r_0$) where the impulse approximation holds, we have

$$\int \{[\phi, f_0] - [\phi_\varkappa, f_\varkappa]\}\, dx = \int \{[\phi, f_0^0] - [\phi_\varkappa, f_\varkappa^0]\}\, dx. \tag{4.20}$$

Therefore, the unified interaction term becomes[5]

$$I = \int [\phi_\varkappa, f_\varkappa]\, dx + \int \{[\phi, f^0] - [\phi_\varkappa, f_\varkappa^0]\}\, dx. \tag{4.21}$$

If we take $\phi_\varkappa$, for example, to be the Debye type potential

$$\phi_\varkappa = e^2\, e^{-\varkappa r}/r, \tag{4.22}$$

then we have

$$I = \int_0^\infty B_\varkappa(b)\, db + \int_0^\infty \{K(k) - A(k^2 + \varkappa^2)^{-2}\, k^3\}\, dk, \tag{4.23}$$

where the first integral is Boltzmann's collision term with the potential (4.22) which has been well investigated[21,22]. This expression was derived by Baldwin[6] from the Liouville equation for an electron plasma; a simple and general proof was given by Kihara[4]. With this expression, various examples have been treated[35–39].

Modification of wave theory

We modify the Eq. (3.9), by replacing the potential ϕ_{12} in the last term on the right side by an auxiliary potential $\phi_\varkappa$ which satisfies the conditions (4.18) and

$$r_0 \ll \varkappa^{-1} \ll l. \tag{4.24}$$

The solution of the modified equation is denoted by $f^\varkappa$, which is linear with respect to $\phi_\varkappa$.

By using $f^\varkappa$, we rewrite the second term on the right side of (4.21) as

$$\int \{[\phi, f^0] - [\phi_\varkappa, f^0_\varkappa]\}\, dx = \int [\phi, f^0 - f^\varkappa]\, dx + \int \{[\phi, f^\varkappa] - [\phi_\varkappa, f^0_\varkappa]\}\, dx \tag{4.25}$$

From the linearity of $f^\varkappa$, we see

$$\int [\phi, f^\varkappa]\, dx = \int [\phi_\varkappa, f^0]\, dx. \tag{4.26}$$

The potential ϕ_k is appreciable only when $r \lesssim \varkappa^{-1} \ll l$ where the impulse approximation holds: $f^0 = f^0_0$. Therefore, we have another expression for the unified interaction term

$$\begin{aligned} I &= \int [\phi_\varkappa, f_\varkappa]\, dx + \int [\phi, f^0 - f^\varkappa]\, dx + \int [\phi_\varkappa, f^0_0 - f^0_\varkappa]\, dx \\ &= \int [\phi_\varkappa, f_\varkappa]\, dx + \int_0^\infty K(k) \frac{\phi(k) - \phi_\varkappa(k)}{\phi(k)}\, dk + \int \frac{A}{k} \frac{\phi(k) - \phi_\varkappa(k)}{\phi(k)} \frac{\phi_\varkappa(k)}{\phi(k)}\, dk, \end{aligned} \tag{4.27}$$

where $\phi(k)$ and $\phi_k(k)$ are the Fourier transform of ϕ and ϕ_k.

Taking ϕ_k as (4.22), for example, we have

$$I = \int_0^\infty B_\varkappa(b)\, db + \int_0^\infty \frac{K(k)\varkappa^2}{k^2 + \varkappa^2}\, dk + \frac{1}{2} A, \tag{4.28}$$

which corresponds to the treatment of Perkins[9]. He investigated the motion of a test particle interacting with a plasma, and obtained the rate of its energy loss in some limiting cases.

We can further transform (4.27) into

$$I = \int [\phi_\varkappa, f_\varkappa]\, dx + \int [\phi - \phi_\varkappa, f^0 - f^\varkappa]\, dx + 2 \int [\phi_\varkappa, f^0_0 - f^0_\varkappa]\, dx. \tag{4.29}$$

If we choose

$$\phi_\varkappa = \begin{cases} e^2\left(\dfrac{1}{r} - \varkappa\right) & \text{for } r < \varkappa^{-1} \\ 0 & \text{otherwise,} \end{cases} \tag{4.30}$$

then (4.29) becomes

$$I = \int [\phi_\varkappa, f_\varkappa]\, dx + \int K(k) \left(\frac{\sin k/\varkappa}{k/\varkappa}\right)^2 dk + 2\left(\ln 2 - \frac{1}{2}\right) A, \tag{4.31}$$

where use has been made of the formula

$$\int_0^\infty \frac{\sin x}{x}\left(1 - \frac{\sin x}{x}\right)\frac{dx}{x} = \ln 2 - \frac{1}{2}. \tag{4.32}$$

Guernsey[8] obtained the expression (4.31) by expanding the hierarchy equations in a double series of the charge and the density. In his paper, the collision cross section for the potential (4.30) has been written, but the last term is missing as pointed out by Gould and DeWitt[11]. They calculated the electrical conductivity using the Maxwell model to show the expressions (4.9), (4.13), (4.23), and (4.31) lead to the same result.

5 STRUCTURE OF KINETIC EQUATION

In order to solve the kinetic equation, we have to know in detail about the interaction term **I**, which is essentially a combination of I_0 given by (2.2) and I^0 given by (2.6). Hereafter, we use the velocity distribution function $f(\mathbf{v})$ instead of the momentum distribution function, as is customary.

The binary collision term I_0 can be expanded in the form

$$I_0 = \sum_{n=1}^{\infty} \frac{1}{n!}\left(-\frac{\partial}{\partial \mathbf{v}}\right)^n \cdot \left(\frac{\langle(\Delta\mathbf{v})^n\rangle_0}{\Delta t} f(\mathbf{v})\right) \tag{5.1}$$

where $\Delta\mathbf{v}$ is the change in particle velocity $\mathbf{v}$ during the time interval Δt, and the notation $\langle\rangle_0$ indicates the average over the binary collision. The tensor $\langle(\Delta\mathbf{v})^n\rangle_0/\Delta t$ of n-th rank is called the n-th moment, and is given by

$$\frac{\langle(\Delta\mathbf{v})^n\rangle_0}{\Delta t} = \sum_s \int \boldsymbol{\Delta}^n f_s(\mathbf{v} + \mathbf{g})\, g\, d\mathbf{g}\, d\mathbf{b}. \tag{5.2}$$

where $\boldsymbol{\Delta}$ is the increment of the velocity per one collision with a particle of the s-th species

$$\boldsymbol{\Delta} = \frac{m_s}{m + m_s} g \left[(1 - \cos\theta)\frac{\mathbf{g}}{g} + \sin\theta \frac{\mathbf{b}}{b}\right], \tag{5.3}$$

the deflection angle being denoted by θ. The first and second moments diverge logarithmically, while the third and higher moments converge. These divergences should be removed by combining the corresponding terms in the wave theory.

The collective interaction term I^0 is composed of two moments

$$I^0 = -\frac{\partial}{\partial \mathbf{v}} \cdot \left(\frac{\langle \Delta \mathbf{v} \rangle^0}{\Delta t} f \right) + \frac{1}{2} \frac{\partial}{\partial \mathbf{v}} \frac{\partial}{\partial \mathbf{v}} : \left(\frac{\langle \Delta \mathbf{v} \Delta \mathbf{v} \rangle^0}{\Delta t} f \right). \tag{5.4}$$

Here the moments can be written in the forms

$$\frac{\langle \Delta \mathbf{v} \rangle^0}{\Delta t} = \frac{q^2}{m} \frac{1}{2\pi^2 i} \int \frac{\mathbf{k}}{k^2 \varepsilon(\mathbf{k}, \mathbf{k} \cdot \mathbf{v})} d\mathbf{k} + \frac{1}{2} \frac{\partial}{\partial \mathbf{v}} \frac{\langle \Delta \mathbf{v} \cdot \Delta \mathbf{v} \rangle^0}{\Delta t}. \tag{5.5}$$

$$\frac{\langle \Delta \mathbf{v} \Delta \mathbf{v} \rangle^0}{\Delta t} = \frac{4q^2}{m^2} \sum_s q_s^2 \iint \frac{\mathbf{k}\mathbf{k}\, \delta(\mathbf{k} \cdot \mathbf{v}_s - \mathbf{k} \cdot \mathbf{v})}{k^4 \, |\varepsilon(\mathbf{k}, \mathbf{k} \cdot \mathbf{v})|^2} f_s(\mathbf{v}_s)\, d\mathbf{v}_s\, d\mathbf{k}, \tag{5.6}$$

where q is the particle charge, and ε is the dielectric permittivity. These moments can be obtained by considering the change in velocity due to the fluctuating forces during a time interval Δt which is sufficiently longer than the continuence of the force correlation but much shorter than the relaxation time of interest.

Hence, the kinetic equation in the unified theory is written as

$$\frac{df(\mathbf{v})}{dt} = \sum_{n=1}^{\infty} \frac{1}{n!} \left(-\frac{\partial}{\partial \mathbf{v}} \right)^n \cdot \left(\frac{\langle (\Delta \mathbf{v})^n \rangle}{\Delta t} f(\mathbf{v}) \right). \tag{5.7}$$

The first and the second moments in the collision and the wave theories are connected into those in the unified theory; the connection formula (4.13) obtained in the previous section, for example, applies to each moment. The other moments are the same as (5.2).

The similarity between (5.1) and (5.7) shows a possibility that the unified kinetic equation can be obtained on the basis of a kind of binary collisions. In fact, several attempts have been made along this line[10,31].

6 GENERALIZATION

Thus far, we considered a classical plasma which was uniform in space and time, and to which no external fields were applied. The unified theory, however, applies to more general cases[4,5]. To show the generalized result, we use the connection formula (4, 23), which is to be denoted by

$$I = I_\varkappa + I^0 - I_\varkappa^0, \tag{6.1}$$

where

$$I_\varkappa = \int B_\varkappa \, db, \quad I^0 = \int K \, dk, \quad \text{and} \quad I_\varkappa^0 = \int A\, k^3/(k^2 + \varkappa^2)^2 \, dk.$$

The non-uniformity or external fields do not change the form (6.1), nor play any role in the difference $I_\varkappa - I_\varkappa^0$, but modify only the term I^0

in the wave theory. The term I^0 or its equivalents have been obtained in case of the presence of a uniform magnetic field[40,41], high-frequency electric field[42,43], or non-uniformity[44,45]; they are not simple enough to be reproduced here. Several examples in the unified theory have been obtained, showing effects of magnetic fields and temporal non-uniformity.[35–37,46–48]. However, effects of instabilities or turbulences, which cause non-uniformity in space and time, have not yet been investigated, because they are not well-known even in ther logarithmic accuracy.

The quantum effect has to be taken into account, when the wave packets of particles overlap each other; the classical theory holds true, when the size of the wave packet is much smaller than the close collision radius. The interaction term including the quantum effect is the same as (6.1) except that the term $I_\varkappa$ should be calculated on quantum mechanics; the difference $I^0 - I^0_\varkappa$ is free from the quantum effect. That is,

$$I^{\text{quantum}} = I_\varkappa^{\text{quantum}} + (I^0 - I^0_\varkappa)^{\text{classical}}. \tag{6.2}$$

In this manner, the quantum effect is separated from the effects of non-uniformity and external fields[4].

The term $I_\varkappa$ can be expanded in the form similar to (5.1); the third and higher moments are independent of $\varkappa$, and are the same as (5.2). Since the quantal differential cross section in a pure Coulomb field is the same as the classical Rutherford formula, the third and higher moments are free from the quantum effect. The first and second moments consist of integrals independent of the quantum effect and the integral of the form

$$Q_\varkappa = \int (1 - \cos\theta)\, d\mathbf{b}, \tag{6.3}$$

where $d\mathbf{b}$ should be regarded as the quantal cross section. This integral has been calculated by Honda[49] as follows:

$$Q_\varkappa = 4\pi\left(\frac{qq_s}{\mu_s g^2}\right)^2 \left\{\ln\frac{2\mu_s g}{\hbar\gamma\varkappa} - \frac{1}{2} - Re\,\psi\left(\frac{i\,|qq_s|}{\hbar g}\right)\right\}. \tag{6.4}$$

Here μ_s is the reduced mass; $2\pi\hbar$ is the Planck constant; $\ln\gamma = 0.577$ is Euler's constant; and ψ is the digamma function. In the limiting cases, it reduces to the well-known results

$$Q_\varkappa = 4\pi\left(\frac{qq_s}{\mu_s g^2}\right)^2 \left(\ln\frac{2\mu_s g^2}{\gamma\,|qq_s|\,\varkappa} - \frac{1}{2}\right), \quad \text{for } \hbar g \ll |qq_s|, \tag{6.5}$$

$$= 4\pi\left(\frac{qq_s}{\mu_s g^2}\right)^2 \left(\ln\frac{2\mu_s g}{\hbar\varkappa} - \frac{1}{2}\right), \quad \text{for } \hbar g \gg |qq_s|. \tag{6.6}$$

When the colliding particles are identical, a special consideration is required.

On the basis of the quantum field theory, Honda[50] derived another expression for I^{quantum} which is equivalent to (6.2). In particular, when the Born approximation holds, his result reduces to

$$I^{\text{Born}} = (I^0)^{\text{Born}}, \tag{6.7}$$

as predicted by Kihara[4]. In this case, the process of unification is not necessary, because the *orbital curveture* at close collisions is completely cevered by the quantal diffraction. In fact, several examples in the Born approximation have been treated along this line[51~54]. The connection formula (6.2) combined with (6.6) leads the same result, as it should.

7 BROWNIAN MOTION OF A SLOW ION

The plasma is composed of several species of charged particles; the velocity distribution functions of the constituents are assumed to be Maxwellian

$$f_s(v_s) = n_s\left(\frac{m_s}{2\pi T}\right)^{3/2} \exp\left(-\frac{m_s v_s^2}{2T}\right). \tag{7.1}$$

where n_s and m_s are the density and the mass of the particle of type s, and T is the temperature in energy units. The ion of interest has the mass m, charge q, and the velocity v which is much smaller than the thermal velocities of the constituents

$$v \ll (T/m_s)^{1/2}, \quad \text{for all } s. \tag{7.2}$$

The slow ion diffuses in the velocity space; the diffusion coefficient is given by the second moment. Since the plasma is isotropic, and the ion is slow, the second moment is diagonal. The diagonal sum of the second moment is denoted by D

$$D = \langle \Delta\mathbf{v} \cdot \Delta\mathbf{v}\rangle / \Delta t. \tag{7.3}$$

In the collision theory, the diagonal sum is

$$D_\varkappa = \sum_s \iint \mathbf{\Delta} \cdot \mathbf{\Delta} f_s(\mathbf{v} + \mathbf{g})\, g\, d\mathbf{g}\, d\mathbf{b}. \tag{7.4}$$

where $\mathbf{\Delta}$ is given by (5.3), and the Debye-type potential (4.22) is used Neglecting $\mathbf{v}$, and using (6.5), we have

$$D_\varkappa = \Sigma A_s\left(\ln \frac{4\mu_s T}{\gamma^2\, |qq_s|\, m_s\varkappa} - \frac{1}{2}\right). \tag{7.5}$$

$$A_s = \frac{16\pi q^2}{m^2} q_s^2 n_s\left(\frac{m_s}{2\pi T}\right)^{1/2}. \tag{7.6}$$

In the wave theory, the moment is given by (5.6), which reduces to

$$D^0 = \Sigma A_s \int \frac{k^3}{(k^2 + k_D^2)^2} dk. \tag{7.7}$$

Here the velocity $\mathbf{v}$ has been neglected; and use has been made of the limit

$$\varepsilon(k, 0) = 1 + k_D^2/k^2, \quad k_D^2 = \sum 4\pi n_s q_s^2/T, \tag{7.8}$$

where k_D is the Debye constant.

The connection formula (4.23) gives

$$D = \sum A_s \left(\ln \frac{4\mu_s T}{\gamma^2 \, |qq_s| \, m_s \varkappa} - \frac{1}{2} + \int_0^\infty \left\{ \frac{k^3}{(k^2 + k_D^2)^2} - \frac{k^3}{(k^2 + \varkappa^2)^2} \right\} dk \right). \tag{7.9}$$

Therefore, the final result[3] is

$$D = \frac{16\pi q^2}{m^2} \sum_s n_s q_s^2 \left(\frac{m_s}{2\pi T} \right)^{1/2} \left(\ln \frac{4Tm}{\gamma^2 \, |qq_s| \, (m + m_s) \, k_D} - \frac{1}{2} \right). \tag{7.10}$$

This result can be obtained on the basis of the binary collision, by using the Debye potential, because the velocity of the ion is sufficiently small compared with the thermal velocities of the plasma constituents. In general, the field around a moving charge is neither of Debye-type nor spherically symmetric.

8 ENERGY LOSS BY A FAST PARTICLE

A charged particle moving fast through a plasma loses its energy, by interacting with the plasma. The rate $\mathbf{P}$ of the energy loss by the particle with the mass m is defined by

$$P = \frac{m}{2} \left(2\mathbf{v} \cdot \frac{\langle \Delta \mathbf{v} \rangle}{\Delta t} + \frac{\langle \Delta \mathbf{v} \cdot \Delta \mathbf{v} \rangle}{\Delta t} \right). \tag{8.1}$$

which is sometimes called the stopping power of the plasma. We consider the same plasma as in the previous section. The velocity v of the particle is assumed to be much larger than the thermal velocity of the constituents

$$v \gg (T/m_s)^{1/2}, \text{ for all } s. \tag{8.2}$$

In this case, the contribution to the energy loss comes only from the electrons in the plasma, because of the mass difference.

From (5.2) and (6.5), we have the loss rate in the modified collision theory

$$P_\varkappa = \frac{4\pi n_e e^2 q^2}{m_e}\left(\ln\frac{2\mu_e v^2}{\gamma\,|qe|\,\varkappa} - \frac{1}{2}\right). \tag{8.3}$$

where e being the electronic charge. In the wave theory, the second moment (5.6) is inversely proportional to v under the condition (8.2), so that it cancels the last member of (5.5). Therefore, we have

$$P^0 = \frac{q^2}{2\pi^2}\int\frac{i\mathbf{k}\cdot\mathbf{v}}{k^2\varepsilon(k, \mathbf{k}\cdot\mathbf{v})}\,d\mathbf{k}. \tag{8.4}$$

Using the fact that $\varepsilon(k, \omega)$ has no zeroes in the upper half-plane of complex ω, we can integrate (8.4) along a large semicircle of radius kv, where ε reduces to

$$\varepsilon(k, \omega) = 1 - \omega_e^2/\omega^2, \quad \text{for} \quad |\omega|/k \gg (T/m_s)^{1/2}. \tag{8.5}$$

ω_e being the electron plasma frequency

$$\omega_e^2 = 4\pi n_e e^2/m_e. \tag{8.6}$$

Then, (8.4) is integrated to be

$$P^0 = \frac{q^2\omega_e^2}{v}\int\limits_{k>\omega_e/v}\frac{dk}{k}. \tag{8.7}$$

The corresponding quantity $P_\varkappa^0$ in the impulse approximation is

$$P_\varkappa^0 = \frac{q^2\omega_e^2}{v}\int\frac{k^3}{(\varkappa^2 + k^2)^2}\,dk. \tag{8.8}$$

The loss rate in the unified theory is given by $P = P_\varkappa + P^0 - P_\varkappa^0$.

Hence we finally obtain[1]

$$P = \frac{q^2\omega_e^2}{v}\ln\frac{2\mu_e v^3}{\gamma\,|qe|\,\omega_e}. \tag{8.9}$$

The close collision radius in this case is $|qe|/\mu_e v^2$, and the screening length corresponds to v/ω_e which is much longer than the Debye length. The result (8.9) is independent of the velocity distribution function of the plasma constituents, even if the plasma is unstable, as fas as the velocity v is large enough[55]. The loss rate for arbitrary velocities has also been calculated[39].

The expression (8.9) was first obtained by Bohr[56] in 1913. He considered a system where the constituent electrons were kept in their positions by a harmonic force with a frequency ω_e; such a system may be regarded as a dielectric medium characterized by the permittivity (8.5). Treatments by Kramers[57], Neufeld and Ritchie[58], and Sitenko[59] are of interest; they also obtained (8.9). However, their methods apply only under the condition (8.2).

The quantum effect is easily obtained, by using (6.4). In particular, when the Born approximation holds, we have[4], from (6.6).

$$P^{\text{Born}} = \frac{q^2\omega_e^2}{v} \ln \frac{2\mu_e v^2}{\hbar\omega_e}, \tag{8.10}$$

in agreement with the result[51] of the quantum field theory, as it should. In this case, the de Broglie wavelength replaces the close collision radius. The loss rate in the Born approximation was first obtained by Bethe[60].

The presence of a magnetic field modifies the expression (8.7) in the wave theory, and hence the result. However, an expression similar to (8.4) holds, if the permittivity is replaced by that including the effect of the magnetic field. Then, the loss rate is calculated to be[46]

$$P = \frac{q^2\omega_e^2}{v}\left(\ln \frac{2\mu_e v^3}{\gamma\,|qe|\,\omega_e} - \frac{1}{2}\ln \frac{\omega_e^2 + \omega_B^2}{\omega_e^2}\right). \tag{8.11}$$

where ω_B is the electron gyrofrequency, and the velocity of the particle is parallel to the magnetic field. For obliquely moving particles, the loss rate is larger than (8.11) but smaller than (8.9). The effect of magnetic fields on the stopping power was first considered by Akhiezer[61], in the Born approximation.

9 RELAXATION BETWEEN ION AND ELECTRON TEMPERATURES

We consider a two-component plasma composed of electrons and one type of ions. The velocity distribution functions are both Maxwellian, but the electron temperature T_e differs from the ion temperature T_i. The energy exchange between the ions and the electrons is much smaller than that among the ions or the electrons, because the mass ratio is far from unity. Hence, the system approach the equilibrium, keeping the distributions Maxwellian.

The rate R of the temperature relaxation is defined by

$$d(T_e - T_i)/dt = -(T_e - T_i)\,R. \tag{9.1}$$

By use of any one of the connection formulas obtained in the section 4, we have[1]

$$R = \frac{8}{3}(n_e + n_i)\frac{m_e}{m_i}\left(\frac{Ze^2}{T_e}\right)^2\left(\frac{2\pi T_e}{m_e}\right)^{1/2}\left(\ln \frac{4T_e}{\gamma^2 Ze^2 k_e} - \frac{1}{2}\right). \tag{9.2}$$

where Z is the charge number of the ion, and $k_e = (4\pi n_e e^2/T_e)^{1/2}$ is the electronic screening constant. It should be noted that the screening length in this case is not the total Debye length but the electronic Debye length; the ions do not play any role in the screening. When a magnetic field is applied, the relaxation rate increases[36].

10 HIGH-FREQUENCY CONDUCTIVITY

When the frequency ω of the applied electric field is much higher than the collision frequency, the interaction between constituent particles can be regarded as a perturbation. The plasma is assumed to be of two-component and in thermal equilibrium. Then, the real part of the conductivity is easily calculated in the form

$$\sigma = \frac{n_\iota}{3\pi}\frac{\omega_e^2}{\omega^2}\left(\frac{Ze^2}{T}\right)^2\left(\frac{2\pi T}{m_e}\right)^{1/2} L. \tag{10.1}$$

where L is the Coulomb logarithm.

When $\omega \ll \omega_e$, we have[35]

$$L = \ln\frac{4T}{\gamma^2 Ze^2 ke} + \frac{k_D^2}{2k_i^2}\ln\frac{k_e^2}{k_D^2}. \tag{10.2}$$

where k_e and k_i are the electronic and the ionic Debye constants, and k_D is the total Debye constant $(k_e^2 + k_i^2)^{1/2}$. This form of the Coulomb logarithm is characteristic to the phenomena caused by the momentum exchange between ions and electrons.

Unless $\omega \ll \omega_e$, we may not use (2.6). However, the conductivity in the wave theory is known[42], by use of which we can calculate[35] the dependence of the Coulomb logarithm L on the frequency ω. The effect of magnetic fields has been obtained by Itikawa[37]. He calculated also the D.C. conductivity[34].

11 RELAXATION OF ANISOTROPIC DISTRIBUTION OF ION VELOCITIES

In order to show some characteristics of the ion-ion interactions, we consider a relaxation of the "spheirodal" velocity distribution of ions

$$f(\mathbf{v}) = n\left(\frac{m}{2\pi T_\parallel}\right)^{1/2}\frac{m}{2\pi T}\exp\left(-\frac{mv_\parallel^2}{2T_\parallel} - \frac{mv_\perp^2}{2T_\perp}\right), \tag{11.1}$$

under the condition

$$|T_\perp - T_\parallel| \ll T \equiv \frac{1}{3} T_\parallel + \frac{2}{3} T_\perp. \tag{11.2}$$

The electron distribution may be assumed to be Maxwellian, because its relaxation is much faster.

The rate of increase of the mean energy $W_\parallel$ associated with the velocity component parallel to the axis of the spheroid is calculated to be[33]

$$\frac{dW_\parallel}{dt} = (T_\perp - T_\parallel) \frac{8}{15} n \left(\frac{Z^2 e^2}{T}\right)^2 \left(\frac{\pi T}{m}\right)^{1/2} \ln \frac{4T}{\gamma^2 Z^2 e^2 [k_e^2 + \eta(Z) k_2^2]^{1/2}}. \tag{11.3}$$

Here $\eta(Z)$ means the effective fraction' of the ionic screening, and its numerical values are

$$\eta(1) = 0.66147, \quad \eta(2) = 0.66098, \quad \eta(3) = 0.66151,$$
$$\eta(4) = 0.66228, \quad \ldots, \quad \eta(\infty) = 0.67137. \tag{11.4}$$

This result indicates that an ion is completely screened by the electrons but partially by the other ions. Of course, the effective fraction of the ionic screening are generally different for different phenomena[33,38].

Acknowledgments

One of the present authors (O.A.) is grateful to Professor A. Isihara of SUNY at Buffalo, where the final manuscript was prepared.

This work was supported partially by the National Science Foundation through GP-9419.

References

1. O. Aono, *J. Phys. Soc. Japan* **17**, 853 (1962).
2. T. Kihara, in *Service de Physique des Plasmas* (École d'été, University of Paris, 1962) Vol. 1, p. 69.
3. T. Kihara and O. Aono, *J. Phys. Soc. Japan* **18**, 837 (1963).
4. T. Kihara, *J. Phys. Soc. Japan* **19**, 108 (1964).
5. O. Aono, *J. Phys. Soc. Japan* **20**, 1250 (1965).
6. D. E. Baldwin, *Phys. Fluids* **5**, 1523 (1962).
7. E. A. Frieman and D. L. Book, *Phys. Fluids* **6**, 1700 (1963).
8. R. L. Guernsey, *Phys. Fluids* **7**, 1600 (1964).
9. F. Perkins, *Phys. Fluids* **8**, 1361 (1965).
10. I. B. Bernstein and J. F. Ahearne, *Ann. Phys.* (*N.Y.*) **49**, 1 (1968).
11. H. A. Gould and H. E. DeWitt, *Phys. Rev.* **155**, 68 (1967).

12. O. Aono, *Phys. Fluids* **11**, 341 (1968).
13. J. H. Jeans, *Problems of Cosmogony and Stellar Dynamics* (Cambridge Univ. Press, 1919) p. 230ff.
14. A. A. Vlasov, *Zh. Eksp. Teor. Fiz.* **8**. 291 (1938).
15. P. Gruner, *Ann. Physik* **35**, 381 (1911).
16. S. Chapman, *Monthly Not. Roy. Astron. Soc.* **82**, 292 (1922).
17. I. Langmuir, *Proc. Natl. Acad. Sci. U.S.* **14**, 627 (1928).
18. L. Landau, *Phys. Z. Sowjetunion* **10**, 154 (1936).
19. R. S. Cohen, L. Spitzer, Jr., and P. McR. Routly, *Phys. Rev.* **80**, 230 (1950).
20. E. Persico, *Monthly Not. Roy. Astron. Soc.* **86**, 93 (1926).
21. T. Kihara, *J. Phys. Soc. Japan* **14**, 402 (1959).
22. R. L. Liboff, *Phys. Fluids* **2**, 40 (1959).
23. R. Balescu, *Phys. Fluids* **3**, 52 (1960).
24. A. Lenard, *Ann. Phys. (N.Y.)* **10**, 390 (1960).
25. N. Rostocker and M. N. Rosenbluth, *Phys. Fluids* **3**, 1 (1960).
26. W. B. Thompson and J. Hubbard, *Rev. Mod. Phys.* **32**, 714 (1960).
27. J. Hubbard, *Proc. Roy. Soc. (London)* **A260**, 114 (1961).
28. T. Y. Wu and R. L. Rosenberg, *Can. J. Phys.* **40**, 463 (1962).
29. for example, T. Y. Wu, *Kinetic Equations of Gases and Plasmas* (Addison-Wesley, 1966).
30. O. Aono, *Phys. Fluids* **11**, 832 (1968).
31. J. Hubbard, *Proc. Roy. Soc. (London)* **A261**, 371 (1961).
32. J. Weinstock, *Phys. Rev.* **133**, A673 (1964).
33. T. Kihara, O. Aono, and Y. Itikawa, *J. Phys. Soc. Japan* **18**, 1043 (1963).
34. Y. Itikawa, *J. Phys. Soc. Japan* **18**, 1499 (1963).
35. O. Aono, *J. Phys. Soc. Japan* **19**, 376 (1964).
36. Y. Itikawa, *J. Phys. Soc. Japan* **19**, 748 (1964).
37. Y. Itikawa, *J. Phys. Soc. Japan* **20**, 1921 (1965).
38. T. Hatori and Y. Itikawa, *J. Phys. Soc. Japan* **21**, 515 (1966).
39. Y. Itikawa and O. Aono, *Phys. Fluids* **9**, 1259 (1966).
40. N. Rostoker, *Phys. Fluids* **3**, 922 (1960).
41. Y. Itikawa, *J. Phys. Soc. Japan* **21**, 1387 (1966).
42. C. Oberman, A. Ron, and J. Dawson, *Phys. Fluids* **5**, 1514 (1962); **7**, 921 (1964).
43. J. Dawson and C. Oberman, *Phys. Fluids* **6**, 394 (1963).
44. R. L. Guernsey, *Phys. Fluids* **5**, 622 (1962).
45. R. Balescu and A. Kuezell, *J. Math. Phys.* **5**. 1140 (1964).
46. N. Honda, O. Aono, and T. Kihara, *J. Phys. Soc. Japan* **18**, 256 (1963).
47. T. Kihara, O. Aono, and R. Sugihara, *J. Phys. Soc. Japan* **19**, 986 (1964).
48. R. Sugihara and Y. Itikawa, *J. Phys. Soc. Japan* **20**, 591 (1965).
49. N. Honda, *J. Phys. Soc. Japan* **19**, 1201 (1964).
50. N. Honda, *J. Phys. Soc. Japan* **19**, 1953 (1964); **21**, 1189 (1966).
51. A. I. Larkin, *Zh. Eksp. Teor. Fiz.* **37**, 264 (1959), [*Sov. Phys.-JETP* **10**, 186 (1960)].
52. O. V. Konstantinov and V. I. Perel, *Zh. Eksp. Teor. Fiz.* **39**, 861 (1961). [*Sov. Phys.-JETP* **12**, 597 (1961)].
53. V. I. Perel and G. M. Eliasberg, *Zh. Eksp. Teor. Fiz.* **41**, 886 (1961). [*Sov. Phys.-JETP* **14**, 633 (1962)].

54. H. W. Wyld and D. Pines, *Phys. Rev.* **127**, 1851 (1962).
55. J. Okutani and O. Aono, *J. Phys. Soc. Japan* **19**, 2207 (1964).
56. N. Bohr, *Phil. Mag.* **25**, 10 (1913).
57. H. A. Kramers, *Physica* **13**, 401 (1947).
58. J. Neufeld and R. H. Ritchie, *Phys. Rev.* **98**, 1632 (1955).
59. A. G. Sitenko, *Electromagnetic Fluctuations in Plasma* (Academic Press, 1967) p. 129.
60. H. Bethe, *Ann. Physik* **5**, 325 (1930).
61. I. A. Akhiezer, *Zh. Eksp. Teor. Fiz.* **40**, 954 (1961). [*Sov. Phys.-JETP* **13**, 667(1961)].

PAPER C–4

Some comments on the structure of Boltzmann's H function

S. HARRIS

College of Engineering, SUNY, Stony Brook, N.Y.

I INTRODUCTION

Boltzmann's H theorem[1] stands as one of the very important results of the kinetic theory. Aside from its important implications with regard to the irreversibility of a given kinetic equation[2], an H theorem also provides the necessary uniqueness proof for the lowest order approximation when solutions are constructed via the Chapman–Enskog procedure, e.g. the Boltzmann equation[2]. Unfortunately no such result yet exists for the Cloh–Uhlenbeck collision term[2], which provides the necessary correction to Boltzmann's equation when moderately dense gases are considered. The formal theory of this equation must therefore still be considered as incomplete. Preliminary considerations[3] indicate that it may be possible to establish an H theorem for the Choh–Uhlenbeck collision term using Boltzmann's original choice for the H function, $H[f] = \int d\mathbf{v}\, f \ln f$ (we will consider only spatially uniform systems here, so that the one particle distribution function $f = f(\mathbf{v}, t)$; further we will define f to be normalized to unity, i.e. $\int d\mathbf{v}\, f = 1$).

The choice of H can not be, in the absence of a general proof, considered to be unique, and we are thus motivated to consider the following question: what are the particular properties of an H function used in conjunction with a given kinetic equation which single it out from all other functionals of that kinetic equation's solution? It seems likely to be most profitable to consider this question in the context of Boltzmann's equation, since so much is already known for this case. Hopefully the answers will prove useful in the establishment of an H theorem for the Choh–Uhlenbeck equation.

II THE *H* FUNCTION AND THE THERMODYNAMIC ENTROPY

It is usually stated (and therefore apparently accepted) that Boltzmann's H function derives much of its significance from the fact that it is not merely a fortuitously chosen test function which demonstrates the approach to equilibrium of solutions to Boltzmann's equation, but rather that this function has a profound physical significance in that it represents an extension, for non-equilibrium states, of the thermodynamic entropy[4]. This last statement is true, but only in a very qualified sense. If the Boltzmann equation is viewed as an *exact* contraction of the Liouville equation, which is only possible by specifying the Boltzmann gas limit (BGL)[5], then the identification of H with the thermodynamic entropy is indeed correct. The reason for this is that the entropy identifiable with the H function is the perfect gas entropy. The BGL is defined by taking $N \to \infty$, $m \to 0$, $\sigma \to 0$, so that $N\sigma^3 \to 0$, and $Nm \to$ const. (N the number of particles in the system, m the mass of a single particle, and σ the range of the inter-particle forces). A consequence of this limit is that $\lambda \propto 1/N\sigma^2$, the mean free path, is finite so that collisions play a role in determining the temporal evolution of the one particle distribution function, but the excluded volume, $N\sigma^3$, is zero, and the inter-particle potential does not therefore contribute to the thermodynamic quantities, which are exactly given by their perfect gas values.

It does not appear to be possible to prescribe a similar limit in which the Liouville equation exactly contracts to the Boltzmann equation plus the Choh–Uhlenbeck collision term. Instead this latter equation must be viewed as an *approximation* in the usual thermodynamic limit, valid at not too high densities. At lower densities the Boltzmann equation again furnishes the (approximate) kinetic description, but now, in contradistinction to the case in the BGL, the thermodynamic properties include a second virial correction term[6]. An H theorem, using $H = \int d\mathbf{v}\, f \ln f$ can also be proved here, but now H is no longer identifiable with the appropriate thermodynamic entropy. Is there any other criterion by which this function is apparently singled out from the class of functionals of the Boltzmann equation solution?

III MCKEAN'S CONJECTURE

In a very elegant paper McKean has considered the problem of the approach to equilibrium in Kac's model of Boltzmann's equation[7], and has conjectured, for technical reasons, that a distinguishing feature of H might be that its successive derivatives alternate in sign, i.e. not only is H monotone decreasing, $\frac{\partial H}{\partial t} \equiv H^{(2)} \leqq 0$, but $H^{(1)} \geqq 0$, and in general $(-)^k H^{(k)} \geqq 0$. In order to give this conjecture substance it must first be demonstrated that it is valid. For Boltzmann's equation no such result has yet been shown[8], however there are some important cases for which a general result can be proved, and we now proceed to consider these.

A Linear, Symmetric, Non-Positive Collision Operators

Consider first the class of kinetic equations which have linear, symmetric, non-positive collision operators,

$$\frac{\partial \phi}{\partial t} = L\phi \tag{1}$$

with

$$(L\phi, \psi) = (\phi, L\psi) \tag{2}$$

$$(L\phi, \phi) \leqq 0 \tag{3}$$

The conjugate linearized H function is

$$H = H[f_0] + \frac{\varepsilon^2}{2}(\phi, \phi) + o(\varepsilon^3) \tag{4}$$

The inner product (,) is with respect to the measure $d\mathbf{u} = f_0\, d\mathbf{v}$, where f_0 is the absolute Maxwellian distribution function which describes the equilibrium state. An equation like (1) often results when an exact kinetic equation

$$\frac{\partial f}{\partial t} = Kf \tag{5}$$

is linearized about the equilibrium value of f, i.e. we set $f = f_0(1 + \varepsilon\phi)$, and only terms linear in the small parameter ε are retained. The kinetic equation (5) need not describe f, e.g. the kinetic equation for F_N, the N particle distribution function, could also be considered within the present discussion. The full collision operator K may or may not contain ε, and may or may not be linear. Examples are the linearized Boltzmann equation (for f)[2] and the Master equation (for f or F_N)[9].

The proof of the alternating property follows directly from the properties of the collision operator. We have, making use of (2), (3), and (4)

$$\frac{\partial^n H}{\partial t^n} \equiv H^{(n)} = \frac{L^n(\phi, \phi)}{2} \tag{6}$$

$$= 2^{n-1}(\phi, L^n\phi) \tag{7-a}$$

$$= 2^{n-1}(L^{n/2}\phi, L^{n/2}\phi) \geqq 0, \quad n \text{ even} \tag{7-b}$$

$$= 2^{n-1}(L[L^{n/2}\phi], L^{n/2-1}\phi) \leqq 0, \quad n \text{ odd} \tag{7-c}$$

with equality only in equilibrium. Thus there exist an infinite set of functions, in the case of (1)–(4), which have the same properties as the proper H function with regard to demonstrating the approach to an unique equilibrium state.

B Collision Operators for which $f^{(1)} = -f^{(2)}$

The linearized system considered in the preceding section offers considerable simplifications in its treatment due to the form of the linearized H function. When the full H function is considered things quickly become more complicated. A general result may be demonstrated however for kinetic equations which possess a certain simplifying feature, namely that

$$f^{(1)} = -f^{(2)} \tag{8}$$

(with an appropriately chosen time scale). This condition is not obeyed by Boltzmann's equation, however it is found in certain models of that equation. The simple BGK model[5], the discrete velocity gas[10], and the two state model[11], proposed by Carleman are examples of model kinetic equations for which (8) is generally satisfied. Further, for certain initial conditions the Boltzmann equation for a test particle moving in a system of randomly situated stationary scatterers (in three dimensions) or through a background fluid which is in equilibrium (in one dimension) also satisfies (8). Finally, we also have (8) satisfied in a flow at constant density and temperature when f is represented by a ten moment approximation[12].

Consider now $H^{(2+k)}$, which we find by direct differentiation of the full H function.

$$H^{(2+k)} = -H^{(k)} + \int d\mathbf{v}[f^{-1}(f^{(1)})^2]^{(k)} \tag{9-a}$$

$$\equiv -H^{(k)} + \int d\mathbf{v}\, A^{(k)} \tag{9-b}$$

Here we have already made use of the condition given by (8). For the case of the discrete velocity gas or Carleman's model the integral appearing in (9) is replaced by a sum over all of the discrete distribution functions, but the proof proceeds exactly as outlined below. A simple inductive argument allows us to conclude that if the $A^{(k)}$ alternate in sign, so that $A^{(2k)} \geqq 0$; $A^{(2k+1)} \leqq 0$, then $(-)^k H^{(2+k)} \geqq 0$, since from the H theorem we already know that $H^{(1)} \leqq 0$. Now since

$$[fA]^{(k)} = \sum_{j=0}^{k} f^{(k-j)} A^{(j)} \binom{k}{j} \tag{10}$$

with $\binom{k}{j} = k!/(k-j)!\,j!/$, we can obtain the following equation for $A^{(k)}$ by again making use of (8):

$$A^{(k)} = f^{-1}\left[(-2)^k (f^{(1)})^2 - \sum_{j=0}^{k-1} f^{(k-j)} A^{(j)} \binom{k}{j}\right] \tag{11}$$

The above equation can be solved by iteration; doing this we find

$$A^{(k)} = (-)^k f^{-1} (f^{(1)})^2 \sum_{j=0}^{k} C_j^k [f^{-1} f^{(1)}]^j \tag{12}$$

with

$$C_0^k = 2^k \tag{13}$$

$$C_{j+1}^k = \sum_{r=0}^{k-1} C_j^k \binom{k}{j} \tag{13a}$$

It is now sufficent to show that the sum on the r.h.s. of (12) is positive to complete out proof. This sum is a polynaomial in $x \equiv f^{-1} f^{(1)}$, and we need only consider values of x which are negative. Further, a simple consideration shows that x is bounded below by -1, so that we need only consider x in the interval $-1 \leqq x \leqq 0$. I have tried to complete the proof at this point by showing that the polynomial $\sum C_j^k x^j$ has no roots in the interval $-1 \leqq x \leqq 0$, but have been unable to prove this and must therefore settle for the more awkward proof that follows (hopefully a better mathematician can come to my rescue). Write

$$P(k) = \sum_{j=0}^{k} C_j^k x^j \equiv \sum_{j=1}^{k+1} P_j(k) \tag{14}$$

where $P_r(k)$ is given explicitly, from (12) and (13), as

$$P_r(k) = \sum_{j=1}^{r} (j+1)^k \binom{r-1}{j-1} (-)^{j+1} (-|x|)^{r-1} \tag{15}$$

Finally, we make use of the binomial theorem to expand $(j+1)^k$, introducing a new dummy index s and *a* term j^s; we change indicies on the j summation to recover a $(j + 1)^s$ which we re-expand, obtaining

$$P(k) = \sum_{r=1}^{k+1} \sum_{s=0}^{k} \sum_{t=0}^{s} |x|^{r-1} \binom{k}{s} \binom{s}{t} \sum_{j=0}^{r-1} \binom{r-1}{j} j^t (-)^{r-j-1} \tag{16}$$

The last sum must be positive; it is a well known combinatorial result[13] which expresses the number of possible arrangements of t objects in $r - 1$ cells. To be even more explicit, we may note that

$$\sum_{t=0}^{\infty} \frac{x^t}{t!} \sum_{j=0}^{r} \binom{r}{j} (-)^{r-j} j^t = (e^x - 1)^r \tag{17}$$

and by differntiating both sides of this equation n times with respect to x and then setting $x = 0$ we obtain the following result

$$\sum_{j=0}^{r} \binom{r}{j} (-)^{r-j} j^n = \frac{\partial}{\partial x^n} (e^{x-1})^r|_{x=0} \geqq 0$$

Therefore $P(k)$ is positive for all x of interest, and consequently we may conclude that the derivatives of the full H function do have the conjectured alternating property, subject to the condition (8).

IV CONCLUDING REMARKS. CONVEXITY OF *H* FOR A MORE REPRESENTATIVE MODEL

The preceding results present a brief review of what has been done up to now. These results lead us to believe that it is extremely likely that some general principle is in operation, however their extension to the full Boltzmann equation appears to be an exceedingly tedious job, and at present even a proof of the convexity of H is lacking[8] for this case[14]. It is toward the establishment of this latter result that we have set as our present goal, since the availible strong mathematical theorem[15] pertinent to the proof of the alternating property requires an explicit knowledge of the time behavior of f, and thus is at present of no use to us.

In the absence of an answer to our original question, it seems profitable to consider additional models which (a) are non-linear, and therefore include $\int d\mathbf{v} f \ln f$ (or its analogue) as the proper conjugate H function and (b) which model the Boltzmann equation more closely than the models considered in III.B, i.e. which do *not* satisfy (8).

A nice example of a model which satisfies the above criteria is the three state model considered by Carleman[11]. For this model we have

$$\begin{aligned} \dot{a} &= b^2 + c^2 - 2a^2 \\ \dot{b} &= a^2 + c^2 - 2b^2 \\ \dot{c} &= a^2 + b^2 - 2c^2 \end{aligned} \tag{18}$$

with

$$a + b + c = 1 \tag{19}$$

It is a simple matter to show that, e.g. $\dot{a} \neq -\ddot{a}$, so that (8) is not valid for this model. The conjugate H function is

$$H = a \ln a + b \ln b + c \ln c \tag{20}$$

and the H theorem is simply demonstrated in the usual manner by computing

$$\begin{aligned} H^{(1)} &= \dot{a} \ln a + \dot{b} \ln b + \dot{c} \ln c \\ &= (b^2 - a^2) \ln (a/b) + (c^2 - a^2) \ln (a/c) + \\ &\quad + (c^2 - b^2) \ln (b/c) \leqq 0 \end{aligned} \tag{21}$$

with equality only for the unique equilibrium state $a = b = c$.

Consider now the proof of the convexity property. By direct differentiation we obtain

$$H^{(2)} = \left[\frac{(\dot{a})^2}{a} + \frac{(\dot{b})^2}{b} + \frac{(\dot{c})^2}{c}\right] + [\ddot{a} \ln a + \ddot{b} \ln b + \ddot{c} \ln c] \equiv H_{\text{I}}^{(2)} + H_{\text{II}}^{(2)} \tag{22}$$

Writing out $H_{\text{II}}^{(2)}$ we obtain

$$\begin{aligned} H_{\text{II}}^{(2)} &= (a^3 - b^3) \ln a/b + (a^3 - c^3) \ln a/c + (b^3 - c^3) \ln b/c \\ &\quad + ba(a - b) \ln a/b + ca(a - c) \ln a/c + bc(b - c) \ln b/c \\ &\quad + b^2(c - a) \ln a/c + c^2(b - a) \ln a/b + a^2(c - b) \ln b/c \end{aligned} \tag{23}$$

Eliminating the third variable in the coefficient of the logrithmic term containing the other two variables, e.g. c in the coefficient of $\ln (a/b)$, by the use of (19) we obtain

$$\begin{aligned} H_{\text{II}}^{(2)} &= (a - b)(2\{a + b\} - 1) \ln a/b + (a - c)(2\{a + c\} - 1) \ln a/c \\ &\quad + (b - c)(2\{b + c\} - 1) \ln b/c \end{aligned} \tag{24}$$

This expression can be shown to be non-negative by the following argument. The first term is only negative if $(a + b) < \frac{1}{2}$, in which case the other two terms are non-negative. Consider this case with $b > a$. Then

$$0 < (a - b) \ln a/b < (a - c) \ln a/c \tag{25}$$

Further, since $a > 0$ it follows that

$$\begin{aligned} 2a + 2 - 2(a + b) - 1 &= 2(a + 1 - a - b) - 1 \\ &= 2(a + c) - 1 > 1 - 2(a + b) = |2(a + b) - 1| \end{aligned} \tag{26}$$

and therefore that the sum of the first two terms on the r.h.s. of (24) are positive, and hence $H_{\mathrm{II}}^{(2)}$ is positive for this case. For $(a + b) < \frac{1}{2}$, $a > b$ we can repeat the same argument using the first and last term on the r.h.s. of (24), and therefore we can conclude that in general $H_{\mathrm{II}}^{(2)} \geqq 0$ with equality only in equilibrium. Since $H_{\mathrm{I}}^{(2)}$ is obviously non-negative, we can conclude that $H^{(2)}$ itself has this property, i.e. H is convex.

There are certain similarities between the structure of $H^{(2)}$ in the case of the model treated above and Boltzmann's equation. For the latter case we have

$$H^{(2)} = \int d\mathbf{v}(f^{(1)})^2 f^{-1} + \int d\mathbf{v} f^{(2)} \ln f \tag{27}$$

If the Boltzmann equation behaves similarly to our model then we will be able to show that the second term on the r.h.s. of (27) is non-negative, avoiding therefore a more complicated argument involving inequalities. In the case of Boltzmann's equation a result similar to (23) can be obtained, expressing the second term on the r.h.s. of (27) as the sum of terms which are always non-positive and non-negative. For the model we have shown that the non-negative terms always dominate – this remains to be done to complete a proof of convexity for Boltzmann's equation. The results presented here indicate that it is not unreasonable to expect that this can be done.

In conclusion we would like to recall some of the early molecular dynamics experiments[16], where results were reported for the H function for solutions to Boltzmann's equation. These results indicate that H is convex! The problem remains of catching up with the computer.

References

1. L. Boltzmann, *Sitzungsberichte Akad. Wiss., Vienna, II* **66**, 275 (1872).
2. G. E. Uhlenbeck and G. W. Ford, *Lectures in Statistical Mechanics*, (American Mathematical Society, Providence, R. I., 1963).
3. S. Harris, *Phys. Fluids* **9**, 1328 (1966).
4. See, e.g., G. H. Wannier, *Statistical Physics* (John Wiley & Sons, Inc., New York, 1966), Chap. 18.

5. H. Grad, in *Handbuch der Physik*, S. Flugge, Ed. (Springer-Verlag, Berlin, 1958), Vol. 12.
6. S. T. Choh, "*The kinetic theory of phenomena in dense gases*", University of Michigan dissertation, unpublished (1958).
7. H. McKean, Jr., *Arch. Rational Mech. Anal.* **21**, 343, (1966).
8. In our talk we gave a preliminary announcement of a proof for the convexity of H. This proof has been shown to be spurious, and the problem remains open.
9. S. Harris, *J. Chem. Phys.* **48**, 4012 (1968).
10. S. Harris, *J. Math. Phys.* **8**, 2407 (1967).
11. T. Carleman, *Problemes Mathematiques dans Theorie Cinetique des Gaz*, (Publications Scientifques de l'Institut Mittag-Leffler, Vol. 2, Uppsala, 1957).
12. H. Grad, *Comm. Pure Appl. Math.* **2**, 331 (1949).
13. See, e.g. W. Feller, *An Introduction to Probability Theory and its Applications*, (John Wiley & Sons, Inc., New York, 1957), Vol. I, Chap. 2.
14. An argument that *H* cannot be convex is given in 12, however no rigorous proof is given. For our model results this argument does not hold, but with regard to the Boltzmann equation the problem is open.
15. G. A. Baker, Jr., *Phys. Rev.* **161**, 434 (1967).
16. B. Alder and T. Wainwright in, *Proceedings of the International Symposium on Transport Processes in Statistical Mechanics*, I. Prigogine ed., (Interscience Publishers, Inc. New York, 1958).

PAPER C–5

*Quantum Boltzmann equation for a laser**

CHARLES R. WILLIS

Boston University, Boston, Mass.

I INTRODUCTION

Kinetic equations enter the description of lasers in three different ways. First the laser system of N atoms per unit volume and the radiation oscillators are in contact with various reservoirs which destroy phase information, dissipate energy, and provide inverted atomic populations by pumping. The various reservoirs are represented by kinetic operators. Second, in the case of gas lasers, the gas atoms collide with other atoms and the collisions are described by the Boltzmann equation of kinetic theory with the additional complexity that the laser atoms are in exited states. The third way that kinetic equations enter the description of lasers is in the treatment of the radiation-matter $j \cdot A$ interaction itself. The details of reservoirs and collisions are not crucial for the understanding of laser phenomena however for most laser phenomena the dissipative processes are larger than the coherence producing $j \cdot A$ radiation-matter interaction. The most important inequality needed to justify the existence of the laser kinetic equation is that $(\gamma^2\omega_0^2N)/\nu_2\nu_D) \ll 1$ where γ^2N is the dimensionless total dipole moment of the gas, $\hbar\omega_0$ is the two-level energy difference, ν_D is the Doppler frequency and ν_2 is the inverse relaxation time of the atomic polarization. The inequality is just the requirement that the collective behavior measured by the laser frequency $\omega_L^2 \equiv \gamma^2 N\omega_0^2$ is small compared with dissipative decay of the atomic polarization, ν_2, and the dephasing atomic motion measured by ν_D. Some recent phenomena involving lasers such as π-pulses and self induced transparency are not dissipation dominated and are not described by the present laser kinetic equations.

* The research reported in this paper was sponsored in part by Air Force Cambridge Research Laboratories, Office of Aerospace Research.

There are two general classes of laser experiments and their corresponding theories. Type I theories treat the radiation only in terms of the first moment[1] $\langle E \rangle$ or the second moment[2], $\langle E^2 \rangle \equiv I$, of the radiation density matrix R where E is the electric field operator. The statistics of the radiation in type I theories are prescribed a priori not determined dynamically by the equations of motion. The self-consistent field approximation, $\langle E \rangle \neq 0$, implies the radiation is practically in a single quantum state[3] known as a Glauber state[4], is very coherent and has only the fluctuation required by the uncertainty principle. The case where the electric field vanished and the radiation is completely determined by the second moment, $\langle E^2 \rangle$, corresponds to Gaussian or "thermal like" statistics because only the second moment is dynamically determined. The two type I theories give the same results[2] for phenomena that depend on only the first or second moments because $\langle E^2 \rangle$ of the "thermal-like" statistics equal to $|\langle E \rangle|^2$ of the self-consistent field approximation. Type I theories are used to study, (a) multimode effects, (b) collisions, (c) nonlinear oscillations, and all phenomena which can be expressed in terms of just the first or second moment of the radiation field.

Type II experiments and theories [5,6,7,8,9] are those in which the complete radiation density matrix R (not just the first or second moment) must be determined in order to explain the experiments. The experiments known as photon counting experiments determine the diagonal matrix elements of R, R_n, as a function of the number of photons in the single laser mode. Below threshold R_n is thermal, $R_n \sim e^{-\beta E_n}$, above threshold R_n is sharply peaked about the average number of photons and is nearly a Poisson distribution. Type I theories are contained in type II theories as approximations. Type I theories are justified by the great deal of labor saved in explaining those experiments which do not measure the full radiation density matrix.

Present laser theories are equivalent to the first Born approximation to the Liouville equation where radiation-matter correlations must be retained in the equations of motion. In this paper we present a derivation of the laser kinetic equations without radiation-matter correlations but with essentially exact single particle dynamics. We show that the radiation and matter distribution functions in a quantum mechanical description of a laser satisfy coupled Boltzmann equations which are generalizations of the coupled Boltzmann equations for two interacting dilute gases. The method of derivation consists of the exact calculation of the probability per unit time of the emission or absorption of a photon by a two-level system in the rotating wave approximation. The resultant transition probability per unit time is time dependent. However, when the finite lifetime of the two-level system is introduced in the calculation, we obtain a transition probability per unit time which is independent of time, and which serves as the kernel in the coupled radiation and matter Boltzmann equations. When we eliminate the matter distribution function adiabatically, we obtain a non-linear equation for the radiation distribution function.

The Boltzmann equation steady state distribution function differs from the usual first Born approximation master equation result only in the replacement of the inversion density $\bar{N}$ produced by the pump in the absence of radiation by $\tilde{N}$ the actual inversion density in the presence of radiation.

Section II contains a review of the Boltzmann equation and Section III contains a derivation of the kernel of the Boltzmann equation for a laser. In Section IV we obtain the Boltzmann equation for the radiation distribution function and solve for the steady state. Section V contains a derivation and solution of the matter Boltzmann equation. In Section VI we compare the Born approximation master equation and the Boltzmann equation.

II THE BOLTZMANN EQUATION

We first express the Boltzmann equation for dilute gases in a form that is easily adaptable to lasers. The Boltzmann equation is

$$\begin{aligned}\frac{\partial f(v,t)}{\partial t} &= \int d\sigma(\Omega)\,|v-\bar{v}|\,\{f(v')f(\bar{v}')-f(v)f(\bar{v})\}\,d^3\bar{v}\\ &= \int k[v',\bar{v}'\,|v,\bar{v}]\,\{f(v')f(\bar{v}')-f(v)f(\bar{v})\}\,d\Omega\,d^3\bar{v}\\ &= \int K[\xi,\eta\,|v,\bar{v}]\,\{f(\xi)f(\eta)-f(v)f(\bar{v})\}\,d\Omega\,d^3\bar{v}\,d^3\xi\,d^3\eta \qquad (2.1)\end{aligned}$$

where

$$K[\xi,\eta\,|v,\bar{v}] = \delta^3(\xi-v')\,\delta^3(\eta-\bar{v}')\,k[v',\bar{v}'|\,v,\bar{v}]$$

and $\sigma(\Omega)$ is the cross section for the collision of two particles with initial velocities $(v, \bar{v})$. The final velocitis $(v', \bar{v}')$ are the solutions of the equations of motion for the two particles with initial velocities $(v, \bar{v})$. The probability per unit time of observing a particle with velocity v at time t is $f(v, t)$. The kernel $K[\xi, \eta \mid v, \bar{v}]$ is the time independent probability per unit of time of a collision between two particles with initial velocities $(v, \bar{v})$ which results in the final velocities (ξ, η). The kernel K is exact to all orders in the interaction potential.

We need a slight generalization Eq. (2.1) for the laser problem because the laser has two different kinds of variables, namely, the radiation and the matter variables. The dilute gas analogue to different kinds of variables is a mixture of two different gases *a* and *b*. Atoms of gas *a* interact with atoms of gas *b* but not with other *a* atoms and *b* atoms interact with *a* atoms but not with other *b* atoms. The coupled Boltzmann equations for the *a* atom distribution, $f_a(v_a, t)$, and the *b* atom distribution, $f_b(v_b, t)$, are

$$\frac{\partial f_a(v_a, t)}{\partial t} = \int K[\xi_a, \eta_b \,|\, v_a, v_b] \{f_a(\xi_a, t) f_b(\eta_b, t) - f_a(v_a, t) f_b(v_b, t)\} \, d\Omega \, d^3 v_b \, d^3 \xi_a \, d^3 \eta_b \tag{2.2}$$

$$\frac{\partial f_b(v_b, t)}{\partial t} = \int K[\xi_a, \eta_b \,|\, v_a, v_b] \{f_a(\xi_a, t) f_b(\eta_b, t) - f_a(v_a, t) f_b(v_b, t)\} \, d\Omega \, d^3 v_a \, d^3 \xi_a \, d^3 \eta_b .$$

There are three crucial requirements necessary for the derivation of Eqs. (2.1) and (2.2). The first requirement is that the gas must be of sufficiently low density so that we can neglect three and more particle effects. Explicitly low density means the dimensionless parameter $\varrho \bar{r}^3$ approaches zero where ϱ is the density of the gas and $\bar{r}$ is the force range. The dimensionless parameter $\varrho \bar{r}^3$ is the ratio of the interaction time or collision time $(\bar{r}/\langle v \rangle)$ to the relaxation time $[\varrho \bar{r}^2 \langle v \rangle]^{-1}$. The second requirement is that $K[\xi, \eta \mid v, \bar{v}] = K[v, \bar{v} \mid \xi, \eta]$; i.e., the transition probability is the same for the collision $(v, \bar{v})$ to (ξ, η) as for the inverse collision (ξ, η) to $(v, \bar{v})$. The third requirement is that the two particles with velocities v and $\bar{v}$ are uncorrelated before collision.

In gas lasers and some solid lasers it is justifiable to neglect multiparticle effects because the density of the two-level systems is sufficiently low. In the next section we show the kernel for the laser Boltzmann equation is symmetric. The relationship of the third requirement to lasers is subtle and we discuss it in Section VII.

III BOLTZMANN KERNEL FOR THE LASER

Our model of a laser consists of N two-level systems in interaction with a cavity mode in the presence of dissipation and pumping. We must calculate the exact probability per unit time that a single atom absorbs or emits a photon in order to obtain the Boltzmann equation for our laser model. The probability per unit time that a two-level system in its excited state at $t = t_0$

in the presence of n photons emits a photon at time t is[10]

$$K[n+1, -; t \mid n, +; t_0] = \frac{4\gamma^2\omega_0^2(n+1)}{[\Delta^2 + 2\gamma^2\omega_0^2(n+1)]^{1/2}}$$

$$\sin \tfrac{1}{2}(t - t_0)\, [\Delta^2 + 2\gamma^2\omega_0^2(n+1)]^{1/2} \cos \tfrac{1}{2}(t - t_0)\, [\Delta^2 + 2\gamma^2\omega_0^2(n+1)]^{1/2} \tag{3.1}$$

where + indicates the excited state and − indicates the ground state of our two-level system. The frequency Δ is equal to $\omega_0 - \Omega$ where $\hbar\omega_0$ is the two-level energy difference, Ω is the cavity frequency, and γ is the dimensionless dipole moment of the two-level system defined in Ref. 9. For convenience we are considering modes with periodic boundary conditions, so there is no spatial dependence to the kernel. If we consider the standing wave case, γ is replaced by $\gamma\Gamma(X)$ where $\Gamma(X)$ is the mode eigenfunction of the cavity evaluated at the center of mass of the two-level system. The integration over the collision cylinder, represented by $d\Omega$ in Eq. (2.1), is replaced in the laser case by the integration of the $j \cdot A$ interaction over the volume which in the dipole approximation leads to the constant γ. The dimensionless dipole moment γ can be written[9] as $\gamma^2 = r_0\lambda^2 V^{-1}$ where x_0 is the classical electron radius, $2\pi\lambda$ equals $c\omega_0^{-1}$, and V is the volume of the system.

Equation (3.1) states that there does not exist a transition probability per unit time independent of time for an isolated permanent two-level system in a cavity without dissipation. This result is reasonable because unlike collisions in a dilute gas we do not have two atoms approach each other, interact, and then go to infinity. Instead the two-level system and the radiation are in a superposition of two exact eigenstates of the Hamiltonian of a single two-level system and the interacting radiation and, consequently the probabilities are periodic functions of the time. However, when we add to the model any interactions which destroy the quantum mechanical phase of the wave-function (such as limiting the lifetime of the radiation or of the two-level system), we obtain a transition probability per unit time independent of time. In a laser both the matter and the radiation have lifetimes, but since the matter lifetime is much shorter, it is more important in determining the transition probability.

In a laser the two-level system is a representation of the two laser levels of a multileveled atom. The relevant excited states decay to the ground state of the atom in 10^{-8} to 10^{-7} seconds. Thus our two-level system has only a transient existence and it has a lifetime of the order of 10^{-8} to 10^{-7} seconds. Therefore, we assume that the probability that a two-level system created between t_0 and $t_0 + dt_0$ decays at time $t > t_0$ is given by $\nu \exp[-\nu(t - t_0)]\, dt_0$. When we average Eq. (3.1) over the creation time t_0 from t to $-\infty$, we obtain

$$K[n + 1, - \mid n, +] = \frac{2\nu\gamma^2\omega_0^2(n + 1)}{\nu^2 + 2\gamma^2\omega_0^2(n + 1)} \tag{3.2}$$

where for convenience we have set $\Delta = 0$. Equation (3.2) is independent of time and is symmetric; i.e.,

$$K[n + 1, - \mid n, +] = K[n, + \mid n + 1, -]. \tag{3.3}$$

IV BOLTZMANN EQUATION FOR R

By using the kernel K of Eq. (3.2) we can immediately obtain the form of the coupled Boltzmann equations, Eq. (2.2), for the laser. The Boltzmann equation for the radiation distribution function R is

$$\frac{dR_n}{dt} = N \sum_{n'} \sum_{s,s'} K[n, s \mid n's'] \{R_{n'}(t)\varrho_{s'}(t) - R_n(t)\varrho_s(t)\} + (\mathscr{K}_r R)_{nn} \tag{4.1}$$

where s and s' each take on the two values $+$ and $-$, N is the number of two-level systems, ϱ_+ and ϱ_- are the probabilities that a two-level system is in its excited state or in its ground state respectively. The operator $\mathscr{K}_r$ represents the effect of the radiative dissipative mechanisms on R. The Boltzmann term is independent of the form of $\mathscr{K}_r$. We discuss the equation for the matter density matrix ϱ in the next section. There are only two transitions that give n photons; namely, the $(n + 1, -)$ to $(n, +)$ absorption of a photon and the $(n - 1, +)$ to $(n, -)$ emission of a photon. The inverse of these two transitions provide the two transitions that destroy the n photon state.

When we perform the sums in Eq. (4.1), we obtain

$$\begin{aligned} \frac{dR_n}{dt} &= (n + 1)\, R_{n+1}(t)\, [\nu_r + C_-(n + 1)] + nR_{n-1}(t)\, C_+(n) \\ &\quad - R_n(t)\, [n(\nu_r + C_-(n)) + (n + 1)\, C_+(n + 1)] \end{aligned} \tag{4.2}$$

where

$$(\mathscr{K}_r R)_{nn} \equiv -\nu_r[nR_n(t) - (n + 1)\, R_{n+1}(t)]$$

$$C_\pm(n) = \frac{2\nu\gamma^2\omega_0^2 N\varrho_\pm(t)}{\nu^2 + 2\gamma^2\omega_0^2 n}.$$

We use a zero temperature radiation reservoir although a finite temperature one can be included easily. We take the geometric mean of the lifetimes of the two laser states for the lifetime ν^{-1} of our two-level system. Thus ν^{-1} is

$$\nu^{-1} = (\nu_+\nu_-)^{1/2} \equiv (\nu_1\nu_2)^{-1/2}$$

where $\nu_\pm^{-1}$ are the lifetimes of the excited and ground states of the two-level systems, ν_1^{-1} is the relaxation time of the diagonal matrix elements of ϱ, and ν_2^{-1} is the relaxation time of the off-diagonal matrix elements of the matter density matrix ϱ. With this choice of ν, we obtain Eq. (10.25) of Ref. 2 for the diagonal matrix elements when we set $\varrho_+ = 1$ and $\varrho_- = 0$.

To solve the time dependent Eq. (4.2) we need to know the time dependence of $\varrho_\pm(t)$ which we obtain in the next section. In the steady state $\varrho_\pm$ are numbers and so we can find the steady state solution for R with $\varrho_\pm$ as parameters. The steady state solution of Eq. (4.2) satisfies the following recursion relation

$$R_{n+1} = \frac{C_+(n)}{\nu_r + C_-(n+1)} R_n \tag{4.3}$$

with the formal solution

$$R_n = \prod_{n'=0}^{n'=n} \frac{C_+(n')}{\nu_r + C_-(n'+1)} R_0. \tag{4.4}$$

The distribution function R has two distinct types of solutions[1,5] depending on whether the numerator is less than or greater than the denominator at $n = 0$ in Eq. (4.3). If the numerator is less than the denominator at $n = 0$, then R_n decays monotonically with n and we have a thermal type of solution. If the numerator exceeds the denominator at $n = 0$, R_n first increases, then decreases, and we have a peaked nonthermal distribution for R. The condition that determines the value of n for which R_n is a maximum is $R_{n+1} = R_n$. When this condition is inserted in Eq. (4.3), we obtain

$$\frac{N(\varrho_+ - \varrho_-)}{1 + 2x_0} = \frac{\nu_r(\nu_1\nu_2)^{1/2}}{\gamma^2\omega_0^2} \equiv N_T \tag{4.5}$$

where N_T is the inversion density of two-level systems required to start laser action and is determined by the Schalow–Townes condition, x is defined as $(\gamma^2\omega_0^2 n/\nu_1\nu_2)$, and n_0 is the value of n where the peak occurs. The solution of Eq. (4.5) is

$$n_0 = \frac{\nu_1\nu_2}{2\gamma^2\omega_0^2}\left[\frac{\tilde{N}_s}{N_T} - 1\right] \tag{4.6}$$

where $\tilde{N}_s \equiv N(\varrho_+ - \varrho_-)_s$ and where the subscript s indicates steady state. If we replace $\tilde{N}_s$ by $\overline{N}$, the inversion density of two-level atoms produced by the pump in the absence of radiation, we obtain the expression n_0 for the maximum photon number of the stationary solution of the first Born approximation master equation [5,6,7,8,9].

We use Fleck's technique for finding the best Gaussian fit of Eq. (4.4) centered around n_0. Near threshold this Gaussian fit is exact[5,9,11,12]. By rewriting Eq. (4.4), we obtain

$$R_n = R_{n_0} \prod_{n'=n_0}^{n} \left[\frac{C_+(n')}{\nu_r + C_-(n'+1)} \right]^{\pm 1} \tag{4.7}$$

where the exponent is positive for $n > n_0$ and negative for $n < n_0$. Next we expand the logarithm of R_n in a Taylor series about n_0. The result is

$$R_n = (\pi/2)^{-1/2}\, \sigma^{-1} \exp -(n - n_0)^2/2\sigma^2 \tag{4.8}$$

where $\sigma^2 = \{(n - n_0)^2\}_{\text{avg}}$. The expression for σ^2 is

$$\sigma^2 = \frac{\nu_1 \nu_2}{4\gamma^2 \omega_0^2} \left(\frac{\tilde{N}_s}{N_T} \right) \left[\frac{\tilde{N}_s}{N_T} + 1 \right], \quad \text{and} \quad \frac{\sigma^2}{n_0} = \frac{\dfrac{\tilde{N}_s}{N_T} + 1}{2\left[1 - \dfrac{N_T}{\tilde{N}_s} \right]} \tag{4.9}$$

The expressions for n_0, σ^2, and σ^2/n_0 have the same functional form as the first Born approximation master equation with $\bar{N}$ replaced by $\tilde{N}_s$.

We need the steady state solution for the matter inversion density $\tilde{N}_s = N(\varrho_+ - \varrho_-)_s$ in order to complete our steady state solution for R. In the next section we obtain and solve the equation of motion for $\varrho_+(t) - \varrho_-(t)$.

V BOLTZMANN EQUATION FOR ϱ

The matter Boltzmann equation is

$$\frac{d\varrho_s(t)}{dt} = \sum_{n,n'} \sum_{s'} K[n, s \mid n's'] \{R_{n'}(t)\, \varrho_{s'}(t) - R_n(t)\, \varrho_s(t)\} + (\mathscr{K}_m \varrho)_{ss} \tag{5.1}$$

where $\mathscr{K}_m$ represents the effects of dissipation and pumping on the matter system. When we perform the sums in Eq. (5.1) and introduce a simple combined matter reservoir pump term that preserves the normalization, we obtain

$$\frac{d\varrho_+(t)}{dt} = \{D(n)\}_{\text{avg}}\, \varrho_-(t) - \{D(n+1)\}_{\text{avg}}\, \varrho_+(t) - \nu_p(\varrho_+(t) - 1) \tag{5.2}$$

where

$$D(n) \equiv \frac{2\gamma^2\omega_0^2 \nu n}{\nu^2 + 2\gamma^2\omega_0^2 n} \qquad \{D(n)\}_{\text{avg}} \equiv 2\gamma^2\omega_0^2\nu \sum_n \frac{nR_n(t)}{\nu^2 + 2\gamma^2\omega_0^2 n}$$

and where we assume the pump creates atoms in only the upper state at a rate ν_p. It is consistent with our model to take $\nu_p = \nu = (\nu_1 \nu_2)^{1/2}$ as we do in the remainder of this paper. We do not need a separate equation for ϱ_- since $\varrho_+(t) + \varrho_-(t) = 1$.

Over the operating range of many lasers the matter distribution function changes much more rapidly than the radiation distribution function and so it is common to assume that the matter can be treated adiabatically; i.e., ϱ synchronizes immediately with the instantaneous value of $R(t)$. Consequently, we set $d\varrho_+/dt$ equal to zero in Eq. (5.2), and solve for $\varrho_+(t)$. The result is

$$\varrho_+(t) - \varrho_-(t) = \frac{1}{1 + \left\{\dfrac{D(n)}{\nu}\right\}_{\text{avg}}} = \frac{1}{1 + 2\left\{\dfrac{x}{1 + 2x}\right\}_{\text{avg}}} \tag{5.3}$$

and

$$\varrho_+(t) = \frac{1}{2}\left[1 + \frac{1}{1 + 2\left\{\dfrac{x}{1 + 2x}\right\}_{\text{avg}}}\right]$$

where we have neglected one compared with n. When we substitute Eq. (5.3) into Eq. (4.2), we completely eliminate the matter and obtain a nonlinear, time dependent integral equation for R because ϱ_+ and ϱ_- depend on $\{x(1 + 2x)^{-1}\}_{\text{avg}}$ where the average is over R. The qualitative behavior of the time dependent nonlinear Boltzmann equation and of the time dependent linear Born approximation master equation for R are very similar to each other and as we see in the next section their stationary states are essentially equivalent.

VI COMPARISON OF THE BOLTZMANN AND MASTER EQUATIONS

The Born approximation master equation in the adiabatic approximation has the same functional form as Eq. (4.2) with different values of the coefficients. The master equation is [5,6,7,8,9]

$$\frac{dR_n}{dt} = (n + 1)\, R_{n+1}(t)\, [\nu_r + A_-(n + 1)] + nR_{n-1}(t)\, A_+(n)$$
$$- R_n(t)\, [n(\nu_r + A_-(n)) + (n + 1)\, A_+(n + 1)] \tag{6.1}$$

where

$$A_+(n) \equiv \frac{\gamma^2\omega_0^2\bar{N}}{\nu_2}\left[\frac{1+x}{1+2x}\right]; \quad A_-(n) \equiv \frac{\gamma^2\omega_0^2\bar{N}}{\nu_2}\left[\frac{x}{1+2x}\right].$$

The formal solution of Eq. (6.1) is

$$R_n = \prod_{n'=0}^{n'=n} \frac{A_+(n')}{\nu_r + A_-(n'+1)} R_0 . \tag{6.2}$$

The solution of Eq. (6.2) is peaked for $\bar{N} > N_T$ where $N_T \equiv \nu_2\nu_r(\gamma\omega)^{-2}$, and the peak occurs at $x_0 = (1/2)\,[(\bar{N}/N_T) - 1]$. The best Gaussian fit of Eq. (6.2), and one that becomes exact as $\bar{N} \to N_T$, is

$$R_n = (\pi/2)^{-1/2}\,\sigma^{-1} \exp\,[-(n - n_0)^2/2\sigma^2] \tag{6.3}$$

where

$$n_0 = \frac{\nu_1\nu_2}{2\gamma^2\omega_0^2}\left[\frac{\bar{N}}{N_T} - 1\right]; \quad \frac{\sigma^2}{n_0} = \frac{[1 + (\bar{N}/N_T)]}{2[1 - (N_T/\bar{N})]} \xrightarrow{\bar{N}\to N_T} \frac{1}{1 - (N_T/\bar{N})}.$$

We have a displaced Gaussian fit for both the master equation and the Boltzmann equation. The only difference is that the number density two-level systems produced in the absence of pumping $\bar{N}$ appears in the master equation steady state and the inversion $N_s = N(\varrho_+ - \varrho_-)_s$ appears in the Boltzmann equation steady state. To compare the two solutions further we need to express the quantity $\varrho_+ - \varrho_-$ in terms of the moments of R. We obtain the equation of motion for $\{n\}_{\text{avg}}$ in the Boltzmann equation by multiplying Eq. (4.2) by n and summing over n

$$\frac{d\{n\}_{\text{avg}}}{dt} + \nu_r\{n\}_{\text{avg}} = \nu_r \frac{\tilde{N}}{N_T}\left\{\frac{n}{1+2x}\right\}_{\text{avg}} \tag{6.4}$$

The master equation for $\{n\}_{\text{avg}}$ is

$$\frac{d\{n\}_{\text{avg}}}{dt} + \nu_r\{n\}_{\text{avg}} = \nu_r \frac{\bar{N}}{N_T}\left\{\frac{n}{1+2x}\right\}_{\text{avg}}. \tag{6.5}$$

When we substitute the steady state solution of Eq. (6.4) into Eq. (5.3), we obtain

$$(\varrho_+ - \varrho_-)_s = \frac{1}{1 + 2\dfrac{N_T\{x\}_{\text{avg}}}{N(\varrho_+ - \varrho_-)}} \tag{6.6}$$

which may be written as

$$\varrho_+ - \varrho_- = 1 - 2\frac{N_T}{N}\{x\}_{\text{avg}} \approx 1 - 2\frac{N_T}{N}x_0 = 1 + \frac{N_T}{N} - (\varrho_+ - \varrho_-)$$

with the solution

$$(\varrho_+ - \varrho_-)_s = \frac{1}{2}\left[1 + \frac{N_T}{N}\right]. \tag{6.7}$$

The approximate equality of $\{x\}_{\text{avg}}$ and x_0 is true for small x because the solution of Eq. (6.4) for small x is $\{x\}_{\text{avg}} = (1/2)\,[\tilde{N}/N_T) - 1]$ which is the same as Eq. (4.6) for x_0. As n increases, x increases, the distribution function becomes more sharply peaked, and $\{x\}_{\text{avg}}$ rigorously approaches x_0. When we multiply Eq. (6.7) by N we obtain the expression $(1/2)\,[N + N_T]$ for $\tilde{N}_s$. For our model of pumping all particles are pumped into the upper state so N is equal to $\bar{N}$ and $\tilde{N}_s = (1/2)\,[\bar{N} + N_T]$. Consequently, the only difference between the first Born approximation master equation and the Boltzmann equation in the steady state is to replace $\bar{N}$ by $(1/2)\,[\bar{N} + N_T]$. A similar analysis shows the time dependent behavior of the Born approximation master equation and the Boltzmann equation approach the steady state monotonically with practically the same time constants.

VII DISCUSSION

In the Born approximation master equation the gas atoms of a gas laser can be traeted as permanent two-level systems with a diagonal relaxation time ν_1^{-1} and an off-diagonal relaxation time ν_2^{-1} even though the two-level systems are excited states of an atom whose lifetimes differ by a factor of ten. In the Boltzmann equation a single two-level system oscillates between the two levels instead of making perturbative transitions to and from the two levels. Consequently, the lifetimes of the two atomic levels enter the Boltzmann equation differently than they enter the master equation. The one ad hoc assumption in our heuristic derivation is that the lifetime of our two-level systems is the geometric mean of the lifetimes of the two single levels; i.e., $\nu^{-1} = (\nu_-\nu_+)^{-1/2}$.

The most difficult condition for the validity of the Boltzmann equation to justify is the low density assumption because it requires an investigation on a more fundamental microscopic level, such as the quantum mechanical Liouville equation for all N particles and the radiation. We have not been able to give a rigorous proof to all orders in γ that the Boltzmann term is the largest term for a given power of γ. However we have been able to prove the Boltzmann term dominates the multiparticle terms to order γ^4 for typical laser parameters. The form of the γ^4 result suggests the Boltzmann terms could dominate to all orders in γ.

To show that the multiparticle effects to order γ^4 are negligible compared with the γ^4 terms of the Boltzmann equation we need the kinetic equation to order γ^4 for N two-level systems and a radiation mode in the presence of dissipation and pumping. The γ^4 kinetic equation is

$$\frac{\partial R}{\partial t} + i[R, H_f] + \gamma^2 K_2 R = \gamma^4 K_4 R. \tag{7.1}$$

The expression for $\gamma^2 K_2 R$ is derived and analyzed in Ref. 9.

There are two types of microscopic processes which contribute to order γ^4. The terms proportional to $\gamma^4 N$ represent a second Born approximation of a single atom with the radiation and is identical to the γ^4 term of our Boltzmann equation, Eq. (4.2). The second type of microscopic process consists of the two particle terms which represent the exchange of a photon between two different atoms directly. These terms cannot be represented as an emission and a subsequent absorption. The Boltzmann term is proportional to $N(\varrho_+ - \varrho_-)$ and the two-particle term is proportional to $N^2(\varrho_+ - \varrho_-)^2$.

The full equation for R is too lengthy to present here so we give the following result for the equation of motion for the average number of photons.

$$\frac{d\{n\}_{\text{avg}}}{dt} = -\nu_r\{n\}_{\text{avg}} + \gamma^2\omega_0^2 N\nu^{-1}(\varrho_+ - \varrho_-)\,\{n\}_{\text{avg}}$$

$$-\gamma^4\omega_0^4 N\nu^{-3}(\varrho_+ - \varrho_-)\,\{n^2\}_{\text{avg}} - 2\gamma^4\omega_0^4 N^2\nu^{-1}(\nu_+ + \nu_-)^{-2}(\varrho_+ - \varrho_-)^2\,\{n\}_{\text{avg}}$$

where the ν_r term is the radiation reservoir dissipation, the γ^2 term is the first Born approximation, the $\gamma^4 N$ term is identical to the γ^4 result for $\{n\}_{\text{avg}}$ obtained from Eq. (4.2) and the $\gamma^4 N^2$ term is the two-particle term. Both γ^4 terms are small compared with the $\gamma^2 N$ term because of the smallness of the dimensionless parameter $\gamma^2\omega_0^2 N\nu^{-2}$ which is the ratio of the time of interaction $\tau_{\text{int}} \equiv \nu^{-1}$ to the relaxation time $\tau_{\text{relax}} \equiv (\gamma^2\omega_0^2 N)^{-1}\nu$.

The possibility that the $\gamma^4 N$ Boltzmann term is larger than the $\gamma^4 N^2$ two particle term is at first surprising. However two factors make up for the extra power of N in the two-particle term. First the Boltzmann term depends on $\{n^2\}_{\text{avg}}$ while the two-particle term depends on only the first power $\{n\}_{\text{avg}}$. Second the two terms depend differently on the lifetimes of the two levels of the atom that make up the two level system. The ratio of the two particle term to the Boltzmann term is

$$\frac{2\gamma^4\omega_0^4N^2(\varrho_+ - \varrho_-)^2\{n\}_{\text{avg}}\,\nu^{-1}(\nu_+ + \nu_-)^{-2}}{\gamma^2\omega_0^2N(\varrho_+ - \varrho_-)\{n^2\}_{\text{avg}}\,\nu^{-3}}$$

$$= \frac{2\gamma^2\omega_0^2N(\varrho_+ - \varrho_-)\{n\}_{\text{avg}}(\nu_+ + \nu_-)^{-2}}{\gamma^2\omega_0^2\{n^2\}_{\text{avg}}\,\nu^{-2}} \approx \frac{2\gamma^2\omega_0^2N_T(\nu_+ + \nu_-)^{-2}}{\gamma^2\omega_0^2\{n_{\text{avg}}\}\,\nu^{-2}}$$

$$= \frac{4\nu\nu_r(\nu_+ + \nu_-)^{-2}}{[(N/N_T) - 1]}$$

where we have used the Born approximation steady state $(\varrho_+ - \varrho_-)_s = N_T/N$, $\{n^2\}_{\text{avg}} \approx \{n\}_{\text{avg}}^2$ which is valid because of the peakedness of the distribution and Eq. (6.2) for $\{n\}_{\text{avg}} \approx n_0$. From the general conditions on relaxation time $(\nu_+ + \nu_-)$ must be greater than 2ν. In most lasers $\nu_r \ll (\nu_+ + \nu_-)$. For the He–Ne laser the dimensionless parameter $4\nu\,\nu_r(\nu_+ + \nu_-)^{-2}$ is approximately 10^{-3}. Consequently in the steady state for He–Ne the Boltzmann term dominates the γ^4 two-particle term as long as $[(N/N_T) - 1]$ is greater than 10^{-3}. For solid lasers the Boltzmann term is even more dominate.

We believe the result we proved to order γ^4 holds to all orders because for a given power γ^{2s}, the Boltzmann term has the highest power of n; i.e. $\{n^s\}_{\text{avg}}$ and as we increase the pump power the Boltzmann term will dominate in each order of γ. However we have been unable to derive an explicit value of the pump power at which the Boltzmann term dominates the multiparticle terms for all orders in γ. We have proved that the γ^4N Boltzmann term dominates the γ^4N^2 two-particle term in the steady state. However if the laser starts with $\{n\}$ very small (which is usually the case) then there is a short time initially when the multiparticle effects might dominate the Boltzmann term. In this short time span both γ^4 terms are negligible compared to the dominant γ^2 term and it is unlikely that the two-particle effects are observable even at short times.

The second assumption required for our derivation of the Boltzmann equation is the product ansatz which states that the radiation and matter are uncorrelated before each interaction. One of the most interesting results of the present work is that the Boltzmann equation which does not include radiation-matter correlations gives as coherent a radiation distribution as the first Born approximation master equation with correlations. The reason for this surprising result is that the crucial requirement for obtaining a peaked nonthermal distribution for R (that the coefficients $C_\pm(n)$ in the kinetic equation for R be nonlinear in n) is satisfied by both the Boltzmann equation and the Born approximation master equation. The Born approximation master equation has nonlinear coefficients $C_{\pm}(n)$ as a direct consequence of the radiation-matter correlations. In the Born approximation master equation (with correlations) the $C_{\pm}(n)$ are expressions for the population of the excited (+) and unexcited (−) states of our two-level systems in the presence of n photons where n is the photon variable, not the average number of photons $\{n\}_{\text{avg}}$. The Boltzmann equation obtains the nonlinear coefficients $C_{\pm}(n)$ as a result of the exact solution for a single two-level system interacting with the radiation. Thus the kernel K, Eq. (3.2) depends on the photon variable n even though we have assumed the radiation matter distribution is a product $R\varrho$.

We have been able to show that if the one particle diagrams are dominant to each order in γ then the product assumption is not a necessary assumption. The resultant new generalized Boltzmann equation includes radiation and matter correlations. However the correlated Boltzmann equation does not lead to anything qualitatively new for most accessible pump powers.

Of the present derivations of the kinetic equation for the radiation density matrix the one closest to the Boltzmann approach is the derivation by Scully and Lamb[8]. They take one particle and treat its interaction with the radiation exactly. We can obtain their results by setting ϱ_+ equal to one and ϱ_- equal to zero in our Eq. (4.2). The resultant equation for R is equivalent to their equation for the diagonal matrix elements of R.

The method of derivation in the present paper does not give the offdiagonal matrix elements of the Boltzmann equation. A general kernel[13] that preserves the normalization, hermiticity, and positive definiteness, and whose diagonal elements are given by Eq. (4.2) is

$$\mathscr{L}R = O_+^\dagger R O_+ + O_- R O_-^\dagger - \tfrac{1}{2}\left[R O_+ O_+^\dagger + R O_-^\dagger O_- + O_+ O_+^\dagger R + O_-^\dagger O_- R\right]$$

where

$$O_+^\dagger \equiv b^\dagger \left(\frac{\gamma^2 \omega_0^2 N}{\nu^2 + \gamma^2 \omega_0^2 b^\dagger b} \right) \quad \text{and} \quad O_-^\dagger \equiv \left(\frac{\gamma^2 \omega_0^2 N}{1 + \gamma^2 \omega_0^2 b b^\dagger} \right) b^\dagger .$$

The operators $b^\dagger$ and b are the usual creation and annihilation operators for the electromagnetic field.

References

1. Willis E. Lamb, Jr., *Phys. Rev.* **134**, A1429 (1964).
2. C. R. Willis, *Phys. Rev.* **156**, 320 (1967).
3. R. Picard and C. R. Willis, *Phys. Rev.* **139**, A10 (1965).
4. R. J. Glauber, *Phys. Rev.* **131**, 2766 (1963).

5. J. A. Fleck, Jr., *Phys. Rev.* **149**, 309 (1966), *Phys. Rev.* **149**, 322 (1966), and *Phys. Rev.* **152**, 278 (1966).
6. M. Lax, *Phys. Rev.* **157**, 213 (1967). This reference contains complete references to the author's earlier work on noise and lasers.
7. V. Artz, H. Haken, H. Risken, H. Sauermann, Ch. Schmid, and W. Weidlich, *Zeits. f. Phys.* **197**, 207 (1966). This reference contains complete references to earlier work of Haken and coworkers.
8. M. Scully and W. E. Lamb, *Phys. Rev. Letters* **16**, 853 (1966), *Phys. Rev.* **159**, 208 (1967).
9. C. R. Willis, *Phys. Rev.* **165**, 420 (1968).
10. W. Louisell, *Radiation and Noise in Quantum Electronics*, (McGraw-Hill Book Co., New York 1964) pp. 212–217.
11. H. Risken and H. D. Volmer, *Zeits. f. Phys.* **201**, 323 (1967).
12. M. Lax and R. D. Hempstead, *Phys. Rev.* **161**, 350 (1967).
13. C. R. Willis, *Phys. Rev.* **127**, 1405 (1962).

PAPER C–6

Kinetic theory of a weakly, unstable plasma

CARL OBERMAN

Plasma Physics Laboratory, Princeton University, New Jersey

Abstract

A brief survey of some recent developments in the kinetic theory of (weakly unstable) plasma is given. These include: the recognition of failure of the Bogoliubov ansatz; the necessity of including wave-particle and wave-wave effects as well as particle-particle effects; the importance of including non-analytic dependences on the smallness parameter; and the importance of spatial inhomogeneity on the notions of transport.

PAPER C–7

Kinetic theory of zero-field gain in a gaseous laser

T. F. MORSE, JOHN CIPOLLA

Division of Engineering, Brown University, Providence, Rhode Island

---

Abstract

Since most gaseous lasers operate in the transition regime where the particle mean free path is of the order of the tube diameter, a basic formulation of the problem must include boundary effects as well as the perturbing effects of collisions. Problems of this genre have been standard in the field of rarefied gas dynamics. The physical model is chosen to be a three-level gas contained between two parallel plates. Input radiation resonant with an internal transition in the gas (as in the case of the cesium laser pumped by He light) provides the energy for inversion.

It is the purpose of this investigation to incorporate these radiative effects into a Boltzmann equation and to solve this equation in conjunction with the equation of radiative transfer in order to obtain the spatial dependence of the population inversion (gain) as a function of internal relaxation times, intensity and profile of the input light (optical excitation is considered), and gas pressure.

The results of this study should specifically illustrate the dependence of the radiation induced anisotropies in the velocity distribution function. Further, it is hoped that such a study will enable one to find the pressures at which population inversion will be maximum for a given light input.

PART D

WAVES AND FLUCTUATIONS

Contributors

R. N. Sudan,
Department of Applied Physics and Electrical Engineering
Cornell University
Ithaca, New York

G. Wannier,
University of Oregon
Eugene, Oregon

F. Claro,
University of Oregon
Eugene, Oregon

A. Ron,
Department of Physics
Technion – Israel Institute of Technology
Haifa, Israel

K. Nishikawa,
Kyoto University
Department of Physics
Kyoto, Japan

K. Matsuda,
Department of Physics
Polytechnic Institute of Brooklyn
Brooklyn, New York

A. H. Kritz,
Aeronautical Research Associates of Princeton, Inc.
Hunter College of the City University of New York
Department of Physics and Astronomy
New York, New York

G. V. Ramanathan,
Courant Institute of Mathematical Science
New York University
Washington Square Campus
New York, New York

G. Sandri,
Aeronautical Research Associates of Princeton, Inc.
Hunter College of the City University of New York
Department of Physics and Astronomy
New York, New York

PAPER D–1

*Kinetic theory of waves in hot, low density plasma**

R. N. SUDAN

Laboratory of Plasma Studies
Cornell University
Ithaca, New York

I INTRODUCTION

In many naturally occuring plasmas and also in laboratory experiments directed towards controlled thermo-nuclear fusion, the effect of two-body particle collisions is negligible because the number of particles contained in a Debye sphere is very large[1]. Often, such a hot low density plasma is far from thermodynamic equilibrium due to a variety of causes e.g. anisotropy of the particle velocity distribution, the presence of currents and streams of particles etc. Collisions are ineffective in relaxing such a plasma to the thermodynamic state in any reasonable time scale, because of their rarity. On the other hand, there is a strong likelihood that the free energy of the plasma is released to a set of unstable oscillating modes which, in the linear approximation, grow exponentially from thermal noise levels. When the amplitudes become sufficiently large the unstable modes begin to interact with each other and also with the stable modes of the system. These oscillating modes or waves can be treated as a gas of interacting quasi-particles in **k**-space provided $\gamma_{\mathbf{k}}/\omega_{\mathbf{k}} \ll 1$, where $\gamma_{\mathbf{k}}$ and $\omega_{\mathbf{k}}/2\pi$ are the growth/damping rate and frequency of the mode with wave vector **k**. The nonlinear interactions of these quasi-particles can be described by a kinetic equation, if the spectrum of excited modes is wide enough and the phases of the individual modes are not correlated to each other so that the random phase approximation is justified. The kinetic equation is easily obtained using the

* Work supported by U.S.A.E.C. Contract No. AT(30-1)-4077.

principle of detailed balance. Such a technique is well established in other areas of physics[2] but there are several features peculiar to a hot plasma because of the possibility of wave-particle interactions in addition to wave-wave interactions.

In what follows we shall confine the discussion to electrostatic waves of a low pressure plasma characterized by $4\pi nT/B^2 \ll 1$ where n is the particle number density, T is the plasma kinetic temperature and B is the confining magnetic field. The generalization to other type of waves is fairly straightforward. Section II gives the derivation of the contribution of the wave-wave interactions to the kinetic equation. Section III discusses the wave-particle contribution.

II WAVE-WAVE INTERACTIONS

1 Elementary 3-wave interaction

In order to develop a kinetic equation, we first define a distribution function for the waves. In other words, we define a quantity $N_\mathbf{k}$ that denotes the "number" of waves occupying the state $\mathbf{k}$, by recalling the well known expression for the wave energy $U_\mathbf{k}$[3],

$$U_\mathbf{k} = \frac{k^2|\phi_\mathbf{k}|^2}{8\pi}\,\omega_\mathbf{k}\frac{\partial\varepsilon'}{\partial\omega_\mathbf{k}} \equiv N_\mathbf{k}\omega_\mathbf{k}S_\mathbf{k} \tag{1}$$

$$N_\mathbf{k} \equiv \frac{k^2|\phi_\mathbf{k}|^2}{8\pi}\left|\frac{\partial\varepsilon'}{\partial\omega_\mathbf{k}}\right|; \quad S_\mathbf{k} = \mathrm{sign}\left(\frac{\partial\varepsilon'}{\partial\omega_\mathbf{k}}\right) \tag{2}$$

where the wave is defined by its potential $\phi(\mathbf{x}, t) = Re\,\phi_\mathbf{k} \exp i[\mathbf{k}\cdot\mathbf{x} - (\omega_\mathbf{k} + i\gamma_\mathbf{k})\,t]$ and $\omega_\mathbf{k}$ is one of the solutions of

$$\varepsilon'(\mathbf{k}, \omega_\mathbf{k}) = 0; \quad \gamma_\mathbf{k} = -\varepsilon''(\mathbf{k}, \omega_\mathbf{k})/\frac{\partial\varepsilon'}{\partial\omega_\mathbf{k}} \tag{3}$$

where ε' and ε'' are the real and imaginary parts of the dielectric constant of the plasma. The wave momentum is given by

$$\mathbf{P}_\mathbf{k} = \frac{k^2|\phi_\mathbf{k}|^2}{8\pi}\,\mathbf{k}\frac{\partial\varepsilon'}{\partial\omega_\mathbf{k}} = N_\mathbf{k}\mathbf{k}S_\mathbf{k}. \tag{4}$$

The correspondence principle of quantum mechanics requires that

$$N_\mathbf{k} = h\tilde{N}_\mathbf{k} \tag{5}$$

where $\tilde{N}_\mathbf{k}$ is now a positive integer.

The simplest case of wave interaction is the one that involves three waves **k**, **k**′, **k**″. For this elementary process we write the resonance conditions in the form

$$\mathbf{k} + \mathbf{k}' + \mathbf{k}'' = 0, \tag{6}$$

and

$$\omega + \omega' + \omega'' = 0. \tag{7}$$

We shall omit the subscripts **k**, **k**′ and **k**″ on occasion to simplify the notation. If $\Delta N, \Delta N'$, and $\Delta N''$ are the changes in the wave amplitudes during such an interaction then the conservation of energy and momentum furnishes us with

$$\Delta N\omega S + \Delta N'\omega' S' + \Delta N''\omega'' S'' = 0, \tag{8}$$

$$\Delta N\mathbf{k}S + \Delta N'\mathbf{k}'S'' + \Delta N''\mathbf{k}''S'' = 0. \tag{9}$$

From Eqs. (6) to (9) we can immediately deduce the relation

$$S\Delta N = S'\Delta N' = S''\Delta N''. \tag{10}$$

If all the three waves have positive energy, then for $\omega' > 0$, $\omega'' > 0$ and $\omega < 0$, we must have $S' = S'' = 1$ and $S = -1$. This means that

$$\Delta N' = \Delta N'' = -\Delta N,$$

i.e., a quantum of the **k**-wave is absorbed emitting a quantum each of **k**′ and **k**″ waves. Therefore

$$|\omega| = |\omega'| + |\omega''|.$$

This is an example of the decay instability (Fig. 1). For the case $\omega' > 0$, $\omega'' < 0$ and $|\omega'| > |\omega''|$ with $\omega < 0$ we must have $S = S'' = -1$, $S' = +1$, i.e.,

$$-\Delta N = -\Delta N'' = \Delta N'.$$

This is an example of wave scattering; a quantum each of the **k** and **k**″ waves is absorbed and emitted as the **k**′-wave. In this case

$$|\omega| = |\omega'| - |\omega''|.$$

From Eq. (10) we can derive a remarkable conclusion viz., if

$$S = S' = S'' \tag{11}$$

then

$$\Delta N = \Delta N' = \Delta N'',$$

i.e. all the interacting waves grow[4,5,6,7]. This result can be interpreted in the following terms. Let $\omega' > 0$ and $\omega'' > 0$ so that $\omega < 0$. In view of (11) the sign of wave energy for the **k** wave will be opposite to that of the **k**′ and **k**″ waves. Thus if $U_{\mathbf{k}'} > 0$ and $U_{\mathbf{k}''} > 0$, then $U_{\mathbf{k}} < 0$ which makes the **k** wave

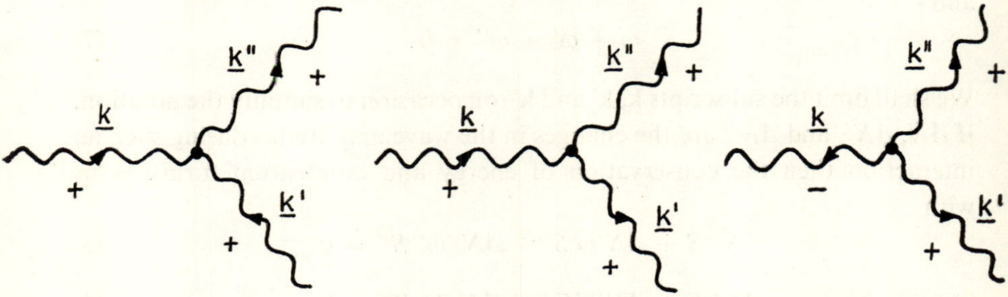

FIG. 1

a negative energy wave. The negative energy wave increases its amplitude by giving up positive energy which is taken up by the other two positive energy waves to increase their amplitudes. Note that (11) is invariant to a Galilean transformation even though wave energy is not invariant. It would appear from the preceding analysis that the waves would increase their amplitudes without limit. However, higher-order effects[8] and more importantly the effect of the waves on particle motion[9] tend to ultimately limit the amplitudes.

The analysis can be easily extended to interactions between more than three waves.

2 Kinetic equation

The principle of detailed balance is expressed as[5,6,10]

$$\frac{d}{dt}\hat{N} = (\text{Emission process} - \text{Absorption process})$$

$$= \sum_{\mathbf{k}',\mathbf{k}''} W_{\mathbf{k},\mathbf{k}',\mathbf{k}''}\{[\hat{N} + \tfrac{1}{2}(1 + \Delta\hat{N})][\hat{N}' + \tfrac{1}{2}(1 + \Delta\hat{N}')]$$

$$\times [\hat{N}'' + \tfrac{1}{2}(1 + \Delta\hat{N}'')] - [\hat{N} + \tfrac{1}{2}(1 - \Delta\hat{N})]$$

$$\times [\hat{N}' + \tfrac{1}{2}(1 - \Delta\hat{N}')][\hat{N}'' + \tfrac{1}{2}(1 - \Delta\hat{N}'')]\}\,\delta_{-\mathbf{k},\mathbf{k}'+\mathbf{k}''}\delta(\omega + \omega' + \omega'') \quad (12)$$

where $W_{\mathbf{k},\mathbf{k}',\mathbf{k}''}$ ir the transition probability of the three-wave interaction, $\Delta\hat{N} = \Delta N/|\Delta N|$, $\hat{N} = N/|\Delta N|$. The factors $\tfrac{1}{2}(1 \pm \Delta\hat{N}) = \pm 1$, result from the Bose–Einstein statistics obeyed by the waves. In view of (10) $\Delta\hat{N}'$ and

$\Delta\hat{N}''$ can be eliminated in favor of $\Delta\hat{N}$. In the classical limit $\hat{N} \gg 1$ we obtain the well known equation[4,5,6,11,12,13]

$$\frac{d}{dt} N_{\mathbf{k}} = \sum_{\mathbf{k}', \mathbf{k}''} |V_{\mathbf{k},\mathbf{k}',\mathbf{k}''}|^2 \, SNN'N'' \left(\frac{S}{N} + \frac{S'}{N'} + \frac{S''}{N''} \right) \times \delta_{-\mathbf{k},\mathbf{k}'+\mathbf{k}''} \, \delta(\omega + \omega' + \omega''). \tag{13}$$

where $|\Delta N| \, W_{\mathbf{k},\mathbf{k}',\mathbf{k}''} \equiv |V_{\mathbf{k},\mathbf{k}',\mathbf{k}''}|^2$. In case a weak space dependence of plasma parameters is allowed then

$$\frac{d}{dt} N_{\mathbf{k}} \rightarrow \left(\frac{\partial}{\partial t} + \frac{\partial \omega_{\mathbf{k}}}{\partial \mathbf{k}} \cdot \frac{\partial}{\partial \mathbf{x}} - \frac{\partial \omega_{\mathbf{k}}}{\partial \mathbf{x}} \cdot \frac{\partial}{\partial \mathbf{k}} \right) N_{\mathbf{k}}(\mathbf{x}, t).$$

A weak damping or growth of the waves can be allowed for by including the term $2\gamma_{\mathbf{k}} N_{\mathbf{k}}$ in the RHS of Eq. (13).

If one defines the entropy of the wave system by

$$H = \sum_{\mathbf{k}} \ln N_{\mathbf{k}}$$

then it is easy to establish from (13) that

$$\frac{d}{dt} H = \sum_{\mathbf{k}} \frac{1}{N_{\mathbf{k}}} \frac{d}{dt} N_{\mathbf{k}}$$

$$= \sum_{\mathbf{k}} 2\gamma_{\mathbf{k}} + \tfrac{1}{3} \sum_{\mathbf{k}, \mathbf{k}', \mathbf{k}''} |V_{\mathbf{k},\mathbf{k}',\mathbf{k}''}|^2 NN'N'' \left(\frac{S}{N} + \frac{S'}{N'} + \frac{S''}{N''} \right)^2 \times \delta_{-\mathbf{k},\mathbf{k}',\mathbf{k}''} \, \delta(\omega + \omega' + \omega'') \tag{14}$$

From (14) we can conclude that equilibrium is only possible if

$$\sum_{\mathbf{k}} \gamma_{\mathbf{k}} \leqq 0.$$

When a system satisfies $\sum_{\mathbf{k}} \gamma_{\mathbf{k}} = 0$, then the equilibrium is determined by

$$\frac{S}{N} + \frac{S'}{N'} + \frac{S''}{N''} = 0 \tag{15}$$

If the system supports a triplet which satisfies (11), then we observe that no equilibrium is possible[4]. If no such triplets exist then one of the S must always be different in sign from the other two, so that (15) becomes

$$\frac{1}{N} = \frac{1}{N'} + \frac{1}{N''}$$

The only solution of this equation keeping the resonance conditions in mind is

$$N_{\mathbf{k}} = \theta/|\omega_{\mathbf{k}} + \mathbf{k}\cdot\mathbf{V}| \tag{16}$$

which is the Rayleigh–Jeans distribution with θ for the wave temperature, viewed from a reference frame moving with velocity $\mathbf{V}$ with respect to the box containing the waves. In practice the wave spectrum rarely approaches the Rayleigh–Jeans distribution because the linear damping and growth of waves become important. Exact stationary solutions for $\Sigma_{\mathbf{k}}\gamma_{\mathbf{k}} < 0$, are very difficult to obtain.

When interacting triplets satisfy (11) over a range of $\mathbf{k}$, comparable to the range of triplets that do not satisfy (11) then it is possible for the entire spectrum to increase in amplitude limited only by effects not included in this calculation. Thus we may have an instability due to nonlinear interaction of linearly stable waves with $\Sigma_{\mathbf{k}}\gamma_{\mathbf{k}} \leqq 0$.

III WAVE-PARTICLE INTERACTIONS

1 Elementary process

The hot plasma differs from other media in the fact that particles can have thermal velocities comparable to the wave phase-velocities. In the linear approximation this gives rise to Landau damping of a wave by particles satisfying the resonant condition,

$$\omega_{\mathbf{k}} - k_{\parallel}V_{\parallel} = n\Omega; \quad n \equiv \text{integer}$$

where $V_{\parallel} = (\mathbf{V}\cdot\mathbf{B})/|B|$, $k_{\parallel} = (\mathbf{k}\cdot\mathbf{B})/|B|$ and $\Omega = \mathrm{e}B/mc$ the particle gyro-frequency. When the nonlinear interaction becomes important we may have a resonant interaction of a particle with the beat wave resulting from two eigenwaves of the plasma. The beat wave need not be an eigenwave. The resonance condition for this case in analogy to the linear situation is

$$\omega' + \omega'' - (\mathbf{k}' + \mathbf{k}'')\cdot\mathbf{V} = n\Omega \tag{17}$$

where of course the frequencies could be of either sign. Conservation of energy and momentum lead to

$$\Delta N'S'\omega' + \Delta N''S''\omega'' + \Delta E_{\perp} + mV_{\parallel}\Delta V_{\parallel} = 0 \tag{18}$$

$$\Delta N'S'\mathbf{k}'_{\parallel} + \Delta N''S''k''_{\parallel} + m\Delta V_{\parallel} = 0 \tag{19}$$

$$\Delta N'S''\mathbf{k}_{\perp} + \Delta N''S''\mathbf{k}''_{\perp} + \Delta\mathbf{p}_{\perp} = 0 \tag{20}$$

where $\Delta N'$ and $\Delta N''$ are the changes in the wave amplitudes during this process, $\Delta E_{\perp}$ is the change in the particle perpendicular energy, $m\Delta V_{\parallel}$ is the change in the particle parallel momentum and $\Delta p_{\perp}$ is the change in the particle perpendicular momentum including any fraction taken up by the magnetic field. From Eq. (17) to (20) we obtain

$$S'\Delta N' = S''\Delta N'' > 0 \tag{21}$$

$$\Delta E_{\perp} \equiv \Delta E_{\perp} + mV_{\parallel}\Delta V_{\parallel} = -S'\Delta N'(\omega' + \omega'') = -S'\Delta N'\left(\frac{U'}{S'N'} + \frac{U''}{S''N''}\right) \tag{22}$$

$$\Delta E_{\perp} = -n\Omega \tag{23}$$

$$\Delta \mathbf{p}_{\perp} = -S'\Delta N'(\mathbf{k}'_{\perp} + \mathbf{k}''_{\perp}) \tag{24}$$

by requiring that $\Delta E_{\perp}/\Omega$ be an integer because of the correspondence principle. The following cases are possible (see Fig. 2):

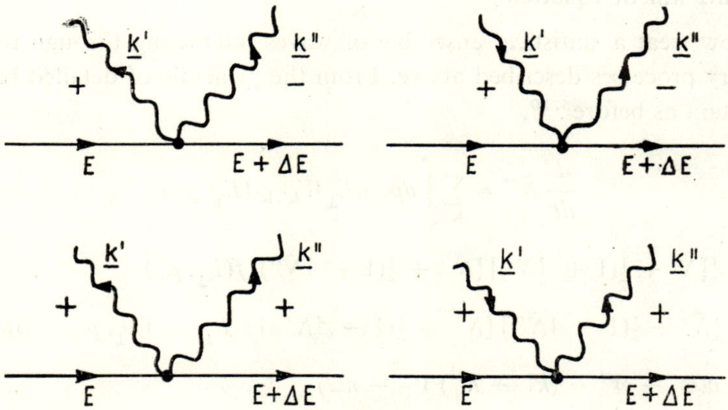

FIG. 2

i) Both waves have positive energy and $S' = S'' = \pm 1$. In this case $\Delta N' = \Delta N''$. A quantum is emitted or absorbed in each of the waves and the particles loses or gains energy of amount $|\Delta N|\,(|\omega'| + |\omega''|)$ according as $S' = +1$ or -1.

ii) Both waves have negative energy and $S' = S'' = \pm 1$. In this case a quantum is emitted or absorbed in each of the waves and the particle also gains or loses energy of amount $|\Delta N|\,(|\omega'| + |\omega''|)$ according as $S' = +1$ or -1.

iii) Both waves have positive energy and $S' = -1$, $S'' = \pm 1$. In this case $-\Delta N' = \Delta N''$. The particle gains or absorbs energy $|\Delta N|(|\omega'| - |\omega''|)$ according as $|\omega'| >$ or $< |\omega''|$. This case describes wave $\mathbf{k}'$ scattering of a particle to produce $\mathbf{k}''$. When both waves have negative energy and $S' = -1$, $S'' = +1$, the particle loses or gains energy according as $|\omega'| <$ or $> |\omega''|$.

iv) Wave $\mathbf{k}'$ has positive energy and $\mathbf{k}''$ has negative energy. When $S' = S'' = 1$ a quantum is emitted in both waves and the particle loses or gains energy $|\Delta N|(|\omega'| - |\omega''|)$ according as $|\omega'| >$ or $< |\omega''|$. When $S' = S'' = -1$ a quantum is absorbed in both waves and the particle gains or loses energy $|\Delta N|(|\omega'| - |\omega''|)$ according as $|\omega'| >$ or $< |\omega''|$. When $S' = -1$ and $S'' = +1$, a quantum is absorbed from the k' wave and emitted into the k'' wave, and the particle loses or gains energy $|\Delta N|(|\omega''| - |\omega'|)$ according as $|\omega'| >$ or $< |\omega''|$.

2 Wave kinetic equation

We now treat a statistical ensemble of waves interacting through the elementary processes described above. From the principle of detailed balance we obtain as before[5,10],

$$\frac{d}{dt}\hat{N}'' = \sum_{\mathbf{k}'}\int dp_\parallel\, dE_\perp W_{\mathbf{k}',\mathbf{k}''}(E_\perp, p_\parallel)$$

$$\times \{[\hat{N}' + \tfrac{1}{2}(1 + \Delta\hat{N}')][\hat{N}'' + \tfrac{1}{2}(1 + \Delta\hat{N}'')]f(E_\perp, p_\parallel)$$

$$- [\hat{N}' + \tfrac{1}{2}(1 - \Delta\hat{N}')][\hat{N}'' + \tfrac{1}{2}(1 - \Delta\hat{N}'')]f(E_\perp + \Delta E_\perp, p_\parallel + \Delta p_\parallel)$$

$$\times \delta(\omega' + \omega'' - (k_\parallel' + k_\parallel'')V_\parallel - n\Omega)$$

In the classical limit, $\Delta E_\perp \ll E_\perp, \Delta p_\parallel \ll p_\parallel, \hat{N} \gg 1$ we obtain after setting $\Delta N'' = S'S''\Delta N'$,

$$\frac{d}{dt}N'' = \sum_{k'}\int dp_\parallel dE_\perp |V_{\mathbf{k}',\mathbf{k}''}(E_\perp, p_\parallel)|^2 \Big\{(S'N'' + S''N')f(E_\perp, p_\parallel)$$

$$+ S''N'N''\left[(\omega' + \omega'' - (k_\parallel' + k_\parallel'')V_\parallel)\frac{\partial f}{\partial E_\perp} + (k_\parallel' + k_\parallel'')\frac{\partial E}{\partial p_\parallel}\right]\Big\}$$

$$\times \delta(\omega' + \omega'' - (k_\parallel' + k_\parallel'')V_\parallel - n\Omega) \tag{25}$$

where

$$\Delta\hat{N}'' W_{\mathbf{k}',\mathbf{k}''}(E_\perp, p_\parallel) = \Delta\hat{N}' W_{\mathbf{k}'',\mathbf{k}'}(E_\perp, p_\parallel) \equiv |V_{\mathbf{k}',\mathbf{k}''}(E_{,\perp} p_\parallel)|^2.$$

Furthermore from (21)

$$S'' \frac{dN''}{dt} = S' \frac{dN'}{dt} \tag{26}$$

The first term of the RHS of (25) is stimulated emission or scattering of waves and the second term constitutes nonlinear Landau damping.

Since the matrix elements for these two processes are identical we need only consider the wave scattering process for computing $V_{\mathbf{k}',\mathbf{k}''}(E_\perp, p_\parallel)$[5,10,14,15]. This process is far simpler to compute than the nonlinear Landau damping which requires going to third order in the conventional method of calculation[16].

As we are operating in the limit $\hat{N} \gg 1$ it might be tempting to neglect the first term in the RHS of (25) with respect to the second term. However, there are certain situations where this term has important physical effects. For example[17], when treating the interaction of a weak electron beam with the tail of an electron thermal distribution it is well known that quasilinear theory (including spontaneous emission terms[18]) describes the flattening of the electron beam and the generation of a narrow wave spectrum in the region where $\partial f/\partial V_\parallel = 0$. The nonlinear coupling of the waves of this primary spectrum interact with particles whose velocities are given by $V_\parallel = (|\omega'| \pm |\omega'|)/(k'_\parallel \pm k^\parallel_\parallel)$. If both $\mathbf{k}'$ and $\mathbf{k}''$ are selected from the primary wave spectrum then $V_\parallel$ falls mostly in the bulk of the electron distribution for wave scattering and nonlinear Landau damping. But if one takes $\mathbf{k}'$ in the primary spectrum and $\mathbf{k}''$ in the tail of the particle distribution, because $N'' \approx 0$, the only term that contributes to dN''/dt is $S''N'f$. Thus, a secondary spectrum can be generated by this term with phase velocities in the tail of the main distribution. It is this process which eventually causes stochastic acceleration of particles by the quasilinear diffusion caused by the secondary spectrum[17].

3 Nonlinear instability

It is clear from (25) and (26) that nonlinear growth of waves occurs when[5,10]

$$S' = S''$$

and

$$S'' \int dp_{\parallel} dE_{\perp} |V|^2 \left\{ [(\omega' + \omega'' - (k'_{\parallel} + k''_{\parallel}) V_{\parallel}] \frac{\partial f}{\partial E_{\perp}} + (k'_{\parallel} + k''_{\parallel}) \frac{\partial f}{\partial p_{\parallel}} \right\}$$

$$\times \delta(\omega' + \omega'' - (k'_{\parallel} + k''_{\parallel}) V_{\parallel} - n\Omega) > 0 \tag{27}$$

When the magnetic field is absent (27) reduces to,

$$S' = S''$$

$$S'' \frac{(k' + k'')}{|k' + k''|} \frac{\partial F}{\partial V_{\parallel}} \bigg|_{V_{\parallel} = \frac{\omega' + \omega''}{k' + k''}} > 0$$

$$F \equiv \int dE_{\perp} |V|^2 f \tag{28}$$

For modes that do not convect along the lines of force of **B** i.e. $k'_{\parallel} = 0 = k''_{\parallel}$, the nonlinear interaction is particularly strong because the entire particle distribution is now resonant with the waves. The instability criterion reduces to

$$S' = S''$$

$$S''(\omega' + \omega'') \int dp_{\parallel} \, dE_{\perp} |V|^2 \frac{\partial f}{\partial E_{\perp}} > 0 \tag{29}$$

$$\omega' + \omega'' = n\Omega .$$

There are four possible cases of this nonlinear instability:

$$(\overline{\partial f / \partial E_{\perp}} \equiv \int dp_{\parallel} \, dE_{\perp} |V|^2 \, \partial f / \partial E_{\perp})$$

1) Two positive energy waves with $|\omega'_{+}| + |\omega''_{+}| = n\Omega$; $\dfrac{\overline{\partial f}}{\partial E_{\perp}} > 0$

2) Two negative energy waves with $|\omega'_{-}| + |\omega''_{-}| = n\Omega$; $\dfrac{\overline{\partial f}}{\partial E_{\perp}} < 0$

3) One negative and one positive energy wave with

$$|\omega'_{+}| - |\omega''_{-}| = |n| \Omega, \frac{\overline{\partial f}}{\partial E_{\perp}} > 0$$

4) One negative and one positive energy wave with

$$|\omega'_{-}| - |\omega''_{+}| = |n| \Omega, \frac{\overline{\partial f}}{\partial E_{\perp}} < 0$$

4 Particle kinetic equation

The evolution of the particle distribution can also be obtained by applying the principle of detailed balance. By summing up the elementary processes represented by the diagrams of Fig. 3 we obtain in the classical limit, $\hat{N} \gg 1, \Delta E_{\perp} \ll E_{\perp}, \Delta p_{\parallel} \ll p_{\parallel}$,

$$\frac{d}{dt} f(E_{\perp}, p_{\parallel}) = \sum_{\mathbf{k}', \mathbf{k}''} \{D|V|^2 N'N''Df - D|V|^2 (S'N'' + S''N')f\} \times \delta(\omega' + \omega'' - (k'_{\parallel} + k''_{\parallel}) V_{\parallel} - n\Omega) \tag{30}$$

where

$$D \equiv (k'_{\parallel} + k''_{\parallel}) \frac{\partial}{\partial p_{\parallel}} + [(\omega' + \omega'') - (k'_{\parallel} + k''_{\parallel}) V_{\parallel}] \frac{\partial}{\partial E_{\perp}}$$

The first term on the RHS of (30) has the structure of a quasilinear diffusion equation in velocity space except that the coefficient of diffusion is propor-

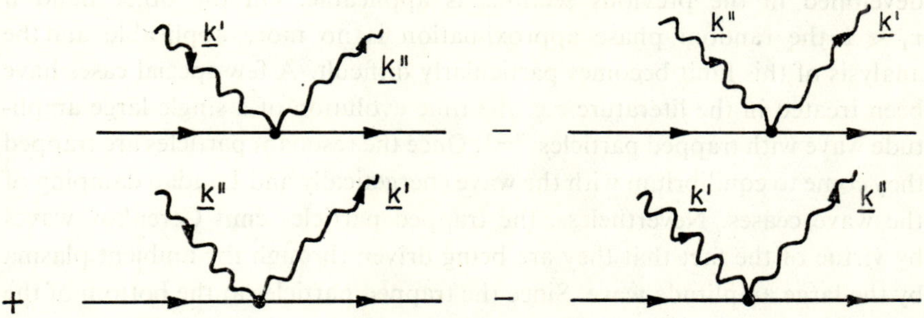

FIG. 3

tional now to $N'N''$. The second term corresponds to a second order frictional term in velocity space, and plays an important role in stochastic acceleration[17].

When the conditions of nonlinear instability whether through wave-particle interactions (25) or through wave-wave interaction (11) are satisfied the diffusion coefficients in (30) increase rapidly. It is clear that the particle distribution would be modified drastically, so that it is possible that the criteria for wave growth are violated at some stage. This would in turn cause wave growth to be arrested. However, an alternative mechanism is available to limit wave growth which is outside the techniques of perturbation theory discussed herein[9,19]. This mechanism involves the effects of the waves on the particles' orbit. In the presence of the wave spectrum the orbit of the particle is no more well defined but indeed is stochastic in nature. In fact the particle diffuses in space and wave damping results because of this diffusion. This damping is dependent on the amplitude of the wave spectrum and can lead to the saturation of nonlinear instabilities[9].

5 Trapped particle effects

A particle travelling initially at the wave phase velocity becomes trapped in the wave potential well in a time approximately given by $\tau_T = (ek^2\phi/m)^{-1/2}$. The time during which the waveform looks roughly stationary to the moving particle i.e. autocorrelation time of the wave spectrum is given by $\tau = 1/\Delta k(V - V_g)$ where Δk is the width of the spectrum, V is the particle velocity and V_g is the group velocity of the wave packet. When $\tau_T \gg \tau$ then the waveform changes before the particle has a chance to be trapped and the random phase approximation together with the perturbation theory developed in the previous sections is applicable. On the other hand if $\tau_T \ll \tau$ the random phase approximation is no more applicable and the analysis of this limit becomes particularly difficult. A few special cases have been treated in the literature e.g. the time evolution of a single large amplitude wave with trapped particles[20,21]. Once the resonant particles are trapped they come to equilibrium with the wave energetically and Landau damping of the wave ceases. Nevertheless, the trapped particles emit Cerenkov waves by virtue of the fact that they are being driven through the ambient plasma by the large amplitude wave. Since the trapped particles at the bottom of the well are isochronous with frequency $\omega_T = 2\pi/\tau_T$ this emission becomes coherent and sidebands at $\omega \approx \omega_0 \pm \omega_T$ are generated. The lower sideband has been observed experimentally[22]. It is possible that the upper sideband is suppressed because it is more heavily Landau damped than the lower sideband. A physically plausible analysis of this situation proceeds as follows[23].

The motion of a particle in the potential well of a large amplitude wave is governed by

$$\ddot{X}_n(t) = -\omega_T^2(X_n - X_{n0} - V_0 t) - \frac{e}{m}\int \frac{dk'}{2\pi}\frac{d\omega'}{2\pi} E(k', \omega') \times \exp i(k'X_n - \omega' t) \tag{31}$$

where V_0 is the phase velocity of the large amplitude wave, $X_n - X_{n0} - V_0 t$ is the position of the nth oscillator relative to the nth trough and $E(k, \omega)$ is the electric field of the emitted wave. Let us define $\zeta_n(t) = X_n(t) - X_{n0} - V_0 t$, then

$$\int dt\, e^{i\omega t}\ddot{\zeta}_n(t) = -\omega^2\zeta_n(\omega) = -\omega_T^2\zeta_n - \frac{e}{m}\int \frac{dk'}{2\pi} E(k', \omega + k'V_0)\, e^{ik'X_{no}} \tag{32}$$

to first order in E. The contribution to the charge density is

$$\varrho_n(X, t) = e\zeta_n(t)\frac{\partial}{\partial X}\delta(X - X_{no} - V_0 t),$$

and summing over all such oscillators

$$\varrho(k, \omega) = ike \sum_n N_n e^{-ik'X_{no}}\zeta_n(\omega - kV_p) \tag{33}$$

where N_n is the number of trapped particles in the nth trough. From Poisson's equation

$$ik\varepsilon(k, \omega)\, E(k, \omega) = 4\pi\varrho(k, \omega) \tag{34}$$

where $\varepsilon(k, \omega)$ is the dielectric function of the background plasma. From (32), (33) and (34) we obtain

$$E(k, \omega) = \frac{\omega_B^2}{\Omega^2 - \omega_T^2}\sum_m \frac{E(k + mk_0, \omega + m\omega_0)}{\varepsilon(k, \omega)} \tag{35}$$

$$\omega_0 = k_0 V_0. \quad \Omega = \omega - kV_0. \quad \omega_B^2 = 4\pi N_T e^2/m$$

where N_T is the mean density of trapped particles. This dispersion relation is solved by recognizing that at $\omega \approx \omega_p$ the two dominant waves are $E(k, \omega)$ and $E(k - 2k_0, \omega - 2\omega_0)$. Neglecting all other terms we obtain

$$1 = \frac{\omega_B^2}{\Omega^2 - \omega_T^2}\left\{\frac{1}{\varepsilon(k, \omega)} + \frac{1}{\varepsilon(k - 2k_0, \omega - 2\omega_0)}\right\} \tag{36}$$

For a warm background plasma $\varepsilon = 1 - \omega_p^2/(\omega^2 - 3k^2V_T^2)$ (V_T is the thermal velocity) it can be shown that (36) has unstable roots which are shifted from $\omega_0 \pm \omega_T$ by a small amount. The upper side band is more heavily damped than the lower side band and consequently is suppressed.

From a quantum mechanical viewpoint one quantum $h\omega_0$ of the large amplitude wave is absorbed to emit one quantum of $\hbar\omega$ of the sideband frequency and one quantum of energy ΔE is delivered to the trapped particle oscillators,

$$\hbar\omega_0 = \hbar\omega + \Delta E. \tag{37}$$

Since the trapped particles are originally in the ground state they can only absorb energy so that from (37) is is clear that only the lower side band $\omega = \omega_0 - \Delta E/\hbar$ can be emitted.

Concluding remarks

Detailed calculations of the matrix elements via the Vlasov equation have been made for many special cases. In particular the electrostatic ion cyclotron modes with $k_{\|} = 0$, of a mirror confined plasma with a loss cone distribution, have been treated in Ref. 5, 6, 10 and shown to suffer from the nonlinear instability discussed in sections II and III. Recent "computer experiments"[24] have verified the three wave nonlinear instability of the electrostatic ion cyclotron modes mentioned above.

The preceding discussion of the kinetic theory of waves in hot plasmas does not pretend to be a complete survey of this rapidly expanding field. Despite much progress in recent years there are still basic problems which have not been resolved satisfactorily e.g. the effect of trapped particles on a wide spectrum of waves. Furthermore, in actual experiments the simplified assumptions of the theory do not always hold, for instance, the number of excited modes in a finite system may be far too few to apply the random phase approximation with any confidence. It has also not been possible to develop the nature of the time asymptotic turbulent spectrum from the kinetic equations, except for some plausible guesses based on dimensional arguments.

Acknowledgements

The bulk of this paper is based on earlier work with B. Coppi and M. N. Rosenbluth, Ref. 5, 6 and 10. The remarks on the trapped particle instability are derived from Ref. 23. The author has benefited from many discussions with Dr. C. T. Dum.

References

1. N. Rostoker and M. N. Rosenbluth, *Phys. Fluids* **3**, 1 (1960).
2. R. E. Peierls, "*The Quantum Theory of Solids*", Oxford University Press, Oxford.
3. See for example T. H. Stix, "*The Theory of Plasma Waves*", McGraw-Hill Book Co., New York, N.Y. (1963).
4. V. M. Dikasov, L. I. Rudakov and D. D. Ryutov, *Zh. Eksperim. i. Teor. Fiz.* **47**, 2266 (1965) trans. Soviet Phys.—*JETP* **21**, 608 (1965).
5. M. N. Rosenbluth, B. Coppi, and R. N. Sudan, Paper: C.N.24/E-13, *Proc. of the 3rd Int. Conf. on Plasma Physics and Controlled Nuclear Fusion Research*, Novosibirsk 1968, I.A.E.A., Vienna.
6. B. Coppi, M. N. Rosenbluth, and R. N. Sudan, *Ann. Phys.* (*N.Y.*) **55**, 207 (1969).
7. R. E. Aamodt and M. L. Sloan, *Phys. Rev. Letters* **19**, 1227 (1967).
8. J. Fukai, S. Krishan, and E. G. Harris, *Phys. Rev. Letters* **23**, 910 (1969).
9. C. T. Dum and R. N. Sudan, *Phys. Rev. Letters* **23**, 1149 (1949).

10. M. N. Rosenbluth, B. Coppi, and R. N. Sudan, *Ann. Phys.* **55**, 248 (1969).
11. M. Camac, A. R. Kantrowitz, M. N. Litvak, R. M. Patrick, and H. E. Petschek, *Nucl. Fusion, Suppl.* Pt. 2, 423 (1962).
12. B. B. Kadomtsev, "*Plasma Turbulence*", Academic Press, London (1965).
13. A. A. Galeev, V. Karpman, and R. Z. Sagdeev, *Nucl. Fusion* **5**, 20 (1965).
14. A. Gailitis and V. N. Tsytovich, *Zh. Eksperim. i. Teor. Fiz.* **47**, 1468 (1964) trans. Soviet Phys.—*JETP* **20**, 987 (1965).
15. L. M. Kovrizhnikh, *Zh. Eksperim. i. Teor. Fiz.* **48**, 114 (1965) *trans. Soviet Phys.—JETP* **21**, 744 (1965).
16. R. E. Aamodt and W. E. Drummond, *Phys. Fluids* **7**, 1816 (1964).
17. C. T. Dum and R. N. Sudan, "*Nonlinear Beam-Plasma Interactions and Stochastic Acceleration*", Laboratory of Plasma Studies, Cornell University, Ithaca, N.Y. Report No. LPS 34.
18. E. G. Harris in "*Advances in Plasma Physics*", John Wiley and Sons, Inc. New York, N.Y. (1969).
19. T. H. Dupree, *Phys. Fluids*, **9**, 1773 (1966).
20. T. M. O'Neill, *Phys. Fluids*, **8**, 2255 (1965).
21. L. M. Altshul and V. I. Karpman, *Soviet Physics—JETP* **22**, 361 (1966).
22. C. B. Wharton, J. H. Malmberg, and T. M. O'Neill, *Phys. Fluids* **11**, 1761 (1958).
23. W. Kruer, J. M. Dawson, and R. N. Sudan, *Phys. Rev. Letters* **23**, 838 (1969).
24. J. A. Byers, in *Proc. of the Third Annual Plasma Simulation Conference*, Calif., Sept. 1969.

PAPER D–2

Relaxation spectrum for phonons

G. WANNIER AND F. CLARO

University of Oregon, Eugene, Oregon

This is a continuation of the work first reported by one of us[1]. The aim is to extend to phonons the analysis of the relaxation rate spectrum which is now well advanced for gases. In this type of work the Boltzmann equation is accepted, and its consequences analyzed. Coherent phenomena are of course lost in such an approach.

We are still studying the simplest model, namely an isotropic sound field without dispersion whose modes are coupled through their cubic anharmonicity. The basic process is three phonon collisions. Because of the assumptions made, the collisions are collinear, and angular relaxation is not possible within this framework. Angular relaxation is thus dependent on some higher order process having a longer relaxation time. For the process under consideration only relaxation in frequency can be investigated. The resultant integral equation is in one dimension only.

Instead of studying departure from a Maxwellian distribution we study here departure from a Bose distribution. Let q be the circular wave number of a particular mode. We then write for the number of quanta $N(q)$ in that mode

$$\frac{1}{N(q)} = \exp\{\beta\hbar cq + \nu(q)\,e^{-\lambda t}\} - 1\,. \tag{1}$$

Here $\nu(q)$ is the function measuring departure from equilibrium and λ the associated decay rate. The other letters have the usual meaning. The advantage of bringing in the departure from equilibrium in this particular way is that the one "relaxation mode" with eigenvalue $\lambda = 0$ takes a simple form. It equals

$$\nu(q) = q\,. \tag{2}$$

It arises from conservation of energy or momentum. Since the process considered does not conserve the particle number and since angular relaxation is inhibited there are no other conserved quantities, and thus no other distribution modes of zero eigenvalue. This particular "mode" acts as a correction to the assumed temperature. There is an orthogonality relation which tells us that the total energy in any of the decaying distribution modes must be zero. The decay of a monochromatic beam can therefore not be a simple exponential process.

We linearize the Boltzmann equation in $\nu(q)$ which results if all possible three phonon processes are handled in the "golden rule" approximation (this brings in irreversibility). The result is an equation of the form

$$\lambda\, \nu(q) = \Gamma(q)\, \nu(q) - \int_0^\infty K(q, q')\, \nu(q')\, dq'. \tag{3}$$

The kernel $K(q, q')$ is compact in all problems considered so far. Because of this $1/\Gamma(q)$ retains at least partly its simple meaning as "phonon life time". By a theorem in operator theory the limit points of the spectrum are the same for two operators which differ by a compact operator. Thus the limit points of the spectrum remain unaltered if the integral term is discarded. Once this is done the solutions of (3) are delta functions $\delta(q - q_0)$, and the associated eigenvalues are $\lambda = \Gamma(q_0)$. This last feature is maintained in the complete Eq. (3). However, the eigenfunctions are more complicated as pointed out earlier. The actual decay of a monochromatic beam must involve complicated transients.

A very tractable form of mathematics results if we assume that the matrix element of all processes compatible with the conservation laws is a constant. This is a regrettable departure from the laws of nature. Even to a rough approximation this quantity is proportional to the product of the square roots of the three wave numbers concerned. This feature makes the damping weaker as the waves get longer. A similar departure from reality is often necessary in the kinetic theory of gases where problems are easily solved under the assumption of a constant mean free time. With the field still practically untouched, the assumption appears useful. Setting

$$\beta hcq = x; \quad \beta hcq' = y \tag{4}$$

and absorbing some constants into λ we get (3) in the form

$$\lambda \nu(x) = \int_{-\infty}^{+\infty} \frac{(x-y)\, y}{x} \frac{\sin h\pi x}{\sin h\pi(x-y) \sin h\pi y} \{\nu(x) - \nu(y) - \nu(x-y)\}\, dy. \tag{5}$$

Here the integral was extended over negative arguments by the stipulation that

$$\nu(-x) = -\nu(x). \tag{6}$$

The two negative terms in the curly bracket are actually identical, and could be rendered as $-2\nu(y)$ in accordance with (3). The chosen form makes (2) trivially a solution. The multiplier of $\nu(x)$ on the right side of (5) equals

$$\Gamma(x) = \tfrac{1}{3}(1 + x^2). \tag{7}$$

We thus know immediately that (5) has one eigenvalue 0, and a continuum extending from 1/3 to $+\infty$.

We shall now show that the problem admits no other eigenvalue, in particular none in the gap between 0 and 1/3. To accomplish this we construct the equation obeyed by the following Fourier transform of $\nu(x)$

$$z(k) = \frac{1}{2\pi}\int_{-\infty}^{+\infty} e^{ikx}\left\{\frac{x}{\sinh \pi x}\,\nu(x)\right\}. \tag{8}$$

We find

$$\frac{d^2z}{dk^2} + \left[3\lambda - 1 + \frac{3}{\cosh^2\left(\frac{k}{2}\right)}\right] z(k) = 0. \tag{9}$$

This is a Schroedinger equation for a shallow potential well, which is explicitly soluble. Because of (6) only odd solutions are admitted. The lowest solution is

$$z_1(k) = \frac{1}{\cosh^3\left(\frac{k}{2}\right)} \sinh\left(\frac{k}{2}\right); \quad \lambda_1 = 0 \tag{10}$$

and the next one

$$z_3(k) = \frac{1}{\cosh^3\left(\frac{k}{2}\right)}\left\{\sinh\left(\frac{k}{2}\right) - \frac{2}{3}\sinh^3\left(\frac{k}{2}\right)\right\}; \quad \lambda_3 = \frac{1}{3}. \tag{11}$$

The other polynomial solutions have the sinh to a degree higher than the third; they are therefore divergent and have to be rejected. The solutions of the continuum have not yet been investigated in detail. They clearly arise for $\lambda \geqq 1/3$, as expected.

The results (10) and (11) constitute proof that our phonon relaxation problem has no discrete relaxation "modes". The solution (10) is simply the solution (2) over again, and (11) is already at the edge of the continuum.

Other discrete solutions there are none. It is remarkable that in spite of its similarity to the hard sphere gas problem the spectral structure is not alike.

An interesting byproduct of this analysis is the solution (11). To our knowledge, this is the first time a continuum solution has been written down for this sort of problem. Writing down these solutions in direct space is known to be very awkward. For the functions still resemble delta functions, but with elaborations arising from their orthogonality to the discrete states. This weird structure apparently does not prevent the functions from having reasonable Fourier transforms. $z_3(k)$ is bounded, but remains finite at infinity with the sign reversed at opposite ends. The weirdness thus takes a form which is easily handled by conventional mathematics. There is a strong possibility that all functions $z(k)$ of the continuum take such a manageable form. For a deeper study of relaxation processes an understanding of the continuum seems essential. Thus the entering wedge achieved here into this field is very hopeful.

References

1. G. H. Wannier, *Bull. Am. Phys. Soc.* **14**, 303 (1969).

PAPER D-3

Kinetic equation methods in quantum fluids: Fermi liquids

AMIRAM RON

Department of Physics Technion—Israel Institute of Technology
Haifa, Israel

I

The theory of low temperature transport phenomena of Fermi liquids was first developed by Abrikosov and Khalatnikov,[1] using Landau's penomenological theory[2]. Later Hone[3] extended their work and calculated the self-diffusion coefficient[4], Recently, due to stimulation by new experimental observations[5], the problem was taken up again by various workers.

The early theories, following Landau's equilibrium studies, were mainly concerned with the asymptotic behavior of the transport coefficients as temperature approaches absolute zero. The phenomenological theory of Fermi liquids is strictly correct only in its asymptotic sense, that is, quantities such as the specific heat, sound velocity, diffusion coefficient, etc., are assumed to exhibit some power series behavior at low temperatures—the Landau theory gives only the leading term of each series (e.g., the specific heat is linear in T, the sound velocity approaches a constant, the diffusion coefficient varies like $1/T^2$ etc.). The theory dos not indicate how to obtain the succeeding terms in the expansions.

The validity of this asymptotic theory is widely accepted both for equilibrium and transport properties. However, the question of when, experimentally, one is in the asymptotic regime is not clearly answered, nor is the related problem of how, theoretically, to obtain the higher order terms. These difficulties are sharply demonstrated, for example, by the specific heat measurements of He^3 below 100 m °K, in which deviations from the asymptotic (linear in T) dependence are observed. Since the degeneracy temperature of He^3 is of the order of a few degrees K°, it is clear that those deviations are due to quasi-particle interactions, which are apparently still important even though the quasi-particle population is quite small. A model calculation by Doniach and Engelsberg[6] indicates that persistent fluctuations (spin fluctuations in their case) can contribute a $T^3 \ln T$ term to the asymptotic γT behavior of the specific heat of Normal Fermi Liquid. A similar result has been obtained by extending the *microscopic* theory of Fermi liquids[7].

There has been a parallel development of low temperature transport theory. Abrikosov and Khalatnikov[1] and Hone[3] have calculated transport coefficients in the asymptotic limit $T \to 0$ using an extended version of the original Landau's quasi-particle theory. Recently, Rice[8] has calculated finite temperature corrections using a model analogous to the one introduced by Doniach and Engelsberg[6]. In Rice's calculations the persistent spin fluctuations are again responsible for the leading corrections, of order T, to the asymptotic limiting forms of the transport coefficients. Subsequently Emery[9] has re-examined the *AK* Boltzman collision term, and was able to demonstrate that the relaxation time τ for quasi-particle collisions may be expanded as[10]

$$\frac{1}{\tau} = aT^2\left(1 + \frac{b}{a}T\right)$$

i.e., the correction to the limiting value of $\dfrac{1}{\tau T^2} \to$ Const is indeed linear in temperature. More recently, Pethnick[11] has presented a variational calculation of the transport coefficients, based on a better ansatz (with respect to *AK*) for the collision integral of Landau's Boltzmann equation. The scattering processes are assumed to be *binary* (as in *AK*) but the quasi-particles are "allowed" to exchange small momenta and energies. Pethnick's work shows that the transport coefficients can be determined in terms of the Landau parameters.

The cornerstone of all transport calculations for Fermi liquids is the *collision integral* of the Boltzmann equation for the quasi-particles. In Landau's papers[2] the collision integral was introduced by making an analogy with a Fermi gas. Without determining the cross section of quasi-particle scatterings, it reads

$$\begin{aligned} I[n_1] = -\int d\tau_2\, d\tau_1'\, d^3p_2'\, w(12;12')\, [n_1 n_2(1-n_1')(1-n_2') \\ -(1-n_1)(1-n_2)n_1'n_2'] \times \delta(\mathbf{p}_1 + \mathbf{p}_2 - \mathbf{p}_1' - \mathbf{p}_2') \\ \times\, \delta(\varepsilon_1 + \varepsilon_2 - \varepsilon_1' - \varepsilon_2')\, \delta_{\sigma_1+\sigma_2,\, \sigma_1'+\sigma_2'} \end{aligned} \qquad (1)$$

where n_i, p_i, ε_i, and σ_i are respectively the distribution function, momentum energy and spin of the quasi-particle, 1 and 2 refer to the incoming particles $1'$ and $2'$ to the outgoing particles. The unknown function $W(1, 2; 1', 2')$ is the transition probability for the $1, 2 \rightarrow 1', 2'$ scattering event. AK[1] used Eq. (1) for their calculations and made the assumption that, for the leading asymptotic terms of the transport coefficients, W may be replaced by f^2, where f is the Landau quasi-particle interaction parameter. Their phase space argument indicates that the most important contributions to the collisions are due to strict forward scatterings and these are given in terms of the Landau parameters. Emery[10] and Pethick[11] have extended this argument to include small angle and energy corrections to the binary transition probability W and were able to account for the deviations of the transport coefficients, from their asymptotic form.

In the present note we describe a method of obtaining the collision integral for Fermi liquids. The Boltzmann collision term [Eq. 1] was previously assumed to be due to binary collisions only, and many quasi-particle correlations were not included. Our approach is basically a correlation approach. First we indicate how the Landau phenomenological description of Fermi liquids in terms of the distribution function, n, and the quasi-particle energy, ε, is generalized to deal with non-ensemble averages of the density matrix and the energy matrix. Landau's kinetic equation is obtained by taking the ensemble average of the equation of motion for the generalized density matrix. However, the collision integral of the quasi-particle kinetic equation is now explicitly given in terms of correlation functions of the quasi-particles. We study the fluctuating components of the density and energy matrices and limit ourselves for clearness of presentation to nearly ferromagnetic Fermi liquids, where only one of Landau's parameters, i.e., the exchange parameter ζ_0, is different from zero. In the kinetic regime, in the Bogoliubov sense, the fluctuating components are expressed in terms of the kinetic ensemble-averaged functions, n and ε, of Landau. The collision integral is then explicitly given by finding the asymptotic behavior of the correlation functions. Our collision integral is given in terms of the Landau parameters, and includes "many body" effects in the dynamics screening of the interaction between quasi-particles. All previous results are obtained (explicitly, in the nearly ferromagnetic case) as special limiting forms of our general result.

II

In Landau's theory of Fermi liquid, the distribution function $n(\mathbf{p}, \mathbf{r}, t)$ and the energy $\varepsilon(\mathbf{p}, \mathbf{r}, t)$ of a quasi-particle play a major role. They essentially replace the many-particle wave functions and Hamiltonian. The distribution function gives the ensemble average value of the density of quasi-particles of momentum $\mathbf{p}$, position $\mathbf{r}$, and time t at some low temperature T. This corresponds to the many-body wave function in the occupation-number representation. The function $\varepsilon(\mathbf{p}, \mathbf{r}, t)$, which replaces the Hamiltonian, gives the ensemble-average value of the energy of a quasi-particle. The total energy of the system measured from its ground state is given by

$$E = \Sigma \varepsilon n, \tag{2}$$

where the summation is over all quasi-particles. The crucial assertion of this phenomenological theory is that ε is a function of n, and for low excitation energies assumes a linear dependence

$$\varepsilon(\mathbf{p}) = \varepsilon^0(p) + \sum_{\mathbf{P}'} f_{\mathbf{pp}'}, n(\mathbf{p}'), \tag{3}$$

where the $f_{pp'}$ are the Landau interaction parameters. We wish to emphasize again, that n and ε are average quantities, and thus one must generalize them in order to include correlations in the phenomenological theory.

To generalize Landau's n and ε, we introduce two quasi-particle (*QP*) operators:

a) the *QP* Density (matrix) *operator* $N_{s_1 s_2}(\mathbf{r}_1, \mathbf{r}_2, t)$ and

b) the *QP* Energy *operator* $E_{s_1, s_2}(\mathbf{r}_1, \mathbf{r}_2; t)$,

where $\mathbf{r}$ and s are the position and spin indices. The ensemble average of N is the usual quantum mechanical density matrix of the quasi-particle states, and its Wigner mixed representation is the Landau distribution function n[12]. The interpretation of $E_{s_1 s_2}(\mathbf{r}_1, \mathbf{r}_2)$ is similar, and E plays the role of the one-particle Hamiltonian; being a functional of N it carries the same statistical interpretation as N.

Using the above definitions, the equation of motion for the density operator is (with $\hbar = 1$)

$$i \frac{\partial}{\partial t} N = [E, N],$$

or explicitly

$$i\frac{\partial}{\partial t}N_{s_1s_2}(\mathbf{r}_1,\mathbf{r}_2;t) = \sum_{s_3}\int d^3r_3\, E_{s_1s_3}(\mathbf{r}_1,\mathbf{r}_2;t)\, N_{s_3s_2}(\mathbf{r}_1,\mathbf{r}_2;t)$$
$$- N_{s_1s_3}(\mathbf{r}_1,\mathbf{r}_3;t)\, E_{s_3s_2}(\mathbf{r}_3,\mathbf{r}_2;t). \tag{4}$$

We now introduce the mixed representation of the operators, e.g.,

$$N_{s_1s_2}(\mathbf{r}_1,\mathbf{r}_2;t) = \sum_p e^{i\mathbf{p}(\mathbf{r}_1-\mathbf{r}_2)}\, N_{s_1s_2}(\mathbf{p},\mathbf{r},t), \tag{5}$$

where $\mathbf{r} = (\mathbf{r}_1 + \mathbf{r}_2)/2$ and $\mathbf{p}$ play the role of position and momentum of the *QP* in analogy with the Wigner representation.

We generalize Landau's ansatz for the energy of low-lying excited states; with explicit spin dependence it reads

$$E_{s_1s_2}(\mathbf{p},\mathbf{r}) = \varepsilon^0(p)\,\delta_{s_1s_2} + \sum_{\mathbf{p}'}\sum_{s_3s_4} f(\mathbf{p},\mathbf{p}')_{s_1s_2;s_3s_4} \times N_{s_3s_4}(\mathbf{p}',\mathbf{r}), \tag{6}$$

wher ε and f have the same interpretation as in the equilibrium Landau theory. Observe that Eq. (6) reduces to Eq. (3) when ensemble averaged. Using spin matrices notation Eq. (6) can be rewritten as

$$E(\mathbf{p},\boldsymbol{\sigma};\mathbf{r}) = \varepsilon^0(p) \times I + \sum_{\mathbf{p}'} Tr_{\sigma'}\, f(\mathbf{p},\boldsymbol{\sigma};\mathbf{p}'\boldsymbol{\sigma}')\, N(\mathbf{p}',\boldsymbol{\sigma}';\mathbf{r}). \tag{6a}$$

We also introduce the Fourier transform in real space

$$N(\mathbf{p},\mathbf{r}) = \sum_{\mathbf{k}} e^{i\mathbf{k}\cdot\mathbf{r}}\, N(\mathbf{p},\mathbf{k}); \tag{7}$$

then Eq. (4) becomes

$$i\frac{\partial}{\partial t}N(\mathbf{p},\mathbf{k}) = \sum_{\mathbf{k}'}\{E[\mathbf{p} + (\mathbf{k}-\mathbf{k}')/2,\mathbf{k}']\, N(\mathbf{p}-\mathbf{k}'/2,\mathbf{k}-\mathbf{k}')$$
$$- N[\mathbf{p} + (\mathbf{k}-\mathbf{k}')/2,\mathbf{k}']\, E(\mathbf{p}-\mathbf{k}'/2,\mathbf{k}-\mathbf{k}')\}, \tag{8}$$

where the time t is suppressed for simplicity, as we deal with all functions at the *same* time.

We now proceed to derive the Landau–Boltzmann equation for the distribution function. We write the operators in terms of their ensemble average; for example

$$N(\mathbf{p},\mathbf{r}) = n(\mathbf{p},\mathbf{r}) + \tilde{N}(\mathbf{p},\mathbf{r}), \tag{9}$$

where $n(\mathbf{p},\mathbf{r})$ is the ensemble average of N, i.e., $\langle N\rangle$, and $\tilde{N}(\mathbf{p},\mathbf{r})$ is the fluctuating part of the density operator with $\langle\tilde{N}\rangle = 0$. It is clear that $n(\mathbf{p},\mathbf{r})$ is nothing but Landau's distribution function for the quasi-particles. The explicit operation of the ensemble average does not have to be indicated, since n and ε are well defined both for equilibrium and nonequilibrium situations. In the same manner

$$E(\mathbf{p}, \mathbf{r}) = \varepsilon(\mathbf{p}, \mathbf{r}) + \tilde{E}(\mathbf{p}, \mathbf{r}) \tag{10}$$

where $\varepsilon(\mathbf{p}, \mathbf{r})$ is the "local" quasi-particle energy in the sense of the Landau theory[4] and again $\langle \tilde{E} \rangle = 0$. Notice that although the ensemble average was taken, n and ε are still matrices in spin space.

Substituting Eqs. (9) and (10) in Eq. (8) and taking the ensemble average of the resulting equation, we obtain

$$\begin{aligned} i\frac{\partial}{\partial t} n(\mathbf{p}, \mathbf{k}) - \sum_{\mathbf{k}'} \{\varepsilon[\mathbf{p} + (\mathbf{k} - \mathbf{k}')/2, \mathbf{k}']\, n(\mathbf{p} - \mathbf{k}', \mathbf{k} - \mathbf{k}') \\ - n[\mathbf{p} + (\mathbf{k} - \mathbf{k}')/2, \mathbf{k}']\, \varepsilon(\mathbf{p} - \mathbf{k}'/2, \mathbf{k} - \mathbf{k}')\} = iI(\mathbf{p}, \mathbf{k}), \end{aligned} \tag{11}$$

where

$$\begin{aligned} I(\mathbf{p}, \mathbf{k}) = \frac{1}{i} \sum_{\mathbf{k}'} \{\langle E[\mathbf{p} + (\mathbf{k} - \mathbf{k}')/2, \mathbf{k}']\, \tilde{N}(\mathbf{p} - \mathbf{k}'/2, \mathbf{k} - \mathbf{k}')\rangle \\ - \langle \tilde{N}[\mathbf{p} + (\mathbf{k} - \mathbf{k}')/2, \mathbf{k}']\, \tilde{E}(\mathbf{p} - \mathbf{k}'/2, \mathbf{k} - \mathbf{k}')\rangle\} \end{aligned} \tag{12}$$

is to be identified with the collision integral. Let us first concentrate on the left hand side of Eq. (11). For simplicity consider the non-magnetic case, i.e., no external magnetic field is applied, and the system has no spontaneous magnetic moment. Thus ε and n are diagonal in the spin space since they are average properties. This is not the case for $\tilde{N}$ and $\tilde{E}$, whose off-diagonal elements can be finite. We expand both ε and n in Eq. (11), e.g.

$$n(\mathbf{p} - \mathbf{k}'/2, \mathbf{k} - \mathbf{k}') \cong n(\mathbf{p}, \mathbf{k} - \mathbf{k}') - \frac{1}{2}\mathbf{k}' \cdot \frac{\partial}{\partial p} n(\mathbf{p}, \mathbf{k} - \mathbf{k}'), \text{ etc.} \tag{13}$$

and obtain

$$\begin{aligned} i\frac{\partial}{\partial t} n(\mathbf{p}, \mathbf{k}) - \sum_{k'} \mathbf{k}' \cdot \frac{\partial \varepsilon(\mathbf{p}, \mathbf{k} - \mathbf{k}')}{\mathbf{p}} n(\mathbf{p}, \mathbf{k}') + \mathbf{k}' \cdot \frac{\partial n(\mathbf{p}, \mathbf{k} - \mathbf{k}')}{\partial \mathbf{p}} \varepsilon(\mathbf{p}, \mathbf{k}') \\ = i\, I(\mathbf{p}, \mathbf{k}), \end{aligned} \tag{14}$$

or, in $\mathbf{r}$ space,

$$\frac{\partial}{\partial t} n(\mathbf{p}, \mathbf{r}) + \frac{\partial \varepsilon(\mathbf{p}, \mathbf{r})}{\partial \mathbf{p}} \cdot \frac{\partial}{\partial \mathbf{r}} n(\mathbf{p}, \mathbf{r}) - \frac{\partial \varepsilon(\mathbf{p}, \mathbf{r})}{\partial \mathbf{r}} \cdot \frac{\partial}{\partial \mathbf{p}} n(\mathbf{p}, \mathbf{r}) = iI(\mathbf{p}, \mathbf{r}). \tag{15}$$

Eq. (15) is the kinetic equation for quasi-particles in Fermi liquids as introduced by Landau. In the more general case, when magnetic fields are present, one obtains

$$\frac{\partial n}{\partial t} + \frac{1}{i}[\varepsilon, n] + \frac{1}{2}\left\{\frac{\partial \varepsilon}{\partial \mathbf{p}}, \frac{\partial n}{\partial \mathbf{r}}\right\} - \frac{1}{2}\left\{\frac{\partial \varepsilon}{\partial \mathbf{r}}, \frac{\partial n}{\partial \mathbf{p}}\right\} = iI \tag{16}$$

where $\{A, B\} = AB + BA$ is the anticommutator. Eq. (16) is Silin's[12] generalization of Landau's kinetic equation.

For homogeneous system we rewrite Eq. (12) as

$$\begin{aligned} I(\mathbf{p}) = \frac{1}{i}\sum_{k} \{&\langle E(\mathbf{p} - \mathbf{k}/2, \mathbf{k})\, \tilde{N}(\mathbf{p} - \mathbf{k}/2, -\mathbf{k})\rangle \\ &- \langle \tilde{N}(\mathbf{p} - \mathbf{k}/2, \mathbf{k})\, \tilde{E}(\mathbf{p} - \mathbf{k}/2, -\mathbf{k})\rangle\}, \end{aligned} \tag{17}$$

that is in terms of the EN correlation function for the homogeneous system.

III

We now return to Eq. (8) and find the equations for the fluctuating parts of N and E. Since we consider only homogeneous ensembles of QP's we may write the Fourier transforms of Eqs. (9) and (10)

$$N(\mathbf{p}, \mathbf{k}) = n(p)\,\delta_{\mathbf{k}_1 0} + \tilde{N}(p, k), \quad E(\mathbf{p}, \mathbf{k}) = \varepsilon(p)\,\delta_{\mathbf{k}, 0} + \tilde{E}(\mathbf{p}, \mathbf{k}), \tag{18}$$

where $n(p)$ and $\varepsilon(p)$ are in general the *time-dependent non-equilibrium* distribution function and energy of the quasi-particles, and are assumed to be diagonal in the spin space. If we substitute Eq. (18) in Eq. (8) and subtract Eq. (11), we readily obtain

$$\begin{aligned} i\frac{\partial}{\partial t}\tilde{N}(\mathbf{p}, \mathbf{k}) &- [\varepsilon(\mathbf{p} + \mathbf{k}/2)\,\tilde{N}(\mathbf{p}, \mathbf{k}) - \tilde{N}(\mathbf{p}, \mathbf{k})\,\varepsilon(\mathbf{p} - \mathbf{k}/2)] \\ &+ [n(\mathbf{p} + \mathbf{k}/2)\,\tilde{E}(p, k) - \tilde{E}(\mathbf{p}, \mathbf{k})\,n(\mathbf{p} - \mathbf{k}/2)] \\ &= i\tilde{J}(p, k), \end{aligned} \tag{19}$$

where $\tilde{J}(\mathbf{p}, \mathbf{k})$ is of the form

$$\tilde{N}\tilde{E} - \langle \tilde{N}\tilde{E}\rangle,$$

i.e., is a higher order fluctuation of the product NE, and can be neglected. Thus, with Eq. (18) we use Eq. (6a) to write

$$\tilde{E}(\mathbf{p}, \mathbf{k}) = \sum_{\mathbf{p}'} Tr_{\sigma'} f(\mathbf{p}, \boldsymbol{\sigma}; \mathbf{p}'\boldsymbol{\sigma}')\,\tilde{N}(\mathbf{p}, \mathbf{k}). \tag{20}$$

Eqs. (19) and (20) now serve as our complete set of equations for the fluctuations.

The program of our calculation is the following. Being interested in the collision integral (Eq. 17), we consider the system in the kinetic regime, i.e. when all internal time processes in the system (e.g. fluctuations) are characterized by a much shorter time scale than that of the distribution function. We thus understand the terms in the collision itegral as being the time asymptotic expressions for correlations between quasi-particles for a given distribution function. In solving Eqs. (19) and (20) for the fluctuations we regard the functions n and ε as independent of time and solve for $\tilde{N}$ and $\tilde{E}$ in the long-time asymptotic limit ($t \to \infty$). In this way we find $\tilde{N}$ and $\tilde{E}$ in terms of the arbitrary n and ε and the initial conditions for $\tilde{N}$ and $\tilde{E}$. In their asymptotic limit the fluctuations follow the slow-time behavior of the distribution function. On the time scale of the fast variations of the fluctuations, the distribution function does not change. Thus, the initial values of the $\langle \tilde{E}\tilde{N} \rangle$ correlations are also determined by the ensemble average values of n and ε. This procedure then yields a collision integral expressed only in terms of the energies and the distribution functions of the quasi-particles of Landau's theory.

We restrict our study here to the simple case of a Fermi liquid with a strong spin-ordering tendency, i.e., when the quasi-particle interactions are *almost* strong enough to render the system ferromagnetic. In the nearly ferromagnetic case one takes

$$f = \zeta_0 \boldsymbol{\sigma} \cdot \boldsymbol{\sigma}' \tag{21}$$

where ζ_0 is a constant equal to the average of $\zeta(\mathbf{p}, \mathbf{p}')$. In this limit Eq. (20) is simplified considerably, and reads

$$\tilde{E}(\mathbf{p}, \mathbf{k}) = \zeta_0 \boldsymbol{\sigma} \cdot \sum_{\mathbf{p}'} Tr_{\sigma'} \tilde{N}(\mathbf{p}, \mathbf{k}) \boldsymbol{\sigma}' \equiv \tilde{E}(\mathbf{k}) \tag{20a}$$

The collision integral, Eq. (17), can be written as

$$I(\mathbf{p}) = 2 \operatorname{Im} \sum_{\mathbf{k}} \langle \tilde{E}(-\mathbf{k})\, \tilde{N}(\mathbf{p} + \mathbf{k}/2, \mathbf{k}) \rangle. \tag{22}$$

In order to find the collision integral expressed in terms of the ensemble averages n and ε (the kinetic functions), we turn to Eq. (19) and indicate explicitly its matrix components. Since the ensemble average functions are diagonal in the spin space, it is convenient to separate Eq. (19) into its diagonal and off-diagonal components. First we rewrite Eq. (19) as

$$i\frac{\partial}{\partial t}\tilde{N}_{\alpha\beta}(\mathbf{p},\mathbf{k},t)-[\varepsilon_\alpha(\mathbf{p}+\mathbf{k}/2)-\varepsilon_\beta(\mathbf{p}-\mathbf{k}/2)]\tilde{N}_{\alpha\beta}(\mathbf{p},\mathbf{k},t)$$
$$+[n_\alpha(\mathbf{p}+\mathbf{k}/2)-n_\beta(\mathbf{p}-\mathbf{k}/2)]\tilde{E}_{\alpha\beta}(\mathbf{k},t)=0. \tag{23}$$

Now, n_α and ε_α are assumed to be time independent (on the time scale of the variation of the fluctuating $\tilde{N}$ and $\tilde{E}$). To find the time asymptotic limit we Laplace transform Eq. (23); and obtain

$$[\omega-\varepsilon_\alpha(\mathbf{p}+\mathbf{k}/2)+\varepsilon_\beta(\mathbf{p}-\mathbf{k}/2)]\tilde{N}_{\alpha\beta}(\mathbf{p},\mathbf{k};\omega)$$
$$+[n_\alpha(\mathbf{p}+\mathbf{k}/2)-n_\beta(\mathbf{p}-\mathbf{k}/2)]\tilde{E}_{\alpha\beta}(\mathbf{k};\omega)=i\overline{N}_{\alpha\beta}(\mathbf{p},\mathbf{k}), \tag{24}$$

where $\overline{N}$ is the initial value of N.

We solve for $\tilde{N}$ and $\tilde{E}$ in terms of $\overline{N}$ and use

$$\langle\overline{N}_{\alpha\beta}(\mathbf{p}_1,-\mathbf{k})\,\overline{N}_{\beta\alpha}(\mathbf{p}_2,\mathbf{k})\rangle=\delta_{p_1p_2}\,n_\beta(\mathbf{p}_1+\mathbf{k}/2)\,(1-n_\alpha(\mathbf{p}_1-\mathbf{k}/2)] \tag{25}$$

Substituting the results into Eq. (22) we obtain the collision integral (for spin up, say)

$$\left.\frac{\partial n_\uparrow(p)}{\partial t}\right|_{\text{coll.}}=[I_\uparrow^{(1)}(p)+I_\uparrow^{(2)}(p)] \tag{26}$$

$$I_\uparrow^{(1)}(\mathbf{p})=2\pi\sum_{\mathbf{k}}\sum_{\mathbf{p}'\sigma'}\left|\frac{\zeta_0}{\chi^+[\mathbf{k},\varepsilon_\uparrow(\mathbf{p}+\mathbf{k})-\varepsilon_\uparrow(\mathbf{p})]}\right|^2\times\{n_\uparrow(\mathbf{p})\,[1-n_\uparrow(\mathbf{p}+\mathbf{k})]$$
$$\times\, n_{\sigma'}(\mathbf{p}'+\mathbf{k})\,[1-n_{\sigma'}(\mathbf{p}')]-n_\uparrow(\mathbf{p}+\mathbf{k})\,[1-n_\uparrow(\mathbf{p})]$$
$$\times\, n_{\sigma'}(\mathbf{p}')\,[1-n_{\sigma'}(\mathbf{p}'+\mathbf{k})]\}\,\delta[\varepsilon_\sigma(\mathbf{p}'+\mathbf{k})-\varepsilon_{\sigma'}(\mathbf{p}')$$
$$-\,\varepsilon_\uparrow(\mathbf{p}+\mathbf{k})+\varepsilon_\uparrow(p)]. \tag{26a}$$

$$I_\uparrow^{(2)}(\mathbf{p})=2\pi\sum_{\mathbf{k}}\sum_{\mathbf{p}'}\left|\frac{2\zeta_0}{\chi^+_{\downarrow\uparrow}[k,\varepsilon_\downarrow(\mathbf{p}+\mathbf{k})-\varepsilon_\uparrow(\mathbf{p})]}\right|^2\times\{n_\uparrow(\mathbf{p})\,[1-n_\downarrow(\mathbf{p}+\mathbf{k})]$$
$$\times\, n_\downarrow(\mathbf{p}'+\mathbf{k})\,[1-n_\uparrow(\mathbf{p}')]-n_\downarrow(\mathbf{p}+\mathbf{k})\,[1-n(\mathbf{p})]\,n(\mathbf{p}')\,[1-n(\mathbf{p}'+\mathbf{k})]\}$$
$$\times\,\delta[\varepsilon_\downarrow(\mathbf{p}'+\mathbf{k})-\varepsilon_\uparrow(\mathbf{p}')-\varepsilon_\downarrow(\mathbf{p}+\mathbf{k})+\varepsilon_\uparrow(\mathbf{p})]. \tag{26b}$$

A similar result can be obtained easily for $I_\downarrow(\mathbf{p})$ by interchanging the "up" and "down" subscripts.

Eqs. (26) for the collision integral are our final results for a nearly ferromagnetic Fermi liquid. The method can be extended to the general Fermi liquid system by following the arguments discussed above. Although this is straighforward, the resulting expressions are much more complicated.

IV

We are now in a position to interpret our result for the collision integral. Although we have obtained the collision integral only for the case of a nearly ferromagnetic Fermi liquid, we maintain that in general the collision integral has the same features as Eqs. (26). The collision integral obtained in the present work has the same general structure as Eq. (1): (i) The Fermion-like distribution functions of the quasi-particles appear explicitly in the order indicated. (ii) The conservation of energy δ-function carries in our theory the proper quasi-particle energy $\varepsilon_\sigma(p)$[13]. Notice that this comes about naturally and is not assumed from the outset. Our main result for the transition probability W is that it is expressed, completely, in terms of the Landau parameters—in our special case only ζ_0 appears. Furthermore, another important feature of the collision integral is the manifestation of dynamic screening as indicated by the susceptibilities in the denominator. While, usually the quasi-particle scattering is assumed to be binary in nature, the screening denominators of Eqs. (26) introduce many-body effects in the form of "quasi-particle clouds" in the vicinity of the colliding quasi-particles.

References

1. A. A. Abrikosov and I. M. Khalatnikov, *Rept. Prog. Phys.* **22**, 329 (1959), [hereafter denoted by AK].
2. L. D. Landau, *J. Exp. Theor. Phys.* **30**, 1058 (1956), *ibid* **32**, 59 (1957). [English translation: *Sovt. Phys.-JETP* **3**, 920 (1957) and **5**, 101 (1957).]
3. D. Hone, *Phys. Rev.* **121**, 669 (1961).
4. For an extensive review of Landau's theory see e.g., D. Pines and P. Nozieres, *The Theory of Quantum Liquids* (W. A. Benjamin, Inc. N.Y., 1966).
5. See e.g., J. C. Wheatley in *Quantum Fluids*, edited by D. F. Brewer (North Holland Pub. Co., Amsterdam, 1966).
6. S. Doniach and S. Engelsberg, *Phys. Rec. Lett.* **17**, 750 (1966).
7. D. J. Amit, J. W. Kane, and H. Wagner, *Phys. Rev.* **175**, 326 (1968), and papers cited there.
8. M. J. Rice, *Phys. Rev.* **159**, 153 (1967), and **162**, 189 (1967).
9. V. J. Emery, *Phys. Rev.* **170**, 205 (1968).
10. The coefficients $a_\pm$ and b for different transport processes were calculated by Brooker and Sykes, *Phys. Rev. Lett.* **21**, 279 (1968). V. J. Emery and C. Cheng, *Phys. Rev. Lett.* **21**, 533 (1968) and others.
11. C. J. Pethick, *Phys. Rev.* **177**, 391 (1969). See also C. J. Pethick, *Phys. Lett.* **27A**, 219 (1968) and K. S. Dy and C. J. Pethick, *Phys. Rev. Lett.* **21**, 876 (1968).
12. See V. P. Silin, *J. Exper. Theor. Phys.* **33**, 1227 (1957) [Soviet Phys.-JETP **6**, 945 (1958)].
13. See Ref. (4) page (58) for a discussion of this point.

PAPER D-4

Wave-wave interactions due to scattering by electrons

KYOJI NISHIKAWA†

Jet Propulsion Laboratory, California Institute of Technology Pasadena, California

---

I INTRODUCTION

In this report we discuss the kinetic wave equation which describes the nonlinear wave-particle interaction in a plasma. The kinetic equations describing a quasilinear wave-particle interaction have been studied by many authors[1-4] and their validity is now well established. This interaction, although in most cases governs the fastest stabilization process in an unstable plasma, is not always sufficient to determine the ultimate turbulent stationary state[5]. In addition, interest is growing in problems, such as the conversion of energy from one type of waves to another[6], which requires a theory that goes beyond the quasilinear approximation. For these problems one needs a theory in which higher order mode-coupling processes are taken into account. Out of the various possible mode-coupling interactions we single out for consideration here only the wave-wave interactions due to scattering by electrons (hereafter to be referred to as the nonlinear wave-particle interaction). This is one of the lowest order mode-coupling interactions and its importance has been pointed out in Refs. 5 and 6.

There have been a number of attempts to obtain a kinetic wave equation suitable for the description of nonlinear wave-particle interactions[5-12].

---

† National Research Council Resident Research Associate, 1968, on leave of absence from the Department of Physics, Faculty of Science, Kyoto University, Kyoto, Japan.

Typical in these methods is the use of the picture of quantized plasmons interacting with particles. In Section II we briefly describe this method. In this method (and in our opinion, in all the others as well) an assumption is made of the Markoffian character of the equation. In the case of *nonlinear* kinetic equations the validity of this assumption has to be carefully examined. We do this in Section III and show that the Markoffian assumption can be justified at least when the plasma is close to a stable stationary state. Restricting ourselves to this case, we derive an explicit expression for the kinetic equation in Section IV. There we use the diagram method developed in Ref. 13. Finally a brief discussion is given in Section V on the application of the result to the case of interaction between an electron-wave and an ion-wave.

II SEMI-QUANTUM MECHANICAL DERIVATION

The method we describe here is a natural extension of the method used by Pines and Schrieffer[14] to derive the quasilinear kinetic equations. A more detailed description of the method can be found in Refs. 9 and 10.

Consider a system which consists of plasmons and electrons. Let $N_{\mathbf{k}}(t)$ and $F_{\mathbf{p}}(t)$ be the time-dependent distributions of plasmons with wavenumber $\mathbf{k}$ and frequency $\omega_{\mathbf{k}}$ and electrons with wave-number $\mathbf{p}$ and energy $E_{\mathbf{p}}$. The rate equation for $N_{\mathbf{k}}(t)$ which takes into account the nonlinear wave-particle interaction can be written, using the folden rule of quantum theory[15], as

$$
\begin{aligned}
\frac{\partial}{\partial t} N_k(t) = \sum\sum \{ & W^{\delta}_{\mathbf{k},\mathbf{k}',p}\,\delta[E_{\mathbf{p}+\mathbf{k}-\mathbf{k}'} - E_{\mathbf{p}} - \hbar(\omega_{\mathbf{k}} - \omega_{\mathbf{k}'})] \\
& \times [(N_{\mathbf{k}}(t) + 1)\, N_{\mathbf{k}'}(t)\, F_{\mathbf{p}+\mathbf{k}-\mathbf{k}'}(t)\,(1 - F_{\mathbf{p}}(t)) \\
& - N_{\mathbf{k}}(t)\,(N_{\mathbf{k}'}(t) + 1)\, F_{\mathbf{p}}(t)\,(1 - F_{\mathbf{p}+\mathbf{k}-\mathbf{k}'}(t))] \\
& + W^{T}_{\mathbf{k},\mathbf{k}',p}\,\delta[E_{\mathbf{p}+\mathbf{k}+\mathbf{k}'} - E_{\mathbf{p}} - \hbar(\omega_{\mathbf{k}} + \omega_{\mathbf{k}'})] \\
& \times [(N_{\mathbf{k}}(t) + 1)\,(N_{\mathbf{k}'}(t) + 1)\, F_{\mathbf{p}+k+k'}(t)\,(1 - F_{\mathbf{p}}(t)) \\
& - N_{\mathbf{k}}(t)\, N_{\mathbf{k}'}(t)\, F_{\mathbf{p}}(t)\,(1 - F_{\mathbf{p}+\mathbf{k}+\mathbf{k}'}(t))]\}.
\end{aligned} \tag{1}
$$

The first term in the sum describes the pure scattering of a plasmon between the two states, $\mathbf{k}$ and $\mathbf{k}'$, and the second term the two-plasmon absorption and emission processes. The coefficients $W^{S}_{\mathbf{kk'p}}$ and $W^{T}_{\mathbf{kk'p}}$ which characterize the two interactions are usually positive definite[16]. We go from a quantum mechanical to a classical description by means of the limit:

$$\hbar\mathbf{k} \to 0, \quad \hbar\mathbf{p} \to m\mathbf{v}, \quad \hbar\omega_{\mathbf{k}} N_{\mathbf{k}}(t) \to U_{\mathbf{k}}(t),$$

$$F_{\mathbf{p}}(t) \to \left(\frac{h}{m}\right)^3 F(\mathbf{v}, t), \quad E_{\mathbf{p}+\mathbf{k}} - E_{\mathbf{p}} \to \hbar\mathbf{k} \cdot \mathbf{v},$$

$$F_{\mathbf{p}+\mathbf{k}}(t) - F_{\mathbf{p}}(t) \to \left(\frac{h}{m}\right)^3 \frac{\hbar\mathbf{k}}{m} \cdot \frac{\partial}{\partial \mathbf{v}} F(\mathbf{v}, t), \tag{2}$$

where $U_{\mathbf{k}}(t)$ and $F(\mathbf{v}, t)$ are the wave energy and electron distribution function. Eq. (1) can then be reduced to the form

$$\frac{\partial}{\partial t}\left(\frac{U_{\mathbf{k}}(t)}{\omega_{\mathbf{k}}}\right) = \sum_{\mathbf{k}'} \int d\mathbf{v} \sum_{\sigma = \pm 1} W^{\sigma}(\mathbf{k}, \mathbf{k}', \mathbf{v})\, \delta[\omega_{\mathbf{k}} - \sigma\omega_{\mathbf{k}'} - (\mathbf{k} - \mathbf{k}') \cdot \mathbf{v}]$$

$$\left\{\frac{U_{\mathbf{k}}(t)}{\omega_{\mathbf{k}}} \frac{U_{\mathbf{k}'}(t)}{\omega_{\mathbf{k}'}} \frac{(\mathbf{k} - \mathbf{k}')}{m} \cdot \frac{\partial}{\partial \mathbf{v}} F(\mathbf{v}, t) + \left(\frac{U_{\mathbf{k}'}(t)}{\omega_{\mathbf{k}'}} - \frac{U_{\mathbf{k}}(t)}{\sigma\omega_{\mathbf{k}}}\right) F(\mathbf{v}, t)\right\}, \tag{3}$$

where $\sigma = 1$ corresponds to the pure scattering case and $\sigma = -1$ to the two-wave absorption and emission case. The first term in the bracket comes from the induced process alone and the second from the coupling between this process with the spontaneous emission process. The second term is often omitted[5,7,8], since it is linear with respect to the wave energy. However, this term is necessary in order that the detailed balancing be satisfied in equilibrium.

The characteristic feature of Eq. (3) is that it describes a Markoffian process. In (3) only one time t appears, so that the process described retains no explicit memory of past states. This assumption was incorporated at the very beginning when we wrote down the quantum mechanical rate Eq. (1).

III DISCUSSION OF THE MARKOFFIAN APPROXIMATION

In order to examine the validity of the Markoffian approximation, we shall avoid the quantum mechanical picture altogether, and derive the kinetic wave equation on a purely classical basis. For simplicity, we assume that $F(\mathbf{v}, t)$ is independent of time and equal to $F(\mathbf{v})$. This may be justified for an open system[13]. We also assume that the wave energy varies only slowly with time.

We start from the standard definition of the wave energy in a dispersive medium[17]:

$$U_{\mathbf{k}}(t) = \frac{1}{4\pi}\left[\frac{\partial}{\partial\omega}\{\omega\varepsilon(\mathbf{k},\omega)\}_{\omega=\omega_{\mathbf{k}}}\right]\langle|\mathscr{P}_{\omega_{\mathbf{k}}}\mathbf{E}_{\mathbf{k}}(t)|^2\rangle$$

$$\fallingdotseq \frac{\mathbf{k}^2}{4\pi}\sigma_{\mathbf{k}}\omega_{\mathbf{k}}\left\{\mathscr{P}_{\omega_{\mathbf{k}}}\int\frac{dp}{2\pi i}e^{pt}\right\}\left\{\mathscr{P}_{-\omega_{\mathbf{k}}}\int\frac{dp'}{2\pi i}e^{p't}\right\}\langle\phi_{\mathbf{k}p}\phi_{-\mathbf{k}p'}\rangle, \tag{4}$$

where $\varepsilon(\mathbf{k}, \omega)$ is the dielectric function, approximated as

$$\varepsilon(\mathbf{k},\omega) \fallingdotseq \sigma_{\mathbf{k}}(\omega-\omega_{\mathbf{k}}) \tag{5}$$

for $(\omega - \omega_{\mathbf{k}})$ small. The symbol P_ω denotes the projection operator which picks out the contribution of the pole at $\mathrm{Im}p = -\omega$, and $\phi_{\mathbf{k}p}$ is the Fourier–Laplace transform of the electrostatic potential:

$$-i\mathbf{k}\,\phi_{\mathbf{k}p} = \mathbf{E}_{\mathbf{k}p} = \int_0^\infty dt\,e^{-pt}\,\mathbf{E}_{\mathbf{k}}(t). \tag{6}$$

The angular bracket $\langle\ \rangle$ denotes the average over initial ensemble of systems.

The basis of our method is to calculate $\langle\phi_{\mathbf{k}p}\phi_{-\mathbf{k}p'}\rangle$ from an appropriate iterative solution of the Klimontovich equation. For the purpose of the present argument, however, we need not appeal to the details of the solution. We need only note that there exist two types of poles which give a significant contribution to $U_{\mathbf{k}}(t)$. The first type of poles satisfy $p + p' = 0$, and in a stable plasma their contribution gives the stationary value of $U_{\mathbf{k}}(t)$, which we designate by $\overline{U}_{\mathbf{k}}$. The other type of poles are those for which

$$p = -i\omega_{\mathbf{k}} - \gamma_{\mathbf{k}}, \quad p' = i\omega_{\mathbf{k}} - \gamma_{\mathbf{k}}, \tag{7}$$

where $\gamma_{\mathbf{k}}$ is the damping constant of the wave, and the "damping constant" includes the effect of both linear and nonlinear contributions. In (4) this pole contributes a factor $\exp\{-2\gamma_{\mathbf{k}}t\}$. Summing contributions of both types of poles and differentiating the result with respect to time, we obtain from (4)

$$\frac{\partial}{\partial t}\{U_{\mathbf{k}}(t) - \overline{U}_{\mathbf{k}}\} = -2\gamma_{\mathbf{k}}\{U_{\mathbf{k}}(t) - \overline{U}_{\mathbf{k}}\}. \tag{8}$$

We note that the method used to obtain this equation is a natural extension of the method used to derive the quasilinear equation for $U_{\mathbf{k}}(t)$[1,13]. The only difference, in fact, is that $\gamma_{\mathbf{k}}$ and $\overline{U}_{\mathbf{k}}$ are here functions of the spectral wave energies. To derive Eq. (3) we must now express $\gamma_{\mathbf{k}}$ and $\overline{U}_{\mathbf{k}}$ in terms of $U_{\mathbf{k}'}(t)$. The problem is that $U_{\mathbf{k}'}(t)$ is truly a function of time, whereas in arriving at (8) we had to assume $\gamma_{\mathbf{k}}$ and $\bar{U}_{\mathbf{k}}$ to be time-independent. This is the problem always encountered when we want to derive a *nonlinear* kinetic equation. Nevertheless, whenever $\gamma_{\mathbf{k}}$ and $\bar{U}_{\mathbf{k}}$, expressed as functions of $U_{\mathbf{k}'}(t)$, are varying slowly in time compared with $(U_{\mathbf{k}}(t) - \bar{U}_{\mathbf{k}})$, the procedure used to derive Eq. (3) is a consistent one, so that the Markoffian approximation is justified. As an example, consider a plasma close to a stable stationary state, either turbulent or quiescent. In this case $|U_{\mathbf{k}}(t) - \bar{U}_{\mathbf{k}}|$ is small compared with $U_{\mathbf{k}}(t)$ itself, so that the logarithmic time rate of change of $[U_{\mathbf{k}}(t) - \bar{U}_{\mathbf{k}}]$ is fast compared with that of $U_{\mathbf{k}'}(t)$ and so ultimately with that of $\gamma_{\mathbf{k}}$ and $\bar{U}_{\mathbf{k}}$, so that the criterion used above is met and the equation is Markoffian.

The criterion is not met when the plasma is nonlinearly unstable in the sense that $\gamma_{\mathbf{k}} < 0$. When this is true $\bar{U}_{\mathbf{k}}$ and $U_{\mathbf{k}}(t)$ usually have different signs as is seen from the quasilinear equation. Then $|U_{\mathbf{k}}(t) - \bar{U}_{\mathbf{k}}|$ can never be smaller than $U_{\mathbf{k}}(t)$ itself, so that the time scale of $[U_{\mathbf{k}}(t) - \bar{U}_{\mathbf{k}}]$ usually becomes comparable to that of $\gamma_{\mathbf{k}}$ and $\bar{U}_{\mathbf{k}}$. When this occurs the kinetic equation will become non-Markoffian which can be seen by the following argument. According to the general theory by Mori[18], in the absence of a time-dependent external field any physical quantity, say $A(t)$, obeys an equation of the following type:

$$\frac{d}{dt}\{A(t) - \bar{A}\} = \int_0^t ds\, K(s)\, \{A(t-s) - \bar{A}\} + F(t),$$

where $\bar{A}$ is the time invariant part of $A(t)$. The first term on the right-hand side contains all the terms linear in $[A(s) - \bar{A}]\ (t \geqq s \geqq 0)$[19] and $F(t)$ denotes the remaining term which is not of interest here. This equation becomes Markoffian only when $\int_0^t ds\, K(s)$ approaches its limit in a time scale much shorter than that of $[A(s) - \bar{A}]$. When this happens the above equation reduces in form to that of (8) where the term corresponding to $F(t)$ does not appear, part of it being absorbed as the contribution of selected poles, the other part being simply neglected. In fact, if the linear term[19] cannot be observed in this way, this in itself implies that the equation should be non-Markoffian.

As a final remark, we point out that in many derivations of nonlinear kinetic equations the non-Markoffian effect is implicitly neglected by decoupling higher-order correlation functions, writing them as the sum of products of lower-order correlation functions, these being evaluated at an arbitrary time t. However, it is not clear that this type of factorization of the higher-order correlation functions persists in the course of time. In fact, we believe that even if this type of factorization is made at an initial time, the equation subsequently becomes non-Markoffian for a nonlinearly unstable plasma at least.

IV EXPLICIT EXPRESSION FOR KINETIC WAVE EQUATION

Assuming that the plasma is close to a stable stationary state, we will now derive an explicit expression for the kinetic wave equation. Since the calculation is lengthy, we only give it in outline.

We begin by calculating $\gamma_{\mathbf{k}}$ and $\bar{U}_{\mathbf{k}}$. The former can be obtained from the imaginary part of the dielectric function:

$$\gamma_{\mathbf{k}} \doteq \frac{1}{\sigma_{\mathbf{k}}} [\operatorname{Im} \varepsilon(\mathbf{k}, \omega)]_{\omega=\omega_{\mathbf{k}}}. \tag{9}$$

Since we are interested in the nonlinear wave-particle interaction, it is sufficient to calculate only those parts of $\gamma_{\mathbf{k}}$ and $\bar{U}_{\mathbf{k}}$ which contain the factor

$$\delta[\omega_{\mathbf{k}} - \sigma\omega_{\mathbf{k}'} - (\mathbf{k} - \mathbf{k}') \cdot \mathbf{v}]. \tag{10}$$

To do this we shall use the diagram method developed in Ref. 13, and for the definition of diagrams we refer to Ref. 13.

We now calculate $\gamma_{\mathbf{k}}$. The generalized dielectric function $\varepsilon_\tau \equiv \varepsilon(\mathbf{k}, \omega)$ ($\tau = [\mathbf{k}, -i\omega]$) has been obtained by a systematic summation of the diagonal part contribution [see Ref. 13, Eq. (79)],

$$\varepsilon_\tau = 1 - \alpha_{\mathbf{k}} \int d\mathbf{v} \int d\mathbf{v}'\, g_\tau(\mathbf{v}, \mathbf{v}')\, Y_\tau(\mathbf{v}'), \tag{11}$$

where $\alpha_{\mathbf{k}} = i4\pi e^2/(mk^2)$ $g_\tau(\mathbf{v}, \mathbf{v}')$ is a renormalized streaming operator defined as the solution of the integral equation

$$g_\tau(\mathbf{v}, \mathbf{v}') = g_\tau^0(\mathbf{v}) \{\delta[\mathbf{v} - \mathbf{v}'] + \int d\mathbf{v}''\, X_\tau(\mathbf{v}, \mathbf{v}'')\, g_\tau(\mathbf{v}'', \mathbf{v}')\} \tag{12}$$

$g_\tau^0(\mathbf{v})$ being the free streaming operator $(= (p + i\mathbf{k} \cdot \mathbf{v})^{-1})$ and $X_\tau(\mathbf{v}, \mathbf{v}')$ and $Y_\tau(\mathbf{v})$ are the contributions of appropriately defined irreducible diagonal parts. In the lowest order we have

$$X_\tau(\mathbf{v}, \mathbf{v}') = 0, \quad Y_\tau(\mathbf{v}) = Y_\tau^0(\mathbf{v}) \equiv \mathbf{k} \cdot \frac{\partial}{\partial \mathbf{v}} F(\mathbf{v}). \tag{13}$$

Substituting (13) into (11), we obtain the dielectric function ε_τ^0 as given by the linear theory. For the present purpose we need only the next order terms. Denothing these terms by the superscript (1), we obtain

$$\varepsilon_\tau^{(1)} = -\alpha_{\mathbf{k}} \int d\mathbf{v}\, g_\tau^0(\mathbf{v}) \{Y_\tau^{(1)}(\mathbf{v}) + \int d\mathbf{v}'\, X_\tau^{(1)}(\mathbf{v}, \mathbf{v}')\, g_\tau^0(\mathbf{v}')\, Y_\tau^0(\mathbf{v}')\}. \quad (14)$$

The diagrams which contribute to $X_\tau^{(1)}(\mathbf{v}, \mathbf{v}')$ and $Y_\tau^{(1)}(\mathbf{v})$ are shown in Fig. 11 and Fig. 13 of Ref. 13. The entire contribution to $\varepsilon_\tau^{(1)}$ can then be written

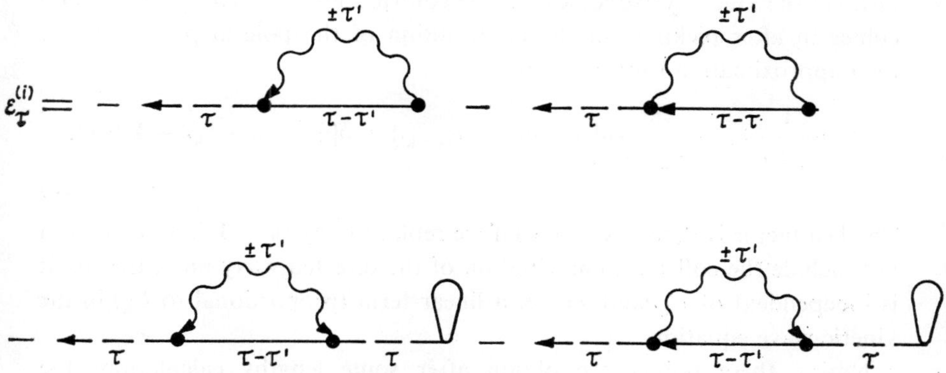

FIG. 1 Diagrams contributing to $\varepsilon_\tau^{(1)}$

diagrammatically in the form of Fig. 1, where we have connected the two wavy lines having the suffixes τ' and $-\tau'$, replacing them by the single wavy line with suffix $\pm\tau'$, and the solid loop denotes the contribution of

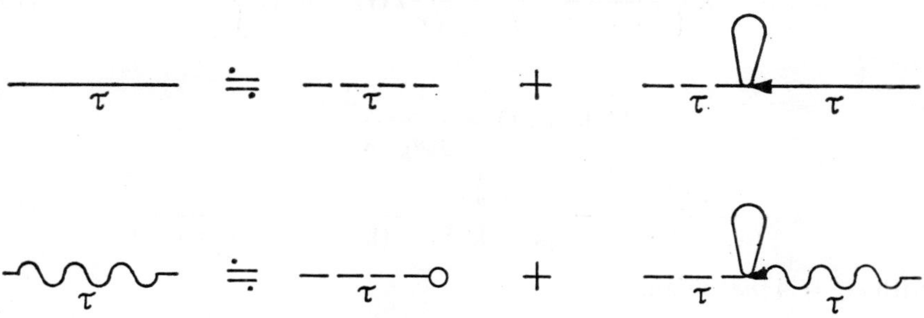

FIG. 2 Schematic representation of the approximation used to calculate the contribution of undirected lines

Y_τ^0. In these diagrams the contribution of undirected lines can be approximated as shown schematically in the diagrams, Fig. 2. We then transform the diagrams in Fig. 1 into mathematical expressions using the prescriptions given in Ref. 13, Table 1.

Now there are two types of terms which contain the δ-function factor (10). One is the contribution from $p' \doteqdot -i\omega_{\mathbf{k}'}$, and the other the contribution from $(p - p') \doteqdot -i\omega_{\mathbf{k}'}$. The contribution from $p' \doteqdot -i\omega_{\mathbf{k}'}$ arises from the collective contribution of all the directed wavy lines with suffixes $\pm\tau'$ (note that a directed wavy line represents the potential). The factor (10) then appears in $\mathrm{Re}\, g^0_{\tau-\tau'}(\mathbf{v})$ and $\mathrm{Im}(^1/\varepsilon_{\tau-\tau'})(\tau = [\mathbf{k}, p], \vartheta mp = -\omega_{\mathbf{k}})$. It is obvious that this contribution results in a nonlinear term (induced process term) in the kinetic wave equation. The contribution from $(p - p') \doteqdot -i\omega_{\mathbf{k}'}$ comes in after picking out the contribution of the pole at $p' = -i\mathbf{k}' \cdot \mathbf{v}$, then approximating $\mathrm{Im}(^1/\varepsilon_{\tau-\tau'})$ as

$$\mathrm{Im}\frac{1}{\varepsilon_{\tau-\tau'}} \doteqdot -\frac{\pi\sigma}{\sigma_{\mathbf{k}-\mathbf{k}'}}\delta[\omega - \omega' - \sigma\omega_{\mathbf{k}-\mathbf{k}'}] \propto \delta[\omega_{\mathbf{k}} - \sigma\omega_{\mathbf{k}-\mathbf{k}'} - \mathbf{k}'\cdot\mathbf{v}]. \tag{15}$$

The last factor is the same as (10) if we replace $\mathbf{k}'$ by $(\mathbf{k} - \mathbf{k}')$. Since we did not include the collective contribution of the directed wavy lines, the result is independent of $U_{\mathbf{k}'}$ and we get a linear term (proportional to $U_{\mathbf{k}}$) in the kinetic wave equation.

Noting these points, we obtain after some lengthy calculations the following expression for $\gamma_{\mathbf{k}}$,

$$\gamma_{\mathbf{k}} = -\frac{1}{2}\sum_{\sigma=\pm 1}\int d\mathbf{v}\sum_{\mathbf{k}'} W^\sigma(\mathbf{k}, \mathbf{k}', \mathbf{v})\,\delta[\omega_{\mathbf{k}} - \sigma\omega_{\mathbf{k}'} - (\mathbf{k} - \mathbf{k}')\cdot\mathbf{v}]$$
$$\times\left\{\frac{U_{\mathbf{k}'}(t)}{\omega_{p'}}\frac{(\mathbf{k} - \mathbf{k}')}{m}\cdot\frac{\partial}{\partial\mathbf{v}}F(\mathbf{v}) - \sigma F(\mathbf{v})\right\} \tag{16}$$

where

$$W^\sigma(\mathbf{k}, \mathbf{k}, \mathbf{v}) = \frac{1}{\sigma_{\mathbf{k}}\sigma_{\mathbf{k}'}}\frac{2\pi\omega_p^4}{k^2k'^2\,n^2}$$
$$\times\left|\frac{\mathbf{k}\cdot\mathbf{k}'}{(\omega_{\mathbf{k}} - \mathbf{k}\cdot\mathbf{v})^2} - \frac{\omega_p^2}{|\mathbf{k} - \mathbf{k}'|^2\,n}\frac{S^\sigma_{\mathbf{k},\mathbf{k}'}}{\varepsilon(\mathbf{k} - \mathbf{k}', \omega_{\mathbf{k}} - \sigma\omega_{\mathbf{k}'})}\right|^2 \tag{17}$$

with $\omega_p^2 = 4\pi ne^2/m$ and

$$S^\sigma_{\mathbf{k},\mathbf{k}'} = -\int d\mathbf{v}\frac{1}{\omega_{\mathbf{k}} - \mathbf{k}\cdot\mathbf{v}}\left\{(\mathbf{k} - \mathbf{k}')\cdot\frac{\partial}{\partial\mathbf{v}}\frac{1}{\sigma\omega_{\mathbf{k}'} - \mathbf{k}'\cdot\mathbf{v}}\mathbf{k}'\cdot\frac{\partial}{\partial\mathbf{v}}\right.$$
$$\left. + \mathbf{k}'\cdot\frac{\partial}{\partial\mathbf{v}}\frac{1}{\omega_{\mathbf{k}} - \sigma\omega_{\mathbf{k}'} - (\mathbf{k} - \mathbf{k}')\cdot\mathbf{v}}(\mathbf{k} - \mathbf{k}')\cdot\frac{\partial}{\partial\mathbf{v}}\right\}F(\mathbf{v}), \tag{18}$$

and we have replaced $\overline{U}_{\mathbf{k}'}$ by $U_{\mathbf{k}'}(t)$, assuming that the plasma is close to a stable stationary state.

Calculation of $\overline{U}_{\mathbf{k}}$ can be carried out in a similar way. The diagrams which contribute to $\overline{U}_{\mathbf{k}}$ can be written in the form of Fig. 3. The fist diagram gives the quasilinear approximation to the spontaneous emission term in the kinetic wave equation. Corrections arise from the next two diagrams.

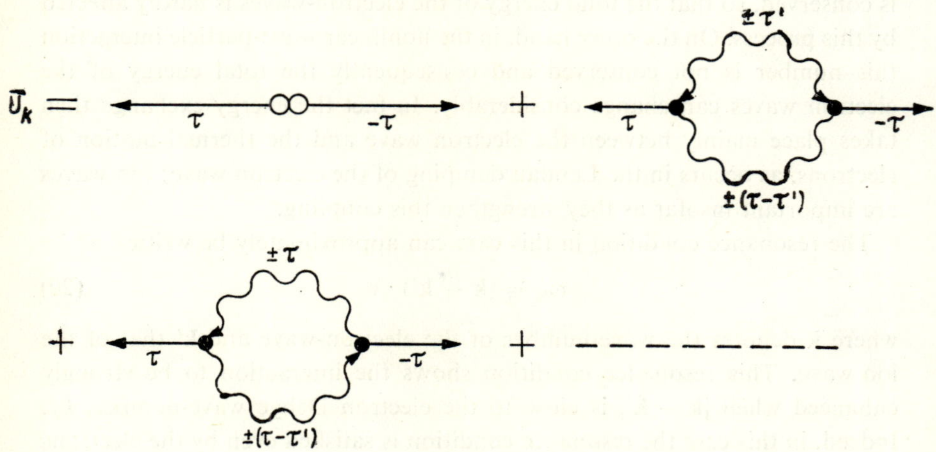

FIG. 3 Diagrams contributing to $\overline{U}_{\mathbf{k}}$

The factor (10) appears in this case either when $p' \doteqdot -i\omega_{\mathbf{k}'}$ or when $(p - p') \doteqdot -i\omega_{\mathbf{k}-\mathbf{k}'}$. We then obtain

$$\overline{U}_{\mathbf{k}} = \sum_{\sigma=\pm 1} \sum_{\mathbf{k}'} \frac{\omega_{\mathbf{k}}\sigma_{\mathbf{k}}^2}{2\pi} \frac{U_{\mathbf{k}'}(t)}{\omega_{\mathbf{k}'}} \int d\mathbf{v} \frac{W^\sigma \mathbf{k}, \mathbf{k}', F(\mathbf{v})}{|\varepsilon(\mathbf{k}, \sigma\omega_{\mathbf{k}} + (\mathbf{k} - \mathbf{k}') \cdot \mathbf{v})|^2}$$

$$\doteqdot \sum_{\sigma=\pm 1} \sum_{\mathbf{k}'} \frac{\omega_{\mathbf{k}}}{2\gamma_{\mathbf{k}}} \frac{U_{\mathbf{k}'}(t)}{\omega_{\mathbf{k}'}} \int d\mathbf{v}\, W^\sigma(\mathbf{k}, \mathbf{k}', \mathbf{v})\, \delta[\omega_{\mathbf{k}} - \sigma\omega_{\mathbf{k}'} - (\mathbf{k} - \mathbf{k}') \cdot \mathbf{v}]\, F(\mathbf{v}). \tag{19}$$

Substituting (16) and (19) into (8), we finally obtain Eq. (3).

Two remarks are in order. First, in the long wave lenght limit when both $\mathbf{k}$ and $\mathbf{k}' \to 0$, the nonlinear term in Eq. (3) agrees exactly with the expression obtained by Kadomtsev[20]. Second, the above result can easily be generalized to a two-component plasma. The only difference is that we now have to consider the full dielectric function of the two-component plasma. We shall use this result in the next section.

V INTERACTION BETWEEN AN ELECTRON-WAVE AND AN ION-WAVE

One interesting example of a nonlinear wave-particle interaction is the coupling between a high-frequency electron-wave and a low-frequency ion-wave. This coupling can also be produced by the three-wave interaction. But then the total number of plasmons associated with the electron-waves is conserved, so that the total energy of the electron-waves is hardly affected by this process. On the other hand, in the nonlinear wave-particle interaction this number is not conserved and consequently the total energy of the electron waves can change considerably. In fact the energy exchange then takes place mainly between the electron wave and the thermal motion of electrons, as occurs in the Landau damping of the electron wave; ion-waves are important insofar as they strengthen this coupling.

The resonance condition in this case can approximately be written as

$$\omega_{\mathbf{k}} \fallingdotseq (\mathbf{k} - \mathbf{k}') \cdot \mathbf{v} \tag{20}$$

where $\mathbf{k}$ denotes the wave-number of the electron-wave and $\mathbf{k}'$ that of the ion-wave. This resonance condition shows the interaction to be strongly enhanced when $|\mathbf{k} - \mathbf{k}'|$ is close to the electron Debye wave-number, k_e. Indeed, in this case the resonance condition is satisfied even by the electrons moving with thermal speed.

Using the kinetic wave equation derived in the preceding section, we have investigated in some detail

1) the effect of an enhancement of short wave-length ion-wave fluctuations on the long wave-length electron waves,

2) the effect of an enhancement of long wave-length electron-wave fluctuations on the short wave-length ion-waves, and

3) the effect of an enhancement of short wave-length electron-waves on the long wave-length ion-waves.

Here we only describe the results. We find for each case:

1) If the electron distribution is Maxwellian and if the electron-wave fluctuations are only thermal, no effect is expected as a result of an enhancement of ion-wave fluctuations. If electron waves are superthermal, they tend to be quenched by an enhancement of ion-waves if $[v\, \partial F/\partial v]_{v=v_r} < 0$, and be excited if $[v\, \partial F/\partial v]_{v=v_r} > 0$ ($v_r (\fallingdotseq \omega_{k/k'})$ is the resonance velocity). The typical damping or growth increment due to this interaction is of the order $\omega_p U_i/(NT_e)$, where U_i is the total energy of the ion-waves.

2) A significant effect appears only when the electron distribution $F(v)$ is not symmetric ($F(v) \neq F(-v)$). If $[\partial F/\partial v]_{v=v_r}$ is different from $[\partial F/\partial v]_{v=-v_r}$, the ion-waves are always excited in one direction of propagation and quenched in the other. The damping or growth increment is of the order $\omega_{pi} U_e/(NT_e)$,

where U_e is the total energy of the electron waves and ω_{pi} the ion plasma frequency.

3) When a relatively slowly moving electron beam is present, electron waves of wave-number close to k_e can be excited. We found that long wavelength ion-acoustic waves can also be excited by the nonlinear wave-particle interaction. The typical growth rate of the ion-waves becomes of the order $\omega_{pi} U_e/(NT_e) N_b/N$, where N_b is the number of electrons constituting the beam. Further details of this work will be reported elsewhere.

In conclusion, I wish to thank Dr. M. Saffren of the Jet Propulsion Laboratory for reading the manuscript. This work presents the results of one phase of research carried out at the Jet Propulsion Laboratory, California Institute of Technology, under Contract No. NAS7–100, sponsored by the National Aeronautics and Space Administration.

References

1. K. Nishikawa and Y. Osaka, *Prog. Theor. Phys.* **33**, 402 (1965).
2. A. Rogister and C. Oberman, *Journ. Plasma Phys.* **2**, 33 (1968).
3. C. S. Wu, *Phys. Fluids* **11**, 1733 (1968).
4. There are many other references which may be found in Refs. 1–3.
5. B. Kadomtsev, *Plasma Turbulence* (Academic Press, London-New York, 1965), Chap. II and IV.
6. V. N. Tsytovich, *Usp. Fiz. Nauk* **90**, 435 (1966) [translation: *Sov. Phys. USPEKHI* **9**, 805(1967)]. Earlier references by C. M. Kovrizhnykh and by V. N. Tsytovich can be found in this paper.
7. E. A. Frieman and P. H. Rutherford, *Ann. of Phys.* **28**, 134 (1964).
8. L. M. Gorbunov and V. P. Silin, *Zh. Eksperim.*; *Teor. Fiz* **47**, 200 (1964) [translation: *Sov Phys. JETP* **20**, 135 (1965)].
9. K. Nishikawa, *Kakuyugo–Kenkyu* (mimeographed circular in Japanese) **16**, 189 (1966).
10. K. Matsuura and T. Kurasawa, *Journ. Phys. Soc. Japan* **22**, 1461 (1967).
11. A. Rogister and C. Oberman, *Journ. Plasma Phys.* **3**, 119 (1969).
12. F. Einaudi and R. N. Sudan, *Plasma Phys.* **11**, 359 (1969).
13. K. Nishikawa, *Prog. Theor. Phys.* **36**, 193 (1966).
14. D. Pines and J. R. Schrieffer, *Phys. Rev.* **125**, 804 (1962).
15. R. E. Peierls, *Quantum Theory of Solids* (Oxford at the Clarendon Press, 1955) Chap. VI.
16. This is true as far as the waves of positive energies are concerned.
17. L. D. Landau and E. M. Lifshitz, *Electrodynamics of Continuous Media* (Pergamon Press, Oxford-London-New York-Paris, 1960) Sec. 61.
18. H. Mori, *Prog. Theor. Phys.* **33**, 423 (1965).
19. The word "linear" used here is different from that used in the other part of this paper where the term $U_{\mathbf{k}} U_{\mathbf{k}'}$, although linear in $U_{\mathbf{k}}$, is called "nonlinear".
20. See Ref. 5, Eq. (IV-5).

PAPER D–5

Test-particle method in kinetic theory of a plasma

KYOKO MATSUDA*

National Aeronautics and Space Administration
Goddard Space Flight Center
Greenbelt, Maryland

Abstract

The superposition principle which has been found by Rostoker is generalized to higher order correlation functions. This principle is useful because the problem of kinetic theory is reduced to determining the conditional probability functions which involves essentially only the Vlasov equation.

Many years have passed since people recognized that the results for the Fokker Plank equation and fluctuations of any physical quantity of plasmas are given in terms of the fields from statistically independent quasi-particles. This became definite when Rostoker[1] found that pair correlation function itself is given in general by a superposition of quasi-particles or suitably dressed test particles.

Although Rostoker's superposition principle is proved also for a system with electromagnetic interactions, we restrict our discussion within coulomb approximation. We consider a system which may be described by the B–B–G–K–Y hierarchy. If we introduce a test particle into the system, the field particles form a cloud surrounding the test particle. The cloud is described by a conditional probability function which obeys to first order

* National Academy of Sciences—National Research Council Resident Research Associate.

in the plasma parameter the following equation

$$\left\{\frac{\partial}{\partial t} + O'(X_1't) + O(X_1t)\right\} P(X_1' \mid X_1t) = \frac{e^2}{m}\frac{\partial f(X_1t)}{\partial \mathbf{v}_1}\cdot\frac{\partial}{\partial \mathbf{x}_1}\frac{1}{|\mathbf{x}_1 - \mathbf{x}_1'|},$$

$$O'(X_1't) = \mathbf{v}_1' \cdot \frac{\partial}{\partial \mathbf{x}_1'} - \frac{e}{m}\mathbf{F}_M(X_1't)\cdot\frac{\partial}{\partial \mathbf{v}_1'},$$

$$O(X_1t) = \mathbf{v}_1 \cdot \frac{\partial}{\partial \mathbf{x}_1} - \frac{e}{m}\mathbf{F}_M(X_1t)\cdot\frac{\partial}{\partial \mathbf{v}_1}$$

$$-\frac{ne^2}{m}\frac{\partial f(X_1t)}{\partial \mathbf{v}_1}\frac{\partial}{\partial \mathbf{x}_1}\cdot\int dX_2 \frac{1}{|\mathbf{x}_1 - \mathbf{x}_2|}\{\cdots\}. \tag{1}$$

X_1 is coordinates of a field particle and X_1' is coordinates of a test particle. $\mathbf{F}_M$ is the macroscopic force. Rostoker found that pair correlation function is given by a superposition of P functions as follows

$$G(X_1X_2t) = f(X_1t)\,P(X_1 \mid X_2t) + f(X_2t)\,P(X_2 \mid X_1t)$$
$$+ n \int dX'\, f(X't)\,P(X' \mid X_1t)\,P(X' \mid X_2t). \tag{2}$$

The equation for P function is essentially the Vlasov equation since it contains only an integral operator, which appears in the Vlasov equation. It was expected that to any higher order the B–B–G–K–Y hierarchy could be reduced to a hierarchy for test particle functions which involves only the Vlasov equation.

An attempt of this problem was made by Rostoker and myself[2] using the procedure of ensemble averages of the singular distribution function. If we assume a system of initially uncorrelated particles, we may derive expressions of correlation functions to any higher order in terms of conditional probability functions which are governed by Vlasov type equations. To second order they are

$$\hat{G}(X_1X_2t) = n \int dX_1'\, F(X_1')\,\tilde{P}(X_1' \mid X_1t)\,\tilde{P}(X_1' \mid X_2t)$$
$$+\frac{n^2}{2}\int dX_1'\,dX_2'\,F(X_1')\,F(X_2')\,\tilde{Q}(X_1'X_2' \mid X_1t)\,\tilde{Q}(X_1'X_2' \mid X_2t), \tag{3}$$

$$\hat{T}(123) = n \int d1'\,F(1')\,\tilde{P}(1' \mid 1)\,\tilde{P}(1' \mid 2)\,\tilde{P}(1' \mid 3)$$
$$+ \sum_P n^2 \int d1'\,d2'\,F(1')\,F(2')\,\tilde{P}(1' \mid 1)\,\tilde{P}(2' \mid 2)\,\tilde{Q}(1'2' \mid 3), \tag{4}$$

$\hat{G}$ and $\hat{T}$ are singular two-particle and three-particle correlation functions defined as

$$\hat{G}(12) = \Delta(1, 2)\, f(1) + G(12)$$
$$\hat{T}(123) = \Delta(1, 2)\,\Delta(2, 3)\, f(1) + \Delta(1, 2)\, G(23) + \Delta(2, 3)\, G(31) + \Delta(3, 1)\, G(12) + T(123)$$

where $\Delta(1, 2) = \dfrac{1}{n}\,\delta(\mathbf{x}_1 - \mathbf{x}_2)\,\delta(\mathbf{v}_1 - \mathbf{v}_2)$. Function $\tilde{P}$ is different from Function P in Reference 1. $\tilde{P}(1' \mid 1)$ characterizes a dressed test particle, but the test particle was introduced into the system nakedly at coordinates X_1' at the initial time. $\tilde{Q}$ expresses a dress shared by a pair of test particles. Those test particles are specified by their initial coordinates. The product of these conditional probability functions are ensemble averaged over initial coordinates. The hierarchy for these conditional probability functions which may be derived from the Klimontovich equation by taking ensemble averages are simple as well as these relationships are simple. However we have derived these relations by assuming a system of initially uncorrelated particles The difficulty of this approach is that this restriction is essential.

To obtain more generalized theory, we extend the approach of Reference 1. To save space, we use singular functions. In terms of singular functions, the Rostoker's superposition principle is reduced to

$$\hat{G}(12) = n \int d1'\, f(1')\, \hat{P}(1' \mid 1)\, \hat{P}(1' \mid 2)$$
$$\hat{P}(1' \mid 1) = P(1' \mid 1) + \Delta(1', 1) \tag{5}$$

To second order in the plasma parameter, the equation for $\hat{P}$ is obtained from the B–B–G–K–Y hierarchy as following

$$\left\{\frac{\partial}{\partial t} + O'(1') + O(1)\right\} \hat{P}(1' \mid 1)$$
$$= \frac{ne^2}{m} \frac{\partial \hat{P}(1' \mid 1)}{\partial \mathbf{v}_1} \cdot \int d2 \frac{\partial}{\partial \mathbf{x}_1} \frac{1}{|\mathbf{x}_1 - \mathbf{x}_2|} \hat{P}(1' \mid 2)$$
$$+ \frac{ne^2}{m} \frac{\partial \hat{P}(1' \mid 1)}{\partial \mathbf{v}_1'} \cdot \int d2 \frac{\partial}{\partial \mathbf{x}_1'} \frac{1}{|\mathbf{x}_1' - \mathbf{x}_2|} \hat{P}(1' \mid 2)$$
$$+ \frac{ne^2}{m} \int d2 \frac{\partial}{\partial \mathbf{x}_1} \frac{1}{|\mathbf{x}_1 - \mathbf{x}_2|} \cdot \frac{\partial}{\partial \mathbf{v}_1} \hat{W}(1' \mid 12)$$
$$+ \frac{ne^2}{m} \int d2 \frac{\partial}{\partial \mathbf{x}_1'} \frac{1}{|\mathbf{x}_1' - \mathbf{x}_2|} \cdot \frac{\partial}{\partial \mathbf{v}_1'} \hat{W}'(1' \mid 12), \tag{6}$$
$$\hat{W}(1' \mid 12) = W(1' \mid 12) + \Delta(1, 2)\, P(1' \mid 1).$$

The new quantity $W(1' \mid 12)$ is the perturbation of the two-particle correlation function $G(12)$ due to the test particle. The first and the third term in the right hand side are clear. A change in the orbit of test particle due to the shielding cloud is taken account by the second term and collision effects on the test particle is taken account by the last term. Since we may use the first order solution for $\hat{P}$ functions in the right hand side, if $\hat{W}$ function is known this equation is a Vlasov type equation. The equation for $\hat{W}$ function is needless to say not Vlasov type. However, $\hat{W}$ function can be given by the perturbation of Eq. (5) as follows

$$\begin{aligned}\hat{W}(1' \mid 12) = n \int d2'\, P(1' \mid 2')\, \hat{P}(2' \mid 1)\, \hat{P}(2' \mid 2) \\ + n \int d2'\, f(2)\, Q(1'2' \mid 1)\, \hat{P}(2' \mid 2) \\ + n \int d2'\, f(2')\, \hat{P}(2' \mid 1)\, Q(1'2' \mid 2).\end{aligned} \tag{7}$$

The new quantity Q characterizes a dress shared by a pair of test particle and obeys the equation

$$\begin{aligned}\left\{\frac{\partial}{\partial t} + O'(1') + O'(2') + O(1)\right\} Q(1'2' \mid 1) \\ = \frac{ne^2}{m} \frac{\partial \hat{P}(1' \mid 1)}{\partial \mathbf{v}_1} \cdot \int d2 \frac{\partial}{\partial \mathbf{x}_1} \frac{1}{|\mathbf{x}_1 - \mathbf{x}_2|} \hat{P}(2' \mid 2) \\ + \frac{ne^2}{m} \frac{\partial \hat{P}(1' \mid 1)}{\partial \mathbf{v}'_1} \cdot \int d2 \frac{\partial}{\partial \mathbf{x}'_1} \frac{1}{|\mathbf{x}'_1 - \mathbf{x}_2|} \hat{P}(2' \mid 2) \\ + \frac{ne^2}{m} \frac{\partial \hat{P}(2' \mid 1)}{\partial \mathbf{v}_1} \cdot \int d2 \frac{\partial}{\partial \mathbf{x}_1} \frac{1}{|\mathbf{x}_1 - \mathbf{x}_2|} \hat{P}(1' \mid 2) \\ + \frac{ne^2}{m} \frac{\partial \hat{P}(2' \mid 1)}{\partial \mathbf{v}'_2} \cdot \int d2 \frac{\partial}{\partial \mathbf{x}'_2} \frac{1}{|\mathbf{x}'_2 - \mathbf{x}_2|} \hat{P}(1' \mid 2).\end{aligned} \tag{8}$$

Since we may use the first order solution for $\hat{P}$ in the right hand side, this equation is Vlasov type.

We may show that to second order in the plasma parameter the first two correlation functions are given in terms of these test particle functions as following

$$\begin{aligned}\hat{G}(12) = n \int d1'\, f(1')\, \hat{P}(1' \mid 1)\, \hat{P}(1' \mid 2) \\ + n^2 \int d1'\, d2'\, P(1' \mid 2')\, P(2' \mid 1')\, \hat{P}(1' \mid 1)\, \hat{P}(2' \mid 2) \\ + \frac{n^2}{2} \int d1'\, d2'\, f(1')\, f(2')\, Q(1'2' \mid 1)\, Q(1'2' \mid 2) \\ + \sum_P n^2 \int d1'\, d2'\, f(1')\, P(1' \mid 2')\, \hat{P}(2' \mid 1)\, Q(1'2' \mid 2),\end{aligned} \tag{9}$$

$$\begin{aligned}\hat{T}(123) = {} & n \int d1'\, f(1')\, \hat{P}(1' \mid 1)\, \hat{P}(1' \mid 2)\, \hat{P}(1' \mid 3) \\ & + \sum_P n^2 \int d1'\, d2'\, f(1')\, P(1' \mid 2')\, \hat{P}(1' \mid 1)\, \hat{P}(2' \mid 2)\, \hat{P}(2' \mid 3) \\ & + \sum_P n^2 \int d1'\, d2'\, f(1')\, f(2')\, \hat{P}(1' \mid 1)\, P(2' \mid 2)\, Q(1'2' \mid 3), \qquad (10)\end{aligned}$$

where $\sum_P$ means the sum over all cyclic permutations of particles involved. The quantities in the right hand sides satisfies the equations for two-particle and three-particle correlation functions. Therefore we may state that these relations always exist provided they are satisfied at the initial time. We also

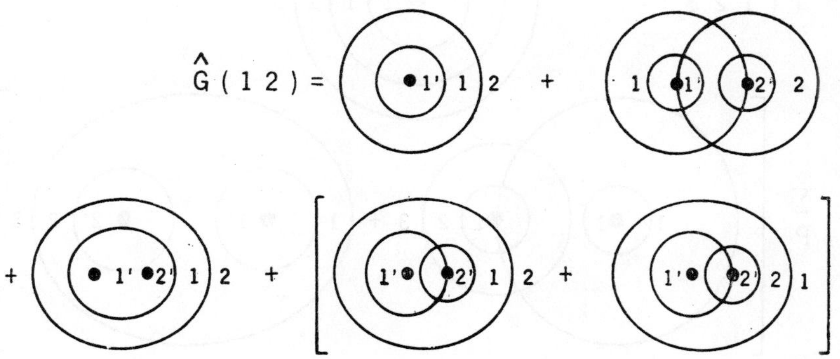

FIG. 1

have checked that by this relationship we may obtain the known correlation functions for an equilibrium system. Each term in Eqs. (9) and (10) is shown schematically in Figures 1 and 2 in the same order.

The rule governing the general relationship between any higher order correlation and the test particle function $P(1' \mid 1)$, $Q(1'2' \mid 1)$, $R(1'2'3' \mid 1)$ etc. can be obtained from these diagrams. A dot indicates a test particle and a circle indicates a shielding cloud. Each dot counts as order $1/\varepsilon$, each circle surrounding one center counts as order ε, each circle surrounding two centers counts as order ε^2, etc., where ε is the plasma parameter. Every center must be surrounded by two circles at least and total order of a diagram is given by the product of orders for circles and dots in the diagram. Therefore, the largest number of centers in a diagram is r if the quantity is considered to order ε^r. It should be noted that beyond first order, correlation functions are given by superpositions of not only statistically independent quasi-particles but also dependent quasi-particles, and the interaction between quasi-particles always comes in as a nonsingular function.

As application of this test particle method, at first we should mention those of the first order superposition principle. Since this principle can be generalized to a system with electromagnetic interactions, the method can be applied to a wide range of problems. We may obtain kinetic equations with the generalized Lenard–Balescu term plus diffusion by waves. On the

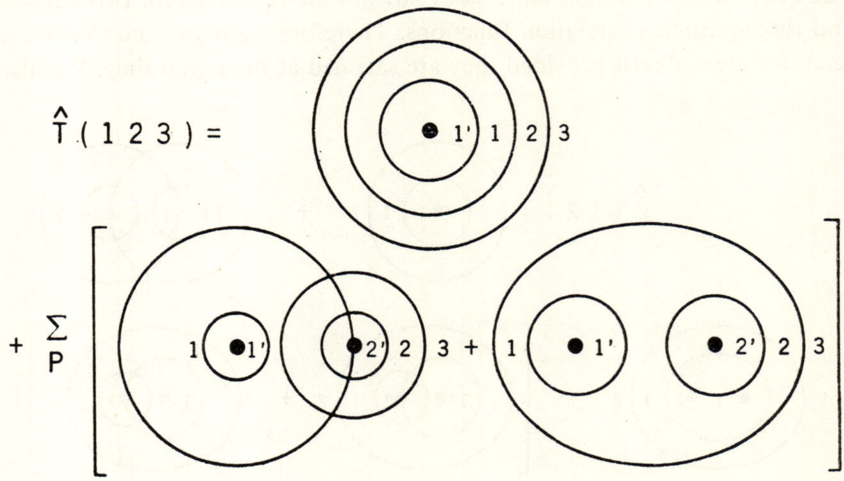

FIG. 2

other hand, we may calculate absorption or emission of waves by particles with respect different kinds of waves.

In the second order theory, the simplest problem is to derive a second order correction to the Lenard–Balescu term for a stable system. But more interesting problems are to derive kinetic equations for weakly unstable systems including the effects of wave-wave interactions. If we intend to obtain the complete description with discrete particle effects, it seems most convenient to treat the long time evolution as a succession of initial value problems which has been suggested by Nishikawa and Osaka[3] and carried out by Price[4] for first order kinetic equations. Now the entire procedure is straightforward although very tedious. The calculation is now in progress. Our problem is how we could approximate the results in transparent forms by making use of only the fact of weakness of instability.

References

1. N. Rostoker, *Phys. Fluids* **7**, 491 (1964).
2. N. Rostoker and K. Matsuda in *Plasma Physics and Controlled Nuclear Fusion Research* (International Atomic Energy Agency, Vienna, 1966), Vol. I, p. 747.
3. J. C. Price, *Phys. Fluids* **10**, 1623 (1967).
4. K. Nishikawa and Y. Osaka, *Progr. Theoret. Phys.* (Kyoto), **33**, 402 (1965).

PAPER D-6

The two-particle correlation function in nonequilibrium statistical mechanics

A. H. KRITZ*

Aeronautical Research Associates of Princeton, Inc. and Hunter College of the City University of New York, Department of Physics and Astronomy

G. V. RAMANATHAN†

Courant Institute of Mathematical Sciences

G. SANDRI*

Aeronautical Research Associates of Princeton, Inc.

Abstract

The two-particle correlation function for systems which are not in thermodynamic equilibrium plays a central role in modern kinetic theory. The theoretical methods currently used to determine the correlation function essentially agree in leading order with Bogolubov's "adiabatic" result. Bogolubov's theory yields a function with an infinite spatial range when the particle separation is parallel to the relative velocity. Previously, the infinite correlation range had been interpreted from a dynamical point of view in terms of the persistence of correlations during the inertial motion that follows a collision. One expects that the range of the two-particle correlation will be limited by three-body collisions.

We show, however, that a proper asymptotic treatment yields a non-adiabatic correlation function whose range is finite (of the order of the mean free path) even with the total neglect of three-body effects. This result demonstrates that the finite spatial range is due primarily to the statistical nature of the correlation function rather than due to three-body collisions. A method of expansion is then developed that yields a

* Sponsored by Air Force Contract No. AF 49(638)1461.

† Sponsored by Air Force Grant AF-AFOSR-815-67.

satisfactory lowest-order collision integral *and* correlation function, opening the possibility of calculating nonequilibrium phenomena beyond lowest order. The sharp contrast between our result and Bogolubov's can be tested with scattering experiments.

I INTRODUCTION

A major tool in the development of the modern kinetic theory of gases and plasmas is the BBGKY hierarchy[1–6]*. The usefulness of the hierarchy in nonequilibrium statistical mechanics stems from the fact that it allows for a systematic decoupling of successively larger clusters of particles. The methods currently used to expand the BBGKY hierarchy utilize the plausible assumption that the one-particle distribution varies slowly compared with the two-particle (and higher) distributions. One then strives to obtain an equation describing the evolution of the one-body distribution in terms of a functional of itself (kinetic equation). This idea is embodied mathematically in three methods: in Bogolubov's synchronized expansion[8], in resumming the strongest singularities of the perturbation expansion[9,10], and in the method of time scales[11,12]. Although these techniques yield the well established lowest-order kinetic equation for each gas regime, they have not led a satisfactory higher-order theory. Several attempts have been made to overcome the difficulties that occur in higher order, but so far no higher-order kinetic equation has been obtained having all the desirable properties, including an entropy theorem[13]. The expansion methods mentioned above yield, essentially, Bogolubov's leading-order results for the kinetic equation and two-particle correlation function. As a consequence, a widely accepted philosophy is that Bogolubov's results are correct in the leading order and that the difficulties that occur in higher order must be resolved by a proper inclusion of many-body effects. For example, the adiabatic two-particle correlation function given by Bogolubov's theory has an infinite spatial range for particle separations parallel to the relative velocity, and the general belief is that one should include three-body effects in order to obtain a finite spatial range†.

In contrast, we take the following different point of view. The fact that the leading-order correlation has an infinite range is a strong indication of

* For the purpose of this paper, Klimontovich's approach does not require a separate discussion[7].

† The adiabatic correlation function depends on time only through the one-body distribution function for times comparable (or larger than) the mean free time.

the breakdown of the approximation, even in leading order. In fact, the infinite range is a result of a singularity of the correlation in Fourier space and the higher-order corrections make the singularity stronger. The singularity indicates a misordering of the correlation in a region of phase space near the singular point. Such misordering shows an inconsistency in the expansion method. A complete asymptotic analysis yields, even *without the inclusion of three-body effects*, a two-particle correlation function which is neither adiabatic nor singular (i.e., the spatial range is finite and, at most, is of the order of the mean free path).

The basic features of the correct correlation function are readily obtained by recordering the variables[14] in the neighborhood of the singularity (Section II). Even though the "reordered" correlation is substantially different from Bogolubov's correlation, both yield the same collision integral and, thus, the same kinetic equation. This can be understood, in view of the following considerations. Bogolubov's correlation is incorrect for large particle separations nearly parallel to the relative velocity, $\mathbf{v}$ (i.e., small $\mathbf{k} \cdot \mathbf{v}$ in Fourier space). However, Bogolubov obtains the correct collision integral as a result of the finite range of the interaction potential. On the other hand, the reordered correlation is not valid for large values of $\mathbf{k} \cdot \mathbf{v}$, but this does not prevent us from obtaining the correct kinetic equation because the main contribution to the collision integral in a Fourier representation comes from small values of $\mathbf{k} \cdot \mathbf{v}$. This is due to the fact that the imaginary part of the correlation (i.e., the only contributing part) is very sharply peaked about $\mathbf{k} \cdot \mathbf{v} = 0$. While we find the leading-order collision integral is rather insensitive to the full applicability of a method, the higher-order collision integrals appear to be far more sensitive. We, thus, adopt the philosophy that singularities and inconsistencies in the leading order must be removed before proceeding to higher order*.

It is of interest to compare, in broad outline, our point of view with the related work of M. Green[18] and of E. Frieman and R. Goldman[19]. A main point suggested by them is that three-body collisions have a damping effect

* We believe that this point of view is necessary for a satisfactory mathematical treatment. Furthermore, it is indicated by the following considerations. (1) The three-body dynamics has been fully developed for the hard sphere model[15] and the relevant surfaces (those bounding the phase space domains corresponding to the allowed collision sequences) have no singularities[16]. (2) The secularities associated with three-body effects have been fully isolated[17]. Therefore, three-body effects are not responsible for the failure of higher-order theory. Rather, it is the method of expansion that fails. This conclusion is illustrated dramatically with the soluble models of Sec. III.

on the spatial and temporal range of the two-body correlation function. Our analysis brings out a new point, namely, that the finite spatial range of the two-particle correlation appears even in lowest order without three-body effects. *This is a consequence of the statistical nature of the correlation function.* We show that peak value of the correlation is umdamped in time in lowest-order theory. We conjecture, however, that three-body effects will damp the peak value within a few mean free times.

On the basis of physical considerations, it is reasonable to expect that the two-particle correlation function exhibits the temporal and spatial properties described below. We discuss two simple cases: (A) a pair of particles moving away from each other and (B) a pair of particles that are on a head-on collision course. We demonstrate, in this paper, that the Bogolubov correlation properly accounts for case (A) but fails to account for case (B). A correct mathematical treatment will yield a two-particle correlation that accounts for both cases.

Consider a system perturbed slightly away from equilibrium and neglect three-body effects in the reasoning. The source of the correlation function is proportional to both the force of the interaction between two uncorrelated particles and to their perturbed velocity distribution functions. Therefore, if two particles, separated by a distance, $\mathbf{x}$, and with relative velocity, $\mathbf{v}$, are uncorrelated initially, no correlation develops between them until they collide. More precisely, we expect the following time evolution of the correlation function.

(A) If $\mathbf{v}$ has a nonpositive component along the line of centers (i.e., the particles move away from each other), the correlation will remain zero forever.

(B) If $\mathbf{v}$ is parallel to the line of centers, the evolution of the correlation has three phases:

(1) The correlation is zero until the collision occurs, ($t < x_{\parallel}/v$).

(2) The correlation rises quickly during the interaction to its peak value, ($t \simeq x_{\parallel}/v$).

(3) The correlation decays within a few mean free times when $t > x_{\parallel}/v$ because the perturbed distribution function itself decays as a consequence of binary collisions.

Thus, we except a finite range for the correlation function because of its statistical properties. We have described the dependence of the correlation function on t for fixed values of $\mathbf{x}$. Similarly, we can discuss the dependence on $\mathbf{x}$ for fixed values of t. We re-examine the two cases above and find that for:

(A) the correlation is zero for all values of $x_{\parallel} < 0$

(B) the correlation function behaves differently in three regions of space if $x_{\parallel} > 0$:

(1) the correlation is zero if $x_{\parallel} > vt$;

(2) the correlation acquires its peak value for $x_{\parallel} \approx vt$;

(3) the correlation damps with decreasing $x_{\parallel}$ when $x_{\parallel} < vt$ within a few mean free paths because the disturbance in the distribution itself damps.

In this paper, we shall demonstrate that the behavior expected on physical grounds is, in fact, borne out. We shall see that the maximum effective range of the correlation function is of the order of a mean free path and *not infinite.* Furthermore, the result does not invoke three-body effects. Thus, the finite spatial range of the two-body correlation function is of statistical origin, rather than of purely dynamical (three-body effects) origin. In addition, we show that in leading order the finite range is propagated along the characteristics of the two-body motion.

It is our belief that the two-particle correlation function for systems which are not in thermodynamic equilibrium is one of the most important quantities that can be calculated with modern kinetic theory. "Modern" theory emphasizes the deduction of nonequilibrium properties from the N-body equations of motion. This deduction is carried out through the calculation of correlation functions, e.g., the collision integral in the kinetic equation is determined by the two-particle correlation function. This approach differs markedly from the "classical" theory of Boltzmann, who used only asymptotic correlation properties (i.e., energy and momentum balance) valid for times large compared to the duration of a collision. As a consequence, the classical theory is insensitive to the details of the two-body phase space behavior. This behavior is of major interest in the modern theory. On the experimental side, there is also a growing interest in the two-particle correlation function. The classical approach has been used primarily to guide the study of transport properties and the nonequilibrium properties of the one-particle velocity distribution. However, with the advances of photon and neutron scattering techniques[20], is has been possible to measure the two-body correlation function directly. The low temperature measurements (below approximately 200°K) are being extended systematically towards higher temperatures, and it is within the foreseeable future that measurements for nonequilibrium conditions can be achieved These measurements will give a crucial experimental test of our theory and, in fact, of "modern" kinetic theory as contrasted to the "classical" theory of Boltzmann.

We now give a brief outline of the contents of this paper. In Section II we derive the reordered correlation function and demonstrate mathematically, the faults of Bogolubov's theory. We treat in detail the weakly coupled gas regime for illustration purposes. The two basic properties of the leading-order correlation function (namely, propagation and finite range) apply to short range and Debye regimes, as well. In Section III, we consider a special case of the Lorentz gas. We present detailed numerical and analytical results which exhibit transparently the propagation and finite range phenomena. A complete explanation is given for the failure of Bogolubov's expansion. We show that Bogolubov's assumptions correspond to choosing a correlation which at time, $t = 0$, grows exponentially with increasing particle separation. This unphysical choice is a consequence of the adiabatic assumption. Finally, in Section IV, we present a method of time and space extension that applies to the coupled equations for the velocity distribution and the two-particle correlation. These results agree with those obtained recently with a method of time extension applied to an integrodifferential approach to the one-particle velocity distribution[14]. In Appendix A, we discuss the connection between the models of Section III and the Lorentz gas. In Appendix B, we prove that the Bogolubov correlation has an infinite range for the short-range regime in the Lorentz gas.

II CRITIQUE OF BOGOLUBOV'S METHOD OF EXPANSION AND REORDERED CORRELATION

When three-body interactions are neglected, the two-particle correlation function satisfies the equation[11]

$$\frac{\partial G}{\partial t} + \mathscr{H} G = (\phi_0/kT)\,\nabla\phi \cdot (\nabla_{v_1} - \nabla_{v_2})\,F_1 F_2 \tag{1}$$

where ϕ is the two-particle potential. The notation used in this paper is summarized in Table I below. Equation (1) is valid in all the regimes of gas theory, providing the operator, $\mathscr{H}$, is suitably chosen. We have, in particular, the following choices for the three-basic regimes:

i) Weak coupling ($\varepsilon = \phi_0/kT \ll 1, nr_0^3 \sim 1$)

$$\mathscr{H} = \mathbf{v} \cdot \nabla$$

ii) Boltzmann gas ($\phi_0/kT \sim 1, \varepsilon = nr_0^3 \ll 1$)

$$\mathscr{H} = \mathbf{v} \cdot \nabla - \nabla\phi \cdot (\nabla_{v_1} - \nabla_{v_2})$$

iii) Plasma ($\varepsilon = \phi_0/kT \sim (nr_0^3)^{-1} \ll 1$)

$$\mathscr{H} = \mathbf{v} \cdot \nabla - \Gamma$$

where, in each case, ϕ_0 is a measure of the strength and r_0 the range of the two-particle potential, and where the operator Γ is defined by

$$\Gamma A(1, 2) = \int d\mathbf{x}_3\, d\mathbf{v}_3[F(\mathbf{v}_1)\, A(2, 3) + F(\mathbf{v}_2)\, A(1, 3)].$$

The effects of three-body interactions would result in additional terms in Eq. (1) that are higher order in ε.

TABLE I Notation

	BBGKY Hierarchy (Sec. II, App. A, App. B)	Reordered Analysis (Sec. II)	Model (Sec. III, App. A)
One-particle distribution function	$F \equiv F^1$	F	$\mathscr{F}$ (complex) f [Eq. (A.22)]
Two-particle distribution function	F^2		
Two-particle correlation function	$G = \begin{cases} F^2 - F \text{ (Lorentz gas)} \\ F^2 - FF \text{ (With recoil)} \end{cases}$	G'	$\mathscr{G}$ (complex) g [Eq. (A.23)] $g^* = g^0 + \varepsilon g^1$ $= g_B + \varepsilon g^1$ (synchronized expansion) g_{IV} (initial value problem)

Bogolubov's "adiabatic" technique is based on the following three assumptions[8].

$$G = \sum \varepsilon^n G^n \tag{2a}$$

$$\partial G/\partial t = \begin{cases} 0(\varepsilon) \text{ Boltzmann gas} \\ 0(\varepsilon^2) \text{ Weak coupling and plasma} \end{cases} \tag{2b}$$

$$G^n(\mathbf{x} - \mathbf{v}\lambda, \mathbf{v}_1, \mathbf{v}_2, t) \xrightarrow[\lambda\to\infty]{} 0 \tag{2c}$$

where the relative position, $\mathbf{x}$, is defined by $x \equiv \mathbf{x}_1 - \mathbf{x}_2$ and the relative velocity, $\mathbf{v}$, by $\mathbf{v} \equiv \mathbf{v}_1 - \mathbf{v}_2$. Condition (2a) requires that the correlation should be a uniformly analytic function of the expansion parameter ε. Condition (2b) states that the time rate of change of the correlation function should be negligible in lowest order, the adiabatic assumption. Finally, the third condition, (2c), is a boundary condition that represents the generalized molecular chaos hypothesis. Equation (2c) has been written for the weakly-coupled regime but is readily generalized to the other regimes.

We observe from Eq. (1) that the adiabatic condition, Eq. (2b), cannot be retained when $\mathscr{H}$ is small, i.e., the neglect of the time derivative of G in Eq. (1) cannot be valid if

$$\mathscr{H} G(1, 2) = \begin{cases} 0(\varepsilon) \text{ Boltzmann gas} \\ 0(\varepsilon^2) \text{ Weak coupling and plasma.} \end{cases}$$

Under these conditions it is the time rate of change of G that balances the source term on the right-hand side of Eq. (1). Quite generally, the domain in which the adiabatic assumption fails is a certain region in the complement of the support of the null space of the operator $\mathscr{H}$. It will be demonstrated that this failure of Eq. (2b) leads to an unphysical behavior of the correlation length. As an example, we discuss, below, the weakly-coupled system in detail[21].

Let us consider a spatially uniform system of particles interacting through a weak spherically symmetric potential. After Fourier transforming, the first two equations of the hierarchy take the form

$$\frac{\partial F}{\partial t} = \varepsilon \iint d^3k\, d^3v_2\, \tilde{\phi}(\mathbf{k})\, (-i\mathbf{k} \cdot \nabla_{v_1})\, \tilde{G}(\mathbf{k}, \mathbf{v}_1, \mathbf{v}_2) \tag{3}$$

$$\frac{\partial G}{\partial t} + i\mathbf{k} \cdot (\mathbf{v}_1 - \mathbf{v}_2)\, \tilde{G} = \varepsilon \tilde{\phi}(\mathbf{k})\, i\mathbf{k} \cdot (\nabla_{v_1} - \nabla_{v_2})\, F(\mathbf{v}_1)\, F(\mathbf{v}_2) + O(\varepsilon^2) \tag{4}$$

where the tilde denotes transformed quantities. We shall consider the simple initial value problem, i.e., $\tilde{G}(t = 0) = 0$[11]. The adiabatic assumption states that F remains constant for finite values of t and that for large t ($t \sim \varepsilon^{-2}$), $\tilde{G}$ will depend on t only through F. The difficulty with this assumption is that it does not hold uniformly for all values of $\mathbf{k}$. In fact, using Bogolubov's assumption, one obtains the following asymptotic solution for G

$$\tilde{G} \sim \varepsilon\tilde{\phi}(\mathbf{k})\,\mathbf{k}\cdot(\nabla_{v_1} - \nabla_{v_2})\,F(\mathbf{v}_1)\,F(\mathbf{v}_2)\cdot\left[\frac{P}{\mathbf{k}\cdot(\mathbf{v}_1 - \mathbf{v}_2)} + \pi i\delta(\mathbf{k}\cdot\mathbf{v}_1 - \mathbf{k}\cdot\mathbf{v}_2)\right].$$

Clearly, when $\mathbf{k}\cdot(\mathbf{v}_1 - \mathbf{v}_2)$ is sufficiently small, $\tilde{G}$ cannot remain of order ε. This is in direct disagreement with Bogolubov's assumption, Eq. (2a). Such an inconsistency cannot remain if we are to obtain a valid theory. Even though the Bogolubov result for $\tilde{G}$ is incorrect, it yields a proper equation for F. The kinetic equation is obtained by substituting the above result for $\tilde{G}$ into the right-hand side of Eq. (3). For a spherically symmetric potential, the principal part integrates to zero in the collision integral and only the δ-function contributes.

If one considers that: (1) the solution for $\tilde{G}$ is singular at $\mathbf{k}\cdot(\mathbf{v}_1 - \mathbf{v}_2) = 0$ and (2) the contribution to the collision integral comes from the neighborhood of $k_{\|} = 0$ ($k_{\|}$ stands for the component of $\mathbf{k}$ along $(\mathbf{v}_1 - \mathbf{v}_2)$), it is clear one should examine the equations more closely for small values of $k_{\|}$ and large values of t. For this purpose, we reorder the terms in the two equations by setting $t = t'/\varepsilon^2$, $k_{\|} = \varepsilon^2 k'_{\|}$ and $\tilde{G} = (1/\varepsilon)\,\tilde{G}'$. We then obtain the following equations for F and $\tilde{G}'$

$$\frac{\partial F}{\partial t'} = \iiint d^2\mathbf{k}_{\perp}\,dk'_{\|}\,d^3v_2\tilde{\phi}(\mathbf{k}_{\perp})\,(-i\mathbf{k}_{\perp}\cdot\nabla_{v_1})\,\tilde{G} + O(\varepsilon^2) \tag{5}$$

$$\frac{\partial\tilde{G}'}{\partial t'} + ik'_{\|}\,|\mathbf{v}_1 - \mathbf{v}_2|\,\tilde{G}' = \tilde{\phi}(\mathbf{k}_{\perp})\,i\mathbf{k}_{\perp}\cdot(\nabla_{v_1} - \nabla_{v_2})\,F(\mathbf{v}_1)\,F(\mathbf{v}_2) + O(\varepsilon^2). \tag{6}$$

Although the kinetic equation obtained by solving Eq. (6) for $\tilde{G}'$ and substituting the result into Eq. (5) is the same as that obtained by using the adiabatic assumption, the Bogolubov and reordered correlation functions are different. The two coupled equations, Eqs. (5) and (6), force $\tilde{G}'$ and F to evolve on the same time scale for small $k_{\|}$. We show, below, that the correlation obtained by reordering is neither adiabatic nor singular.

Neglecting the three-body terms of the order ε^2, we solve Eq. (6) for $\tilde{G}'$, i.e.,

$$\tilde{G}' = \tilde{\phi}(\mathbf{k}_{\perp})\,i\mathbf{k}_{\perp}\cdot(\nabla_{v_1} - \nabla_{v_2})\int_0^{t'} d\lambda\, e^{-ik'_{\|}|\mathbf{v}_1 - \mathbf{v}_2|\lambda}\,F(\mathbf{v}_1, t' - \lambda)\,F(\mathbf{v}_2, t' - \lambda).$$

Returning to the original variables, we have

$$\widetilde{G} = \varepsilon\widetilde{\phi}(k_\perp)\, i\mathbf{k}_\perp \cdot (\nabla_{v_1} - \nabla_{v_2}) \int_0^t d\mu\, e^{-ik_\parallel |\mathbf{v}_1 - \mathbf{v}_2| \mu} F(\mathbf{v}_1, t-\mu) F(\mathbf{v}_2, t-\mu). \quad (7a)$$

Inverse Fourier transforming, we obtain

$$G = \frac{\varepsilon}{|\mathbf{v}_1 - \mathbf{v}_2|}\left[\int_{-\infty}^{+\infty} dx_\parallel \nabla\phi\right]$$

$$\cdot (\nabla_{v_1} - \nabla_{v_2}) F\left(\mathbf{v}_1, t - \frac{x_\parallel}{|\mathbf{v}_1 - \mathbf{v}_2|}\right) F\left(\mathbf{v}_2, t - \frac{x_\parallel}{|\mathbf{v}_1 - \mathbf{v}_2|}\right) \cdot \theta\left(t - \frac{x_\parallel}{|\mathbf{v}_1 - \mathbf{v}_2|}\right) \quad (7b)$$

where θ is the Heaviside function with value unity for positive argument and zero, otherwise. The finite spatial range and propagation properties discussed in the introduction can be seen by inspection of Eqs. (7). The fact that there is no singularity for $k_\parallel = 0$ assures that G has a finite spatial range. As a consequence of the step function, $\theta\left(t - \frac{x_\parallel}{|\mathbf{v}_1 - \mathbf{v}_2|}\right)$, and the dependence of F on the argument, $t - \frac{x_\parallel}{|\mathbf{v}_1 - \mathbf{v}_2|}$ we observe the propagation and finite range properties of the correlation function. Thus, even with the neglect of three-body effects, G has a finite spatial range because of its statistical nature. We have shown that using the reordered correlation, given in Eq. (7), one readily recovers the Fokker–Planck kinetic equation[14]. This result is verified explicitly with the special Lorentz models discussed in the nect section. With these models, the propagation and finite range properties of G are clearly illustrated.

III SPECIAL CASES OF THE LORENTZ GAS

From the single-particle and two-particle distribution functions for a spatially homogeneous "Lorentz gas", under some restrictions we can deduce a linear pair of exactly soluble equations, Appendix A. The Lorentz gas consists of noninteracting light particles embedded in a medium of randomly distributed fixed scattering centers. We decompose the distribution and the correlation function in velocity space spherical harmonics. Retaining only sectoral harmonics, we find that each f_n and g_n satisfies the equations (omitting the harmonic subscript):

$$\frac{\partial f(t)}{\partial t} = -\varepsilon \int d\mathbf{x}\, I(\mathbf{x})\, g(\mathbf{x}, t) \tag{8}$$

$$\frac{\partial g}{\partial t} + \mathbf{v} \cdot \nabla g = J(\mathbf{x})\, f(t) \tag{9}$$

where the interaction functions, I and J, are defined by

$$I = \mathbf{w}_1 \cdot \nabla\phi \tag{10}$$

$$J = \mathbf{w}_2 \cdot \nabla\phi. \tag{11}$$

The two complex vectors, $\mathbf{w}_1$ and $\mathbf{w}_2$, are given by

$$\mathbf{w}_1 = -in\hat{u}, \quad \mathbf{w}_2 = \mathbf{w}_1 - \hat{v} \tag{12}$$

where $\hat{v}$ is the unit vector in the direction of the relative velocity, $\mathbf{v}$, and $\hat{u}$ is a unit vector in the scattering plane that is perpendicular to $\mathbf{v}$.* In order to make our demonstrations particularly simple, we study model situations in which the interaction functions are real quantities with spherical symmetry in the relative position vector.

For a first example, we choose

$$I = J = \delta(\mathbf{x}). \tag{13}$$

Under these conditions, it is possible to solve explicitly Eqs. (8) and (9). For this simple choice, it is readily shown that

$$f(t) = e^{-\varepsilon t} f(0). \tag{14}$$

We assume vanishing initial correlations and normalize the initial distribution so that

$$g(0) = 0$$
$$f(0) = 1. \tag{15}$$

With these conditions, the correlation is given by

$$g_{\mathrm{IV}} = \theta(x_{\parallel}) \exp\left[-\varepsilon(t - x_{\parallel})\right] \theta(t - x_{\parallel})\, \delta(x_{\perp}). \tag{16}$$

where the Heaviside θ-function is defined

$$\theta(x) = \begin{cases} 1, & x > 0 \\ 0, & x < 0. \end{cases} \tag{17}$$

* It is interesting to note that Eqs. (8) and (9) are the classical analogue of the Weisskopf–Wigner equations for the two-level radiative decay.

We use the subscript IV for g to emphasize that this solution is based on the initial value assumption $g(0) = 0$. We have normalized $|\mathbf{v}|$ to unity and $x_\perp$ and $x_\parallel$ are the components of $\mathbf{x}$ perpendicular and parallel to $\mathbf{v}$, respectively. If a nonvanishing initial correlation $A(\mathbf{x})$ is given, then the solution is simply given by Eq. (16) with the term $A(x_\parallel - t, \mathbf{x}_\perp)$ added to it.

For the choice for I and J given in Eq. (13), we find that Bogolubov's correlation function, $g_B \equiv g^0$ (i.e., the first term in the series expansion in Eq. (2a) is given by

$$g_B = \theta(x_\parallel) f(t)\, \delta(x_\perp) \tag{18}$$

and that the correction g^1 is

$$g^1 = \varepsilon x_\parallel g_B. \tag{19}$$

Clearly, the series expansion is not uniformly valid, since if $x_\parallel = 0(1/\varepsilon)$, then $g^1 \approx g_B$. Thus, we find that we cannot expect the Bogolubov correlation given in Eq. (18) to have the physical properties exhibited by the simple initial value correlation given in Eq. (16). If the correlation has finite range,

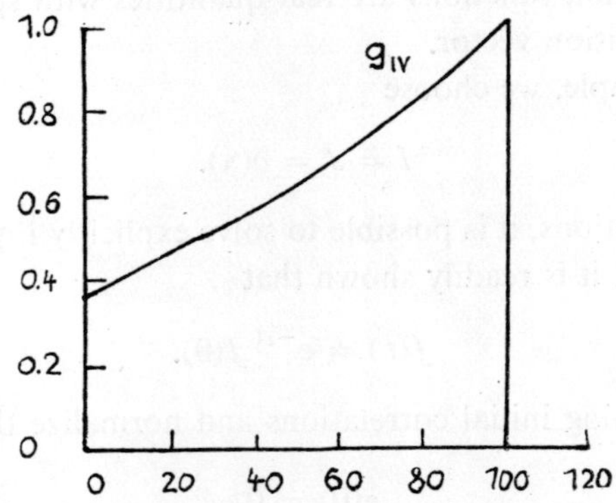

FIG. 1a The initial value correlation is plotted as a function of particle separation for the delta function interaction

it will always have a finite range, as seen in Eq. (16) and Fig. 1a. The Bogolubov correlation, Fig. 1b, however, has infinite range. The Fourier transforms of g^0 and g^1 are singular at $\mathbf{k} \cdot \mathbf{v} = 0$ corresponding to their undamped behavior for large $x_\parallel$. These singularities accumulate in higher order, giving rise to the divergences of higher-order kinetic theory.

For our simple model, the entire series, indicated in Eq. (2a), can be calculated and, in fact, resummed. We use the notation g^* for the summed correlation when the Bogolubov conditions Eqs. (2b) and (2c), are used in obtaining the results. The resulting function, g^*, is given by

$$g^* = s(x_{\parallel})\, f(t)\, \delta(x_{\perp}) \tag{20}$$

where

$$s(x_{\perp}) = \theta(x_{\parallel}) \exp(+\varepsilon x_{\parallel}) \tag{21}$$

We see clearly the reason for the accumulation of singularities in Fourier space. The function g^* cannot be Fourier analyzed since it grows exponentially. This result is in sharp contrast with the simple initial value result given in Eq. (16), and corresponds to the fact that Bogolubov's assumptions, (2a),

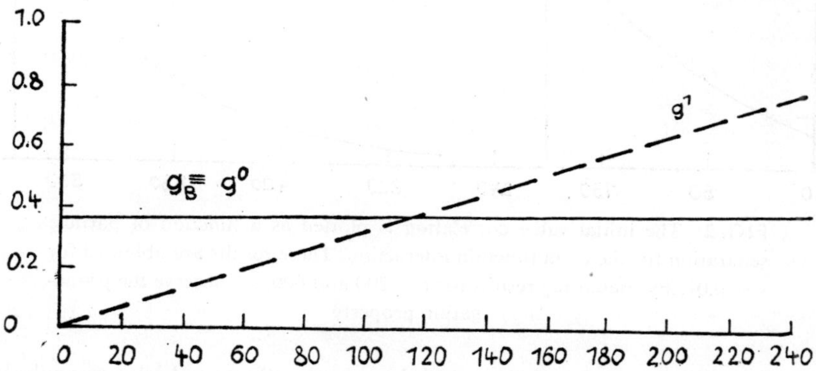

FIG. 1b The Bogolubov correlation, g^0, and the first-order correction, g^1, are presented for $t = 100$ and $\varepsilon = .01$. This figure shows the unphysical behavior of the range of the Bogolubov correlation and the nonuniformity of the Bogolubov expansion

(2b), and (2c), imply a nonphysical choice of initial conditions. Particle pairs are more strongly correlated as their spatial separation increases.

It is interesting to observe that the Bogolubov ansatz, Eq. (2), leads to a separable solution, as shown by Eq. (20). We can obtain another separable solution of the form given by Eq. (20) with the streaming function

$$s = -\theta(-x) \exp(+\varepsilon x). \tag{22}$$

For this solution g^* has a finite spatial range. This corresponds to having assumed an initial correlation in which all two-particle interactions have been completed, and describes the damping of these correlations due to the decay of the velocity distribution. However, the expansion in ε is still not uniform.

Another important property of g_{IV} given by Eq. (16) is its "propagation" character (see Fig. 2). This is seen in the $t - x_{\parallel}$ dependence of g_{IV}. The physical meaning of propagation corresponds to allowing the particle pairs to be uncorrelated prior to interaction, to become correlated because of the interaction, and then become decorrelated because of the decay of the one-particle distribution function. We thus see that the decay of the correlation

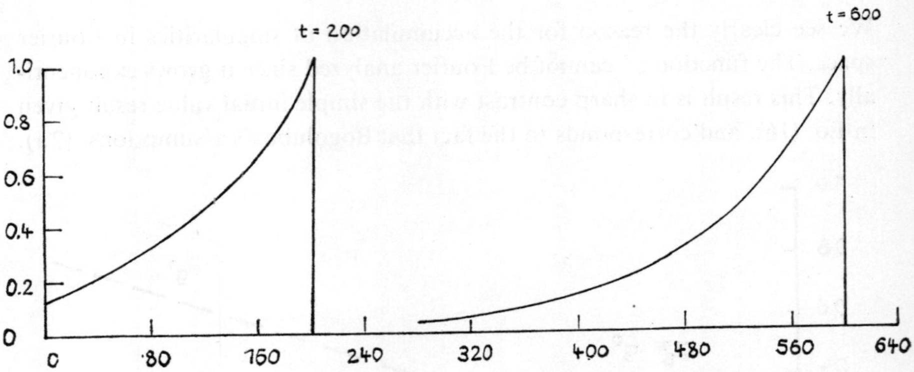

FIG. 2 The initial value correlation is plotted as a function of particle separation for the delta function interaction. These results are obtained for $\varepsilon = 0.01$. By presenting results for $t = 200$ and 600, we observe the propagation property

is not dependent on the presence of three-body effects (which are excluded in the construction of the model) but that the finite range is due to the decay of f and is of the order of a mean free path. We see that the full evolution of the correlation, as expressed by Eq. (16), is nonadiabatic and does not satisfy the assumptions in Eq. (2). This is due to the fact that the time dependence of g_{IV} is not simply given in terms of $f(t)$ but also through $\theta(t - x_{\|})$.

The properties of g_{IV}, g_B, and g^*, demonstrated above, hold for quite general interaction functions, $I(\mathbf{x})$ and $J(\mathbf{x})$, and not simply for the special case discussed above, $(I = J = \delta(\mathbf{x}))$. We have considered in detail the following models (g_{IV} and f were computed numerically):

1) Debye interaction:

$$I(\mathbf{x}) = J(\mathbf{x}) = \frac{e^{-\mu|\mathbf{x}|}}{|\mathbf{x}|} \tag{23}$$

2) Square barrier interaction:

$$I(\mathbf{x}) = \frac{3}{2\pi} \quad J(\mathbf{x}) = \begin{cases} 1 & |\mathbf{x}| < a/2 \\ 0 & |\mathbf{x}| > a/2 \end{cases} \tag{24}$$

Figs. 3 and 4 compare the three relevent correlations and show the unphysical correlation lengths of g_B and g^* for the Debye and square barrier models. In Figs. 5 and 6 we see the propagation property of g_{IV}. Bogolubov's

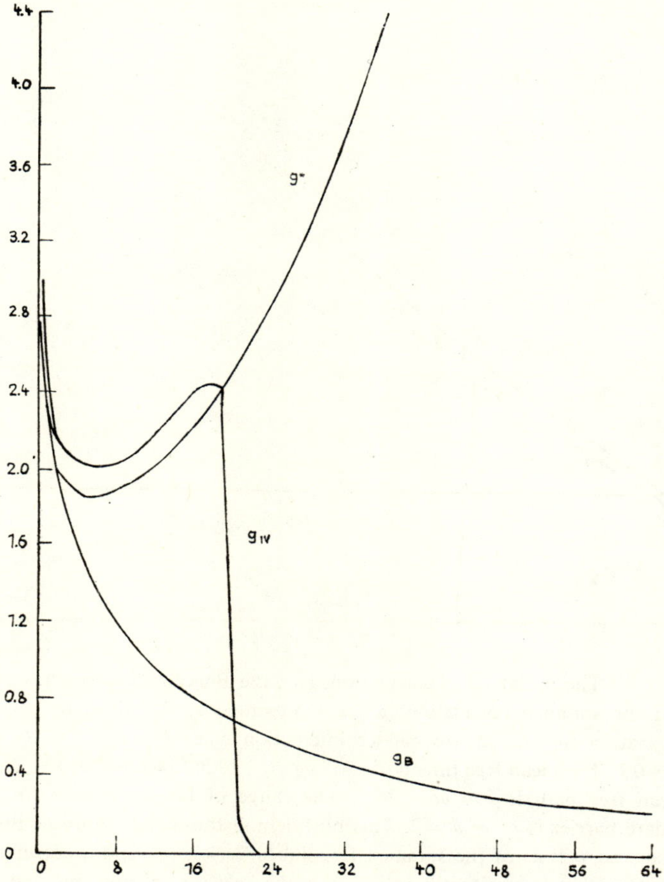

FIG. 3 The initial value correlation, g_{IV}, the Bogolubov correlation, g_B, and the summed correlation, g^*, are presented as a function of particle separation for the Debye interaction when $\theta = 1°$, $t = 20$, and $\varepsilon = .01$. The mean free time is $\tau = (2\pi\varepsilon)^{-1} = 15.9$. Since we set $v = 1$, the mean free path is $\lambda = v\tau = 15.9$. The decay constant, μ, for the Debye interaction is taken to be equal to one (i.e., the Debye length is equal to unity), so that the effective range of interaction is $r_0 = \mu^{-1} = 1$. This plot demonstrates the failure of the Bogolubov and the summed Bogolubov correlation to yield a uniformly valid result when the relative velocity and separation are nearly parallel, i.e., $\theta = \cos^{-1}\left(\frac{\mathbf{x} \cdot \mathbf{v}}{|\mathbf{x}|\,|\mathbf{v}|}\right)$ is small

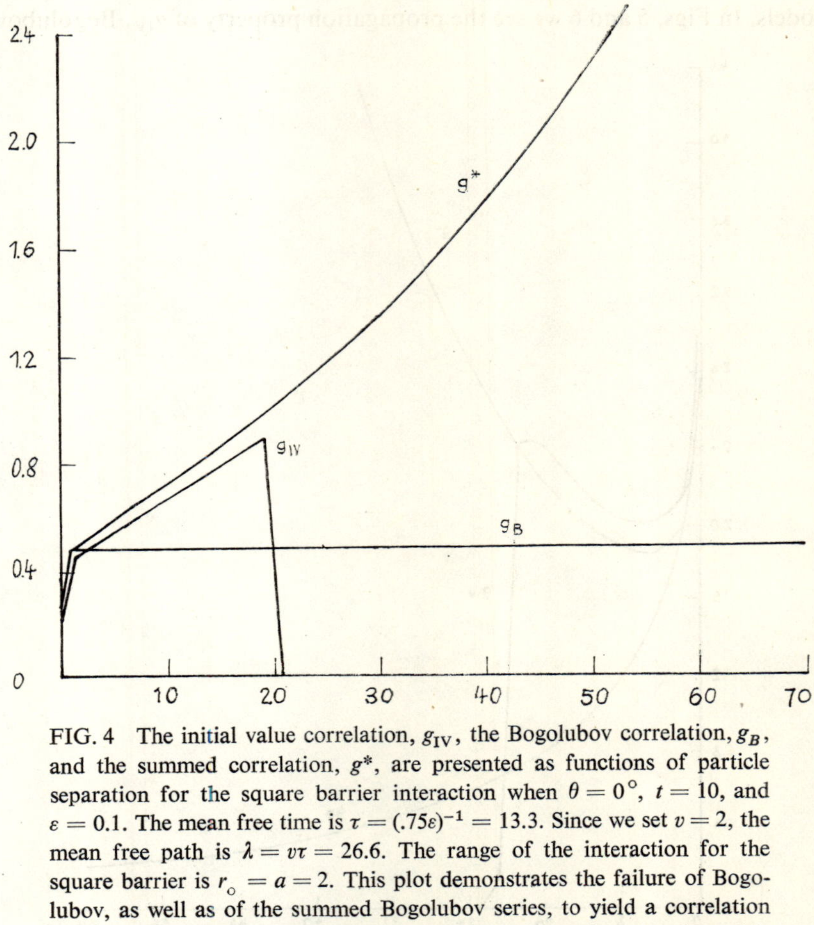

FIG. 4 The initial value correlation, g_{IV}, the Bogolubov correlation, g_B, and the summed correlation, g^*, are presented as functions of particle separation for the square barrier interaction when $\theta = 0°$, $t = 10$, and $\varepsilon = 0.1$. The mean free time is $\tau = (.75\varepsilon)^{-1} = 13.3$. Since we set $v = 2$, the mean free path is $\lambda = v\tau = 26.6$. The range of the interaction for the square barrier is $r_0 = a = 2$. This plot demonstrates the failure of Bogolubov, as well as of the summed Bogolubov series, to yield a correlation function that is uniformly valid when the relative velocity and particle separation are nearly parallel, i.e., when $\theta = \cos^{-1}\left(\frac{\mathbf{x} \cdot \mathbf{v}}{|\mathbf{x}|\,|\mathbf{v}|}\right)$ is small

result fails not only when $\theta = 0$, where $\cos\theta = \mathbf{x} \cdot \mathbf{v}/|\mathbf{x}|\,|\mathbf{v}|$, but fails in a substantial angular region. This is seen from the plot of relative error in Fig. 7.

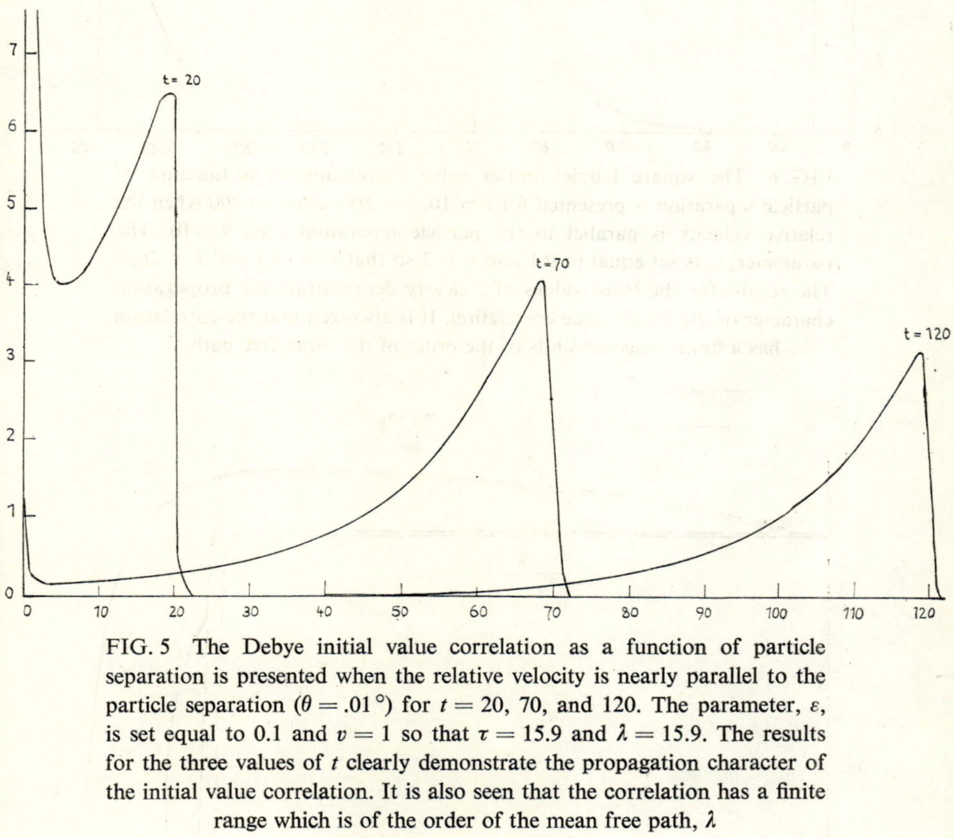

FIG. 5 The Debye initial value correlation as a function of particle separation is presented when the relative velocity is nearly parallel to the particle separation ($\theta = .01°$) for $t = 20$, 70, and 120. The parameter, ε, is set equal to 0.1 and $v = 1$ so that $\tau = 15.9$ and $\lambda = 15.9$. The results for the three values of t clearly demonstrate the propagation character of the initial value correlation. It is also seen that the correlation has a finite range which is of the order of the mean free path, λ

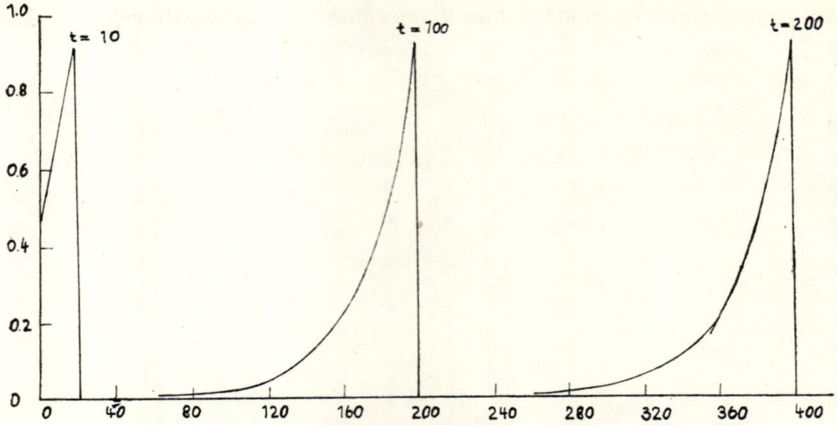

FIG. 6 The square barrier initial value correlation as a function of particle separation is presented for $t = 10$, $t = 100$, and $t = 200$ when the relative velocity is parallel to the particle separation (i.e., $\theta = 0$). The parameter, ε, is set equal to 0.1 and $v = 2$ so that $\tau = 13.3$ and $\lambda = 26.6$. The results for the three values of t clearly demonstrate the propagation character of the initial value correlation. It is also seen that the correlation has a finite range which is of the order of the mean free path, λ

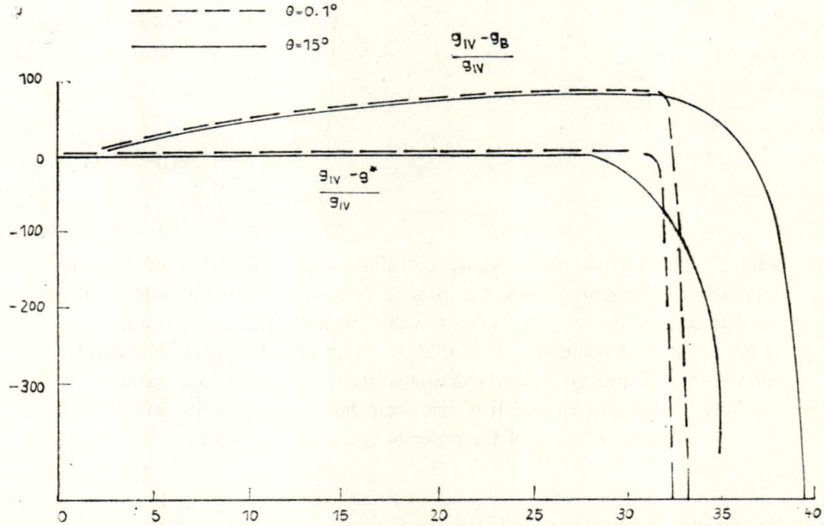

FIG. 7 The percent deviation of g_B and g^* from g_{IV} is plotted as a function of particle separation for the Debye interaction when $t = 32$ and $\varepsilon = .01$ ($\tau = 15.9$, $\lambda = 15.9$). Results are presented for two different values of the angle between the relative direction and the relative velocity, i.e., for $\theta = 0.1°$ and $15°$. Once again, these results demonstrate that Bogolubov and the summed Bogolubov series fail to give a uniformly valid approximation to the initial value correlation function in a finite cone about the direction where the relative velocity and relative separation are parallel

IV IMPROVED METHOD OF EXPANSION

We now describe a method that will yield a good approximation to the lowest-order correlation. This method is motivated by the direct perturbation expansion of Eqs. (8) and (9). We will show that from the behavior of f the introduction of time scales, as suggested by Bogolobov, is necessary. However, because of the nonuniform behavior of the correlation function, it is essential to introduce more freedom by way of extension of the space variable*. This was omitted in Bogolubov's treatment. Taylor expansion of Eqs. (8) and (9) gives the following results. For the one-particle distribution function, we have

$$f^0 = \text{constant}, \quad f^1 \sim \text{constant} \cdot t \tag{25}$$

while for the spatial Fourier transform of the two-particle correlation, we obtain

$$\tilde{g}^0 = \tilde{J}(k) \int_0^t e^{-ik\lambda}\, d\lambda\, f^0 = \frac{\tilde{J}(k)}{ik}(1 - e^{-ikt})\, f^0. \tag{26}$$

We notice that the lowest-order distribution function is approximately constant in time, with a correction that grows linearly with time. We, thus, expect that a correct representation of the distribution function will exhibit variation on a slow time scale in agreement with Bogolubov's suggestion of time scales. We notice, from Eq. (26), an additional feature which motivates the introduction of space scales. The direct perturbation expansion yields a correlation that is not of order one for all the values of k and, in fact, grows to order $1/\varepsilon$ for values of k which are of order ε. Under this condition, one needs a stretching of k in Fourier space[16]. This removes the singularity for small values of k. We have shown that, correspondingly, extension of the x-space removes the unphysical long-range behavior of the correlation function.

We now develop a general method that enables us to determine a valid lowest-order correlation, as well as higher-order corrections. For simplicity, we write the equations for f and g in one dimension as follows:

$$\frac{\partial f}{\partial t} = -\varepsilon \int dx\, I(x)\, g(x, t) = F \tag{25a}$$

$$\frac{\partial g}{\partial t} + \frac{\partial g}{\partial x} = J(x)\, f(t). \tag{25b}$$

* The extension we employ is fundamentally different from the space scale approch of Ref. 19. In contrast, we extend the phase mixing integral and, in addition, make use of an "extension of unity" in the source term of the g equation. This procedure is equivalent to stretching $k_{||}$ in Fourier space.

We first extend the right-hand side of Eq. (25a) as follows, ($\xi_n = \varepsilon^n x$, $\tau_n = \varepsilon^n t$)

$$F = -\varepsilon \iint d\xi_0\, d\xi_1 I(\xi_0)\, g(\xi_0 \xi_1, \tau_0 \tau_1)\, \delta(\xi_1 - \varepsilon\xi_0). \tag{26}$$

We expand in powers of ε. To first order in ε we have

$$\frac{\partial f^0}{\partial \tau_0} + \varepsilon\left(\frac{\partial f^0}{\partial \tau_1} + \frac{\partial f^1}{\partial \tau_0}\right) = -\varepsilon \int d\xi_0 I g^0(\xi_0, 0)\, \varepsilon^2 \int d\xi_0 \left[I g^1(\xi_0, 0) + \xi_0 I \frac{\partial g^0}{\partial \xi_1}(\xi_0, 0)\right]. \tag{27}$$

For the correlation function, we require a special device; namely, we introduce in the source term on the right-hand side of Eq. (25b) an extension of unity, which is denoted by $U(\xi_1 - \varepsilon\xi_0)$ where $U(0) = 1$*. Again, correct to first order in ε, the g equation reads

$$\frac{D}{Dt}(g^0 + \varepsilon g^1) = J(\xi_0)\,(f^0 + \varepsilon f^1)\left(U - \varepsilon\xi_0 \frac{\partial U}{\partial \xi_1}\right)$$

$$= J f^0 U + \varepsilon\left(J F^1 U - f^0 J \xi_0 \frac{\partial U}{\partial \xi_1}\right). \tag{28}$$

(Note that U has been expanded in a Taylor series in ε with only first-order terms retained.)

The lowest-order theory is now developed as usual[11]. We obtain, for the one-particle distribution function,

$$\frac{\partial f^0}{\partial \tau_0} = 0 \tag{29}$$

and for the correlation function

$$\frac{\partial g^0}{\partial \tau_0} + \frac{\partial g^0}{\partial x_0} = J f^0 U. \tag{30}$$

We see that this procedure is in agreement with Bogolubov's time scale approach; however, there is a novelty—namely, the factor U on the right-hand side of Eq. (30) will allow us to eliminate suitably the long range misbehavior of the correlation. Integration of Eq. (30) yields

$$g^0 = f^0 U \int_0^{\tau_0} d\lambda\, J(\xi_0 - \lambda). \tag{31}$$

* In general, U will be a velocity-dependent operator.

We now discuss the first-order theory. The one-particle distribution function satisfies

$$\frac{\partial f^0}{\partial \tau_1} + \frac{\partial f^1}{\partial \tau_0} = -f^0 \int d\xi_0 I \int_0^{\tau_0} d\lambda\, J(\xi_0 - \lambda) = -f^0 \int_0^{\tau_0} d\lambda\, K(\lambda) \tag{32}$$

where we have introduced the notation

$$K(\lambda) = \int d\xi_0\, I(\xi_0)\, J(\xi_0 - \lambda). \tag{33}$$

Elimination of the secularity in Eq. (32) immediately yields

$$\frac{\partial f^0}{\partial \tau_1} = \gamma f^0 = -f^0 \int_0^{\infty} K(\lambda)\, d\lambda \tag{34}$$

with the first-order distribution given by

$$\frac{\partial f^1}{\partial \tau_0} = +f^0 \int_{\tau_0}^{\infty} d\lambda\, K(\lambda). \tag{35}$$

Equation (34) is the kinetic equation. It coincides with Bogolubov's result. The mechanism of the derivation is, however, substantially different. In particular, τ_0 is allowed to tend to infinity only after phase mixing, thus preventing a mixing of the two scales. Therefore, the adiabatic assumption is not used.

The equation for the correlation function in first order is given by

$$\frac{\partial g^1}{\partial \tau_0} + \frac{\partial g^1}{\partial \xi_0} = JUf^1 - J\xi_0 \frac{\partial U}{\partial \xi_1} f^0 - \frac{\partial g^0}{\partial \tau_1} - \frac{\partial g^0}{\partial \xi_1}. \tag{36}$$

The first two terms on the right-hand side of Eq. (36) have finite spatial range for any given value of τ_0, while the last two terms do not. We thus require that the third and fourth terms balance each other

$$\frac{\partial g^0}{\partial \tau_1} + \frac{\partial g^0}{\partial \xi_1} = 0. \tag{37}$$

Inserting g^0, given in Eq. (31), into Eq. (37), we conclude that

$$\frac{\partial U}{\partial \xi_1} = -\gamma U \tag{38}$$

or, alternatively,

$$U = e^{-\gamma \xi_1}. \tag{39}$$

This result for U, inserted into Eq. (31), yields the correct lowest-order behavior of the correlation function, *including the propagation property and the finite spatial range.*

The correlation function, g^1, which is the correction to g^0, is now obtained from Eq. (36).

$$g^1 = e^{-\gamma\xi_1} \int_0^{\tau_0} d\lambda\, J(\xi_0 - \lambda)\, f(\tau_0 - \lambda) + \gamma e^{-\gamma\xi_1} f^0 \int_0^{\tau_0} d\lambda(\xi_0 - \lambda)\, J(\xi_0 - \lambda). \quad (40)$$

It is readily verified that no improper behavior for large separations occurs. The corrections to the velocity distribution function can also be obtained. We have

$$\frac{\partial f^0}{\partial \tau_2} + \frac{\partial f^2}{\partial \tau_0} = -\gamma f^0 \int_{\tau_0}^{\infty} d\lambda\, K(\lambda) - \int_0^{\tau_0} f^1(\tau_0 - \lambda)\, K(\lambda)$$

$$- f^0 \int_0^{\tau_0} d\lambda \int d\xi_0\, I(\xi_0)\, (\xi_0 - \lambda)\, J(\xi_0 - \lambda) + \gamma f^0 \int_0^{\tau_0} d\lambda \int d\xi_0\, \xi_0 I(\xi_0)\, J(\xi_0 - \lambda). \quad (41)$$

With a slight rearrangement of the terms:

$$\frac{\partial f^0}{\partial \tau_2} + \frac{\partial f^2}{\partial \tau_0} = -\gamma f^0 \int_{\tau_0}^{\infty} d\lambda\, K(\lambda) \int_0^{\tau_0} f^1(\tau_0 - \lambda)\, K(\lambda)\, d\lambda + \gamma f^0 \int_0^{\tau_0} d\lambda\, \lambda K(\lambda). \quad (42)$$

It is readily verified that this equation coincides essentially with the one derived by Ramanathan and Sandri[14] and shown by them to give the correct second-order approximation to the one-particle distribution function. Our improved method allows us, therefore, to obtain not only a correct lowest-order correlation but also systematic corrections to *both* f and g. Thus, a concrete program has been outlined for a kinetic theory of dense gases.

V CONCLUSIONS

We summarize briefly our point of view and the main conclusions reached in this paper. Our philosophy is quite different from that usually adopted in nonequilibrium statistical mechanics. It is widely believed that Bogolubov's results are correct in the leading order and that the difficulties of higher-order theory have to be attributed to intrinsic complexities associated with three-body effects. To the contrary, we show that it is the leading-order theory which has been miscalculated. Singularities and inconsistencies in the leading order must be removed, before proceeding to higher-order theory, for a satisfactory mathematical treatment. Furthermore, we believe that three-body effects cannot account for the difficulties of higher-order theory because of the following considerations. (1) Three-body dynamics has been fully developed for hard-sphere models[15] and the surfaces of the phase space domains corresponding to the allowed collision sequences have no singularities[16]. (2) The secularities associated with three-body effects have been fully isolated[17]. In spite of the control that we have on the relevant ingredients with regard to three-body effects, a satisfactory higher-order theory is not available. Thus, three-body effects are not the main cause of the difficulty. Rather, it is the method of expansion that fails. We have demonstrated that the failure occurs in lowest-order theory. Fourier space analysis, in fact, shows the presence of a singularity in the lowest-order correlation that implies an inconsistency in the expansion method. This singularity manifests itself in configuration space by an infinite correlation length for particle separation parallel to the direction of relative velocity. We have shown that this unphysical behavior is not confined to the weakly coupled regime, but is a general feature of Bogolubov's theory (Section II and Appendix B).

The main conclusions of our asymptotic analysis are:

1) The adiabatic assumptions of Bogolubov's method [Eq. (2)] are physically and mathematically inconsistent, even in the leading order.

2) A proper treatment yields a two-particle correlation function which is nonadiabatic and has a finite spatial range even with the neglect of three-body effects.

With exactly soluble models which, by construction, do not contain three-body effects (and which are related to the Lorentz gas), we illustrate in detail the conclusions 1 and 2 above. With the soluble models, we can resum completely the Bogolubov series. This allows us to establish precisely the reason for the failure of the Bogolubov expansion in higher-order theory. Bogolubov's adiabatic assumption forces a solution for the two-particle correlation function which has a nonphysical initial condition. In particular, the greater the particle separation, the stronger the correlation for relative position and velocity parallel to each other.

The two main features of the correlation function that our analysis has uncovered are the propagation property and the finite spatial range property. These properties are exhibited in detail by means of our soluble models. In particular, we show that:

1) The results are independent of the interaction potential. We exhibit this by studying, in detail, δ-function, Debye, and square barrier interactions. These three widely different interactions give essentially the same behavior for sufficiently large times.

2) The failure of the adiabatic assumption is not confined to a region of zero measure; rather, it occurs in a finite domain. The failure does not affect the lowest-order kinetic equation, but does significantly affect the lowest-order correlation function.

Having shown that the reasons for failure in the higher-order theory are not to be attributed to three-body effects, but rather to a miscalculation of the lowest-order theory, we develop an improved expansion method which involves extension of both time and space. This method yields the correct kinetic equation *and* the correct leading-order correlation. This method gives the same results that have been obtained previously[14] with the integrodifferential approach that combines the f and g equations. Application of the new expansion technique to the soluble models allows us to calculate systematically to any order.

We reiterate that it is our belief that the two-particle correlation function for systems which are not in thermodynamic equilibrium is one of the most important quantities that can be calculated with modern kinetic theory. Our result for the two-particle correlation function differs sharply from the results of Bogolubov. The contrast between our results and his can be tested with scattering experiments which would constitute a crucial test for modern kinetic theory.

APPAENDIX DERIVATION OF THE MODEL EQUATIONAS FOR TWO-DIMENSIONAL LORENTZ GAS

The BBGKY hierarchy for a Lorentz gas[22] can be written as

$$\frac{\partial F^s}{\partial t} + \mathbf{v} \cdot \nabla F^s = \varepsilon \nabla \phi^s \cdot \nabla_v F^s + \varepsilon \int d\mathbf{x}_s \, \nabla \phi_s \cdot \nabla_v F^{s+1} \tag{A.1}$$

where F^s is the probability distribution for one electron and $s - 1$ heavy "ions" which are fixed at the locations $\{\mathbf{x}_i\}$. The position and velocity of the electron are denoted by $\mathbf{x}$ and $\mathbf{v}$. The absence of ion recoil introduces three major simplifications: (1) free streaming occurs only in the electron variables, (2) the interaction operators can be expressed in terms of potentials for any s and, most important, (3) the phase mixing does not involve velocity space. The interaction potentials are given by

$$\phi_s = \phi(x - x_s), \quad \phi^s = \sum_{i=1}^{s-1} \phi_i.$$

Consider the spatially homogeneous two-dimensional problem. Polar coordinates in the electron configuration (r, θ) and velocity space (v, θ) are used. We then introduce "relative" polar coordinates by

$$\theta' - \theta = \psi \tag{A.2}$$

$$\theta' = \alpha \tag{A.3}$$

We write for the electron variables in two dimensions

$$x = r\cos(\alpha - \psi) \tag{A.4a}$$

$$y = r\sin(\alpha - \psi) \tag{A.4b}$$

$$v_x = v\cos\alpha \tag{A.4c}$$

$$v_y = v\sin\alpha. \tag{A.4d}$$

The geometry is shown in Fig. A.1. In terms of the "relative" polar variables, r, v, α, ψ, the partial derivatives in the electron phase space are

$$\frac{\partial}{\partial x} = \cos(\alpha - \psi)\frac{\partial}{\partial r} + \frac{\sin(\alpha - \psi)}{r}\frac{\partial}{\partial \psi} \tag{A.5}$$

$$\frac{\partial}{\partial y} = \sin(\alpha - \psi)\frac{\partial}{\partial r} - \frac{\cos(\alpha - \psi)}{r}\frac{\partial}{\partial \psi} \tag{A.6}$$

$$\frac{\partial}{\partial v_x} = \cos\alpha\frac{\partial}{\partial v} - \frac{\sin\alpha}{v}\left(\frac{\partial}{\partial \alpha} + \frac{\partial}{\partial \psi}\right) \tag{A.7}$$

$$\frac{\partial}{\partial v_y} = \sin\alpha\frac{\partial}{\partial v} + \frac{\cos\alpha}{v}\left(\frac{\partial}{\partial \alpha} + \frac{\partial}{\partial \psi}\right). \tag{A.8}$$

In terms of these variables, the spatial homogeneity of the electron distribution, F, reads

$$\frac{\partial F}{\partial r} = 0, \quad \frac{\partial F}{\partial \psi} = 0.$$

The electron-ion correlation, g, is defined by

$$G \equiv F^2 - F \tag{A.9}$$

if we take the ions to have a statistically homogeneous distribution. The F and G equations in these variables are

$$\frac{\partial F}{\partial t} = \varepsilon \int_0^\infty r\, dr\phi' \int_0^{2\pi} d\psi \left\{ \cos\psi \frac{\partial G}{\partial v} - \frac{\sin\psi}{v}\left(\frac{\partial G}{\partial \alpha} + \frac{\partial G}{\partial \psi}\right)\right\} \tag{A.10}$$

$$\frac{\partial G}{\partial t} + v\cos\psi \frac{\partial G}{\partial r} - \frac{v\sin\psi}{r}\frac{\partial G}{\partial \psi} = \phi' \left\{ \cos\psi \frac{\partial F}{\partial v} - \frac{\sin\psi}{v}\frac{\partial F}{\partial \alpha}\right\}. \tag{A.11}$$

We now introduce the following expansions

$$F = \sum_n (a_n \cos n\alpha + b_n \sin n\alpha) \tag{A.12}$$

$$G = \sum_n (A_n \cos n\alpha + B_n \sin n\alpha). \tag{A.13}$$

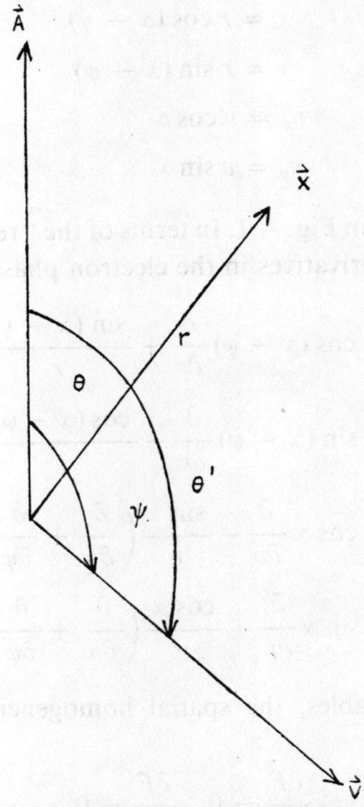

FIG. A.1 Geometry for a Two-dimensional Lorentz Gas. The variables **x** and **v** are the relative position and velocity, respectively. The vector **A** is an arbitrary reference vector in the **x**, **v** scattering plane. Note that in Appendix A the angle θ is not the angle between **x** and **v**. The angle between **x** and **v** is ψ

These correspond to a Fourier decomposition of F and G. This is suitable for two-dimensional situations. It is easy to prove that the equations corresponding to the index n decouple from those for the index m. In fact, dropping the harmonic subscript, we obtain the following equations. For the one particle distribution, we have

$$\dot{a} = \varepsilon \int r\, dr\, \phi' \int d\psi \left\{ \cos\psi \frac{\partial A}{\partial v} - \frac{\sin\psi}{v} \frac{\partial A}{\partial \psi} - \frac{n \sin\psi}{v} B \right\} \tag{A.14}$$

$$\dot{b} = \varepsilon \int r\, dr\, \phi' \int d\psi \left\{ \cos\psi \frac{\partial B}{\partial v} - \frac{\sin\psi}{v} \frac{\partial B}{\partial \psi} + \frac{n \sin\psi}{v} A \right\}. \tag{A.15}$$

For the two-particle correlation function, we obtain the pair of equations

$$\dot{A} + v \cos\psi \frac{\partial A}{\partial r} - \frac{v}{r} \sin\psi \frac{\partial A}{\partial \psi} = \phi' \left(\cos\psi \frac{\partial a}{\partial b} - \frac{n \sin\psi}{v} b \right) \tag{A.16}$$

$$\dot{B} + v \cos\psi \frac{\partial B}{\partial r} - \frac{v}{r} \sin\psi \frac{\partial B}{\partial \psi} = \phi' \left(\cos\psi \frac{\partial b}{\partial v} + \frac{n \sin\psi}{v} a \right) \tag{A.17}$$

We now introduce complex combinations

$$\mathscr{F} = a + ib \tag{A.18}$$

$$\mathscr{G} = A + ib \tag{A.19}$$

The equations for the single-particle distribution and for the correlation function then acquire a fairly simple form. For the one-particle distribution, we have

$$\dot{\mathscr{F}} = \varepsilon \int r\, dr\, \phi' \int d\psi \left\{ \cos\psi \frac{\partial \mathscr{G}}{\partial v} - \frac{\sin\psi}{v} \frac{\partial \mathscr{G}}{\partial \psi} + in \sin\psi \mathscr{G} \right\}. \tag{A.20}$$

For the correlation function, we find

$$\frac{\partial \mathscr{G}}{\partial t} + v \cos\psi \frac{\partial \mathscr{G}}{\partial r} - \frac{v}{r} \sin\psi \frac{\partial \mathscr{G}}{\partial \psi} = \phi' \left\{ \cos\psi \frac{\partial \mathscr{F}}{\partial v} + in \sin\psi \mathscr{F} \right\}. \tag{A.21}$$

We now assume a special form for the velocity dependence of both the velocity distribution and the correlation functions. In particular, we assume for large times

$$\mathscr{F} = e^{-v^2/2} f \tag{A.22}$$

$$\mathscr{G} = e^{-v^2/2} g. \tag{A.23}$$

Inserting (A.22) and (A.23) into (A.21) and (A.22), we obtain

$$\dot{f} = ni\varepsilon \int r\,dr\,\phi' \int d\psi \sin\psi g \tag{A.24}$$

$$\frac{\partial g}{\partial t} + \cos\psi \frac{\partial g}{\partial r} - \frac{\sin\psi}{r}\frac{\partial g}{\partial \psi} = \phi'\{-\cos\psi + in\sin\psi\}. \tag{A.25}$$

In particular, if we choose $n = 1$, we can rewrite the equations in a compact form. We find

$$\dot{f} = i\varepsilon \int r\,dr\,\phi' \int d\psi \sin\psi g \tag{A.26}$$

$$\frac{\partial g}{\partial t} + \cos\psi \frac{\partial g}{\partial r} - \frac{\sin\psi}{r}\frac{\partial g}{\partial \psi} = -\phi'\, e^{-i\psi} f. \tag{A.27}$$

We now notice that (A.26) and (A.27) can be written as

$$\dot{f} = e \int d^2x\, Ig \tag{A.28}$$

$$\frac{\partial g}{\partial t} + v \cdot \nabla g = Jf \tag{A.29}$$

where we have made use of the following symbols

$$I = in\phi' \sin\psi \tag{A.30}$$

$$J = \phi'(-\cos\psi + in\sin\psi). \tag{A.31}$$

We can further simplify (A.30) and (A.31). We can write

$$I = -in\hat{u} \cdot \nabla\phi \tag{A.32}$$

The J function can be written as

$$J = I - \hat{v} \cdot \nabla\phi = -in\hat{u} \cdot \nabla\phi - \hat{v} \cdot \nabla\phi$$

$$= -(in\hat{u} \cdot \hat{v}) \cdot \nabla\phi. \tag{A.33}$$

We now introduce two complex vectors which are defined as follows

$$\mathbf{w}_1 = -in\hat{u} \tag{A.34}$$

$$\mathbf{w}_2 = -(in\hat{u} + \hat{v}). \tag{A.35}$$

In terms of these two complex vectors, we can rewrite the interaction functions in a particularly simple way, namely,

$$I = \mathbf{w}_1 \cdot \nabla\phi \tag{A.36}$$

$$J = \mathbf{w}_2 \cdot \nabla\phi \tag{A.37}$$

which correspond to the results quoted in the text.

We thus notice that the Lorentz gas is, in fact, very closely related to the model discussed in Ref. 4. Specifically, the Lorentz gas, when excited from equilibrium with a suitable harmonic, corresponds to our model. The harmonics that have been used are harmonics that preserve the θ-symmetry, namely, the functions are initially latitude independent. This property is then preserved in time. The asymmetry that is assumed initially is an azimuthal asymmetry (tangerine distribution).

APPENDIX B CORRELATION FUNCTION FOR THE BOLTZMANN REGIME IN THE LORENTZ GAS

Below, we discuss the situation which occurs in the short-range regime of the Lorentz gas. The main result of the analysis is to prove that the two-particle correlation function computed with Bogolubov's method is independent of $x_{\parallel}$ when $x_{\parallel}$ is large compared to the range of interaction. Thus,

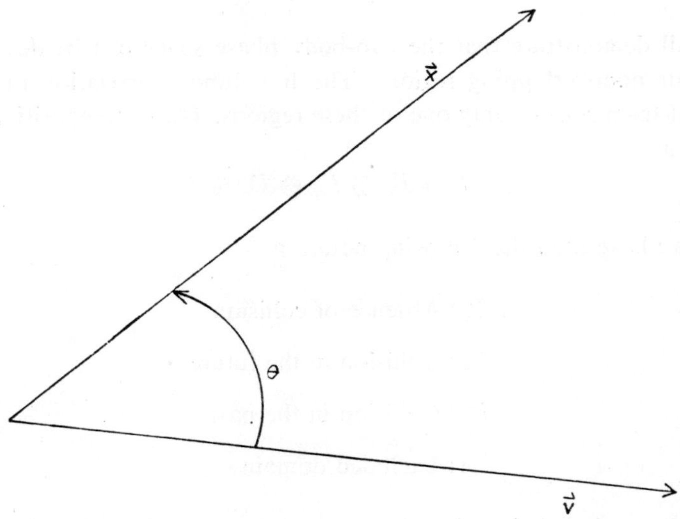

FIG. B.1 The scattering plane. The vectors **x** and **v** represent the position and the velocity of the light particles

it will be shown that the correlation length given by Bogolubov's theory is infinite when **x** is parallel to **v**, exactly as in the weakly coupled regime.

We start by introducing suitable variables in the phase space. We are concerned with the phase space which represents the support of the two-particle distribution function—one particle having finite mass and the other infinite mass. We *call* the light particle an electron and the heavy particle an ion, although the interaction is that of elastic impenetrable spheres. The two-particle phase space for a spatially homogeneous system will, therefore, consist simply of a velocity vector and a position vector. The velocity vector is the electron velocity; the position vector is the relative position of the electron to the fixed ion. These two vectors define a plane (see Fig. B.1). In terms of this plane, we can reduce the needed variables to three scalar variables. Thus, for example, we can completely characterize the phase space by means of the following choice:

$$\Gamma = \{x, v, \theta\}. \tag{B.1}$$

The range of the variables is

$$\begin{aligned} 0 &< x < \infty \\ 0 &< v < \infty \\ 0 &< \theta < \pi . \end{aligned} \tag{B.2}$$

We shall demonstrate that the two-body phase space can be decomposed into four nonoverlapping regions. The Bogolubov correlation function is different from zero in only one of these regions. The decomposition can be written as

$$\Gamma = \Gamma_0 \oplus \Gamma_+ \oplus \Gamma_- \oplus E \tag{B.3}$$

where we have used the following notation

$$\begin{aligned} \Gamma_0 &: \text{Absence of collision} \\ \Gamma_+ &: \text{Collision in the future} \\ \Gamma_- &: \text{Collision in the past} \\ E &: \text{Excluded domain} \end{aligned} \tag{B.4}$$

where the excluded domain, E, corresponds to the fact that the electron should not be inside the ion's core. Within Γ_+ and Γ_- we can determine the time, τ, at which a collision occurs, through the equation

$$|\mathbf{x} - \mathbf{v}\tau| = r_0 . \tag{B.5}$$

This condition, Eq. (B.5), yields a quadratic equation for

$$x^2 - 2\mathbf{x} \cdot \mathbf{v}\tau + v^2\tau^2 = r_0^2 . \tag{B.6}$$

We can examine the solution for τ written in the following alternative forms

$$\begin{aligned}\tau &= \frac{\mathbf{x}\cdot\mathbf{v}}{v^2} \pm \frac{1}{v^2}\sqrt{(\mathbf{x}\cdot\mathbf{v})^2 - x^2v^2 + r_0^2v^2} \\ &= \frac{x}{v}\cos\theta \pm \frac{1}{v}\sqrt{x^2(\cos^2\theta - 1) + r_0^2} \\ &= \frac{x_{\parallel}}{v} \pm \frac{1}{v}\sqrt{r_0^2 - x_{\perp}^2}. \end{aligned} \tag{B.7}$$

For a collision to occur, the square root must be real; for future collisions, τ is positive and for past, negative. Furthermore, if a collision is to occur in the future, we must choose the positive solution, which is the smaller one of the two. The other solution corresponds to the nonphysical geometric contact, corresponding to particles that penetrate one another. Thus, for a collision in the future, we must choose

$$\begin{aligned}\tau_+ &= \frac{x_{\parallel}}{v} - \frac{1}{v}\sqrt{r_0^2 - x_{\perp}^2} \\ &= \frac{x\cos\theta}{v} - \frac{1}{v}\sqrt{r_0^2 - x^2\sin^2\theta}. \end{aligned} \tag{B.8}$$

By a similar reasoning, we can determine the time at which a past collision occurred

$$\tau_- = \frac{x_{\parallel}}{v} + \frac{1}{v}\sqrt{r_0^2 - x_{\perp}^2} \tag{B.9}$$

(where $x_{\parallel}$ is negative). The region that corresponds to no collisions is then characterized by a complex solution for τ, i.e.,

$$\Gamma_0 : r_0^2 < x_{\perp}^2 = x^2\sin^2\theta. \tag{B.10}$$

The phase space can be parameterized with alternative variables to those given in Eq. (B.1). For example, Eqs. (B.8) and (B.9) suggest the following set of variables

$$\Gamma = \{v, x_{\parallel}, x_{\perp}\}. \tag{B.11}$$

The range of these variables is

$$\begin{aligned} 0 &< v < \infty \\ -\infty &< x_{\parallel} < \infty \\ 0 &< x_{\perp} < \infty. \end{aligned} \tag{B.12}$$

Notice that the parallel component of the velocity ranges over the full line, in contrast to the perpendicular component. In terms of these variables, the phase space decomposition is particularly simple, as shown in Fig. B.2.

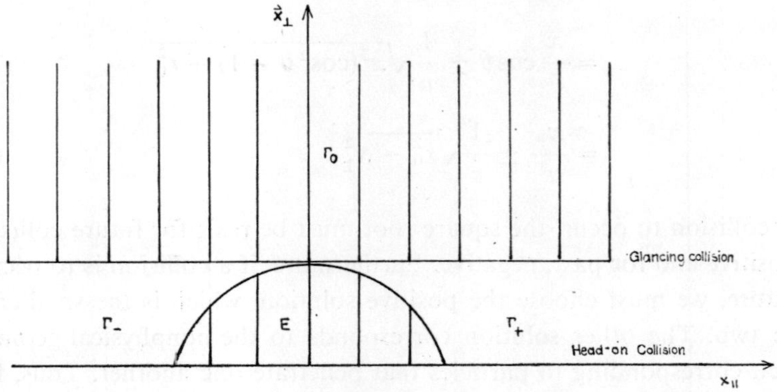

FIG. B.2 The regions of phase space on the $x_\parallel$, $x_\perp$ plane that correspond to the absence of collisions, Γ_0; collisions in the past, Γ_-; collisions in the future, Γ_+, and the excluded region, E

We can now characterize the domains as follows. For the domain in which no collision occurs, we have

$$\Gamma_0 : x_\perp > r_0. \tag{B.13}$$

For the domain in which collisions occur in the future, we can write

$$\Gamma_+ : x_\perp < r_0; \quad x_\parallel > \sqrt{r_0^2 - x_\perp^2}. \tag{B.14}$$

For the domain corresponding to collisions in the past, we have

$$\Gamma_- : x_\perp < r_0; \quad x_\parallel < \sqrt{r_0^2 - x_\perp^2}. \tag{B.15}$$

Finally, the domain of exclusion is simply given by

$$E: x_\parallel^2 + x_\perp^2 < r_0^2. \tag{B.16}$$

Figure B.2 is clearly a summary of the situation*.

The Bogolubov two-particle distribution function is given by[11]

$$F_B^2 = S^2 F \tag{B.17}$$

* An alternative set of variables which is particularly suitable for the discussion of the null space of the Hamiltonian operator consists of total energy, angular momentum, and $x_\parallel$.

where we have used the statistical homogeneities of the ion distribution. Therefore, the Bogolubov correlation is, using Eq. (A.9),

$$G_B = F_B(\mathbf{v}^*) - F_B(\mathbf{v}). \tag{B.18}$$

We can now, without difficulty, construct explicitly $\mathbf{v}^*$. We have, in terms of Bogolubov's two-particle streaming operator, S^2,

$$\mathbf{v}^* = S^2\mathbf{v}. \tag{B.19}$$

It is clear that we will not have a modification in the velocity vector unless a collision occurs in the future. Therefore, the velocity vector is completely unaffected by S everywhere except in the Γ_+ domain. We can therefore state:

$$\mathbf{v}^* = v\varepsilon\Gamma_+. \tag{B.20}$$

Within Γ_+ we have

$$\mathbf{v}^* = \mathbf{v} - 2n(\mathbf{n}\cdot\mathbf{v}) \tag{B.21}$$

where $\mathbf{n}$ is the unit vector along the line of centers at the moment of impact. A simple computation gives

$$\mathbf{n} = \frac{1}{r_0}\left(\mathbf{x} - \mathbf{v}\left[\frac{x_\parallel}{v} - \frac{\sqrt{r_0^2 + x_\perp^2}}{v}\right]\right). \tag{B.22}$$

Thus, within the Γ_+ domain, i.e., $x_\perp < r_0$ and $x_\parallel > (r_0^2 - x_\perp^2)^{1/2}$, the Bogolubov correlation is

$$G_B = F\left(\mathbf{v}\left[\frac{2x_\perp^2 - r_0^2}{r_0^2}\right] - \frac{\mathbf{x}_\perp\cdot}{r_0^2}2\mathrm{v}\sqrt{r_0^2 - x_\perp^2}\right) - F(\mathbf{v}). \tag{B.23}$$

The correlation is zero in Γ_- and Γ_0. We see from Eq. (B.23) that the resulting correlation function is independent of $x_\parallel$ when $x_\parallel$ lies within the domain Γ_+. Therefore, the correlation function has an infinite range in the $x_\parallel$ direction because no matter how large the value of $x_\parallel$, Eq. (B.23) yields a finite value for the correlation function. Furthermore, if we ask under what conditions the correlation vanishes because of the symmetry of the F function, we find the satisfactory result that $G_B = 0$ if, and only if, F is spherically symmetric. Since all spherical solutions are stationary solutions of a linearized Boltzmann equation, we conclude that the Bogolubov correlation will have finite range only in equilibrium. We have thus demonstrated a full analogy between the Boltzmann and the Fokker–Planck regimes for the Lorentz gas. In particular, the infinite range of the Bogolubov correlation function is not dependent on the gas regime.

References

1. Yvon, J., *Actualities Scientifiques et Industrielles*, **I**, Hermann, Paris (1935).
2. Kirkwood, J. G., *J. Chem. Phys.* **14**, 180 (1946).
3. Born, M. and Green, H. S., *A General Kinetic Theory of Liquids* (Cambridge Univ. Press, England, 1949).
4. Bogoliubov, N. N., *J. Phys. U.S.S.R.* **10**, 265 (1946).
5. Wu, T. Y., *Kinetic Equations of Gases and Plasmas* (Addison-Wesley Pub. Co., Reading, Mass, 1966).
6. Liboff, R., *Introduction to the Theory of Kinetic Equations* (Wiley Pub. Co., Somerset, N. J., 1969).
7. Klimontovich, Yu. L., *The Statistical Theory of Nonequilibrium Processes in a Plasma* (MIT Press, Cambridge, Mass. (trans.), 1967).
8. Bogoliubov, N. N., *Studies in Statistical Mechanics* (J. de Boer and Uhlenbeck, eds., North-Holland Pub. Co., Amsterdam, 1962).
9. Prigogine, I., *Nonequilibrium Statistical Mechanics* (Interscience, N. Y., 1962).
10. Kawasaki, K. and Oppenheim, I., *Phys. Rev.* **139A**, 1763 (1965).
11. Sandri, G., *Ann. Phys.* (N.Y.), **24**, 332, 380 (1963).
12. Frieman, E., *J. Math. Phys.* **4**, 410 (1963); McCune, J., Morse, T., and Sandri, G., *Rarefied Gas Dynamics*, I (Laurmann, J., ed., Academic Press, N.Y., 1963); *Book, D., MATT Rept.* 274, Princeton University, Princeton, N. J., 1964; Frank, D., Pfirsch, D., Priess, S., *Z. Naturforsch.* **20a**, 147 (1965); Montgomery, D. and Tidman, D., *Phys. Fluids* **2**, 242 (1964); Montgomery, D., "*Connection between the Bogolyubov and the Frieman and Sandri Methods*", preprint (1965); Schram, P., Euratom Rept. EUR. 1905.e (1964).
13. Kritz, A. H. and Sandri, G., *Phys. Today* **19**, 57 (1966).
14. Ramanathan, G. V. and Sandri, G., *J. Math. Phys.* **10**, 1763 (1969).
15. Thurston, W. and Sandri, G., "*Classical Hard-sphere Three-body Problem*", *Bull. Am. Phys. Soc.* **9**, 386 (1964); Sandri, G., Sullivan, R., and Norem, P., *Phys. Rev. Letters* **13**, 743 (1964); Sandri, G. and Sullivan, R., *Not. Am. Math. Soc.* **12**, 215 (1965); Foch, D., private communication.
16. Sandri, G. and Kritz, A. H., *Phys. Rev.* **150**, 92 (1966).
17. Goldberg, P. and Sandri, G., *Phys. Rev.* **154**, 188 (1967); **154**, 199 (1967); Kritz, A. H., Radin, S., and Sandri, G., *Phys. Rev.* **155**, 101 (1967); *J. Math. Phys.* **9**, 325 (1968).
18. Green, M. S., *Physica* **24**, 393 (1958).
19. Frieman, E. A. and Goldman, R. *J. Math. Phys.* **7**, 2153 (1966); Goldman, R. and Frieman, E. A., *J. Math. Phys.* **8**, 1410 (1967). See also the interesting paper by Green, M. quoted here.
20. Frisch, H. L. and Salsburg, Z. W., *Simple Dense Fluids*, (Academic Press, N.Y., 1968).
21. Lewis, M., has investigated the possibility of a logarithmic behavior in the higher-order correlations for this system, private communication.
22. Fibich, M. and Sandri, G., *ARAP Rept.* **82**, Aeronautical Research Associates of Princeton, Inc., Princeton, N. J., March 1966.

Group Photograph

Symposium on Kinetic Equations, June, 1969, Cornell University

1. R. Sudan
2. I. Freeman
3. R. Liboff
4. G. Sandri
5. K. Nishikawa
6. J. Cipolla
7. N. Rostoker
8. R. Snyder
9. I. Lerch
10. C. Oberman
11. C. Myerscough
12. O. Aono
13. J. Hieronymus
14. J. Snow
15. K. Matsuda
16. S. Harris
17. J. Dorfman
18. M. Bixon
19. E. Gross
20. Unidentified
21. A. Ron
22. D. Gillespie
23. H. Gould
24. A. Kritz
25. M. Leiser
26. H. Hsuan
27. R. Hakim
28. J. Dardis
29. H. Berk
30. H. Smith
31. J. Sengers
32. Unidentified
33. N. Pope
34. C. Dum
35. P. Orteleva
36. L. Caroff
37. G. Ramanathan
38. G. Wannier

Index

Aamodt, R. E. I. 270, 271
Abraham, A. 133
Abrikosov, A. A. 286
Ahearne, J. F. 221
Alder, B. J. 193, 233
Altshul, L. M. 271
Akhiezer, I. A. 223
Amit, D. J. 286
Aono, O. xiv, 221, 222, 223
Akcasu, A. Z. 67, 68
Artz 249
Auto Correlation Formula 17

Baldwin, D. E. 221
Balescu-Lenard Equation, or Balescu-Lenard-Guernsey Equation, or BLG Equation vii, 27
Balescu, R. xiv, 43, 222
Baker, G. A., Jr. 233
Banwell, C. N. 133
Baranger, M. 108, 109
Barut, A. O. 74
BBGKY Equation 130
Benedek, G. B. 67
Bekefi, G. 108
Berne, B. 68
Bernstein, I. B. 221
Bethe, H. 223
Bezzerides, B. 108
Bird, R. B. 192
Birmingham, T. J. 109
Bixon, M. 68
Bludman, S. A. 89
Bogoliubov, N. N. 18, 27, xiv, 192, 340
Bogoliubov Theory xii, 22, 307, 308, 309, 312–316, 329
  Form of Boltzmann Equation 144
  -ansatz 251, 319
  Correlation Function 318, 321, 322, 339
Bohr, N. 223
Bohr radius 14
Boltzmann Equation Reference to the Boltzmann Equation may be found on almost every page of the text
Boltzmann H-Theorem 225
  H-Function 225–233
Boltzmann, L. xiv, 232
Boltzmann-Langevin Equation 75
Boltzmann-Like Operator 57
Boltzmann Operator 59
Book, D. C. xiv, 221
Boon, J. 68
Borckmans, P. 97
Born, M. 340
Brittin, W. E. 74
Brout-Prigogine Equation 30, 35
Brout, R. 43, 67, 98
Brownian Motion 216
Buneman, O. 200
Burnett, D. 193
Buti, B. 89
Byers, J. A. 271

Callen, H. B. 67
Camac, M. 271
Carleman, T. 233
Case, K. M. xv, 200
Chandrasekhar, S. 67
Chapman, S. 19, 192, 222
Chapman-Enskog Theory 17, 75, 138
Chappell, N. 108
Chen, F. M. 133
Choh, S. T. 193, 233

Choh-Uhlenbeck Equation 26, 139, 150–153, 174, 178, 180, 225
Chung, C. H. 67
Clemmow, P. C. 89
Cohen, E. G. D. 27, 28, xiv, 192, 193
Cohen, R. S. 222
Cohen-Tannoudji, C. 122
Collins, M. 98
Conservation Equations 3
 of energy 12
Cooper, J. 108
Cooper, W. 109
Coppi, B. xiv, 270, 271
Cowling, T. G. 19, 192
Crawford, J. A. 133
Curtis, C. F. 192
Cyclic Collisions 156, 162, 164

Daniels, E. 67
Dawson, J. 109, 200, 222, 271
Debye-Hückel Approximation 14, 17
Debye Length 104
DeHaan, M. 122
DeLeener, M. 97
DeWitt, H. E. 221
Dikasov, V. M. 249
Distribution functions
 pair 5, 21
 higher order 9
Doniach, S. 286
Doppler Broadening (of spectral lines) 100
Dorfman, J. R. 27, 28, xiv, 192
Drummond, W. E. 271
Duderstadt, J. J. 68
Dum, C. T. 270, 271
Dupree, T. H. 28, xiv, 109, 271
Dy, K. S. 286

Einaudi, F. 297
Eliasberg, G. M. 222
Emery, J. T. 286
Engelsberg, S. 286
Enskog Theory 138, 140, 142, 147–149
Erasure, Method of 14
Ernst, M. H. 28

Fano, J. 108, 133
Felderhof, B. U. 200
Feller, W. 233
Fermi, Theorem 30
Fibich, M. 340
Fleck, J. A., Jr. 249
Fleck's Technique (for Gaussian Fit) 242
Flynn, G. P. 192
Fokker-Planck
 Equation 299, 316, 339
 Form of Damping Theory 57
 Operator 59
Ford, G. W. 232
Fowler, T. K. 200
Frank, D. 340
Fried, B. D. xv
Frieman, E. A. 27, xiv, 192, 221, 297, 340
Frisch, H. L. 340
Fujita, S. xiv
Fukai, J. 270

Gailitis, A. 271
Galeev, A. A. xiv, 271
Gallavoti, G. 28
George, C. 122
Gibson, A. G. 133
Glauber, R. J. 248
Glauber State 236
Goldberg, P. 340
Golden, S. 43
Goldman, R. 27, 28, 192, 340
Gorbunov, L. M. 297
Gould, H. A. 221
Grad, H. 19, xv, 233
Green, H. S. 18, 19, 340
Green, M. S. 27, 193, 340
Greytak, J. J. 67
Griem, H. 108, 109
Gross, E. P. xv
Gruner, P. 222
Guernsey, R. 28, 221, 222

Haber, S. 193
Haken, H. 249
Hakim, R. 74
Halperin, B. 98
Harris, E. G. 270, 271
Harris, S. 232, 233
Haselgrove, C. B. 193
Haselgrove-Method 174
Hatori, T. 222
Hauge, E. 27
Hempstead, R. D. 249
Henin, F. 43, 122
Hilbert, D. 19
Hirschfelder, J. O. 192
Hoegy, W. 27, 28
Hojgaard, Jensen H. 128
Hoffman, D. K. 19, 192
Hohenberg, P. 98
Honda, N. 222
Hone, D. 286
Hooper, C. F. 108
Hubbard, J. xiv, 89, 222
Hypothetical Collision 155, 156, 162, 164

Inönü, E. xv
Irving, J. H. 18
Itikawa, Y. 222

Jauch, J. M. 133
Jeans, J. H. 222

Kadanoff, L. P. 67
Kadomtsev, B. xiv, 271, 297
Kane, J. W. 289
Kanonisches Normal System 30
Kantrowitz, A. R. 271
Karpman, V. 271
Kawasaki, K. 28, 192, 193, 340
Khalatnikov, I. M. 286
Kihara, T. xiv, 221, 222
Kim, S. K. 192
Kirkwood, J. G. 340
Klimontovich Theory xiii, 73
  Equation 290, 301
Klimontovich Method 73
Klimontovich, Y. L. xiv, 74, 340
Kohler Transport Theory 123
Kolb, A. 108
Konstantinov, O. V. 222
Kovrizhnikh, L. M. 271
Kramers, H. A. 223
Krishan, S. 270
Kritz, A. H. 193, 340
Krook, M. xv
Krver, W. 271
Kuezall, A. 222
Kunze, H. J. 109
Kurasawa, T. 297

Lamb Shift 120
Lamb, W. E., Jr. 248, 249
Landau
  Equation 38, 278, 281
  Damping 74, 262, 265, 268, 296
  Theory of Low Temperature Transport Phenomina 277–286
Landau, L. D. 67, 222, 286, 297
Langer, J. S. 27
Langevin Equation 45, 51, 55, 66, 75
Langmuir, I. 222
Larking, A. I. 222
Lawrence, W. E. 128
Lax, M. 249
Leaf, B. 74
Lebowitz, J. L. 27
Leipnik, R. B. 18
Lenard, A. xiv, 89, 222
Lewis, M. 109, 340
Liboff, R. L. xiv, 222, 340
Lifshitz, E. M. 67, 297
Liouville Theorem or Equation 8, 10, 21, 29, 72, 112, 113, 211, 226, 236, 245
Lippmann-Schwinger Equation 131
Litvak, M. N. 271
Longuet-Higgins, H. C. 43
Lorentz-Dirac Equations 70
Louisell, W. 249
Lowe, I. 98

McCone, J. 340
McElroy, M. B. 192
McLennan, J. A. xv
McKean, H. 233
McR. Routly, P. 222

Malmberg, J. H. 271
Mangeney, A. 74
Markoffian
  Diffusion Equation 95
  Approximation 289–292
Markovian Behavior 103, 105, 288
  Non-Markovian Behavior 103
Marshall, W. 97
Martin, P. C. 67
Matsuda, K. 305
Matsuura, K. 297
Mathiessen's rule 126, 127
Mazur, P. 97
Menzinger, F. 98
Minkiewiez, V. 98
Misra, B. 133
Modeled Collision Operators 66
Møller Wave Operator 130, 131
Monte Carlo Quadratore 174
Montgomery, D. 27, xiv, 340
Montroll, E. A. 43
Mori, H. 67, 297
Morse, T. F. xv, 340
Mountain, R. 67
Mozer, B. 109
Murase, C. 27

Nathans, R. 98
Neal, T. 27
Nelkin, M. 67
Neufeld, J. 223
Nijboer, B. R. A. 43
Nishikawa, K. 297, 305
Norberg, R. 98
Norem, P. 193, 340
Nozieres, P. 286

Oberman, C. 89, 109, 200, 222, 297
Okutani, J. 223
O'Neil, T. M. 271
Oppenheim, J. 28, 192, 193, 340
Ornstein-Zernicke Factor 96
Ortoleva, P. J. 67
Osaka, Y. 297, 305

Partition function 8
Passel, L. 98
Patrick, R. M. 271
Pauli Equation 111, 114, 116, 119
Peierls, R. E. 270, 297
Percus, J. K. 18
Perel, V. I. 222
Perisco, E. 222
Perkins, F. 221
Pethick, C. J. 286
Petschek, H. E. 271
Picard, R. 248
Piccirelli, R. A. 27, 67, 193
Pickart, J. 98
Pines, D. xiv, 223, 286, 297
Pfirsch, D. 340
Plasma 17
  oscillations xi
Poincaire-Recurrence time 29
Pomeau, Y. 28
Price, J. C. 305
Prigogine, I. xiv, 43, 68, 74, 97, 122, 340
Prigogine-Résibois
  Master Equation 30
Priess, S. 340
Primas, H. 133
"Pseudo-Spin Wave" 97

Radin, S. 340
Rahman, A. 67
Ramanathan, G. V. 340
Raman Scattering 120
Ranganathan, S. 67
Rayleight-Schrödinger
  Perturbation Expansion 116, 119
Recollisions 156, 163
Résibois, P. 43, 67, 97, 122
Rice, M. J. 286
Rice, S. A. 68

Risken, H. 249
Ritchie, R. H. 223
Robertson, B. 67
Rohrlich, F. 74
Rogister, A. 89, 297
Romig, J. 89
Ron, A. 222
Rosenberg, R. L. 222
Rosenbluth, M. xiv, 89, 222, 270, 271
Ross, J. 192
Rostoker, N. 89, 109, 222, 270, 305
Rostoker's Superposition Principle 299
Rudakov, L. I. 270
Rutherford, P. H. 297
Ryton, S. M. 67
Ryutov, D. D. 270

Sagdeev, R. Z. xiv, 271
Salsburg, Z. W. 340
Sandri, G. 43, 193, 340
Sauermann 249
Schawlow-Townes Condition 241
Schmid, C. 249
Schofield, P. 67
Schrieffer, J. R. xiv, 297
Scully, M. 249
Second Quantization xiii
Seeger, R. J. 18
Sengers, J. V. 27, xiv, 192, 193
Shen, K. Y. 108
Shirane, G. 98
Silin, V. P. 286, 297
Sitenko, A. G. 223
Sloan, M. L. 270
Smith, E. W. 108
Smith, H. 128
Snider, R. F. 133
Sonine Polynomials 146, 151, 152, 180
Spitzer, L., Jr. 222
Stark Broadening (of spectal lines) 99, 100, 108
Stogryn, D. E. 192
Sudan, R. xiv, 270, 271, 297
Sullivan, R. D. 193, 340
Superoperator 129

Sykes, J. 27
Synge, J. L. 89

Temple, G. 18
ter Haar, D. 43
Terwiel, R. 97
Tidman, D. A. 27, xiv, 109, 340
Titchmarsh, E. C. 43
Thompson, W. B. 89, 222
Thurston, W. 340
Transport Coefficients 5, 9
Triple Collision Transport Integral 167–177
Tsytovich, V. N. 271, 297
Turner, W. 122

Uhlenbeck, G. E. 193, 232
Umklapp-scattering 128

van Kampen, N. G. 43
van Leeuwen, J. M. J. 27, 28, 67
Vidal, R. 108
Vlasov, A. A. 222
Vlasov
  Equation 43, 73, 74, 81, 195–200, 299
  Like (or type) Equation 57, 300, 302
  Kinetic Equation for a Liquid 61
  Relativistic Equation 72, 79
Volk, H. J. 200
Volmer, H. D. 249
Von Neumann Equation (Von Neumann-Lionville Equation) 112, 113, 117, 129

Wagner, H. 286
Wainwright, T. E. 193, 233
Waldmann, L. 133
Walgraef, D. 97
Wang Chang, C. S. 133
Wang-Chang-Uhlenbeck Equation 132
Wannier, G. H. 232, 276
Watson, K. M. 89
Weidlich, W. 249

Weinstock, J. 27, 192, 193, 222
Welton, T. A. 67
Weyland, A. 27
Wharton, C. B. 271
Wheatley, J. C. 286
Wigner, E. P. 18
Wigner-Weisskopf Model 120
Willis, C. R. 248, 249
Wilson, A. 89
Wilkins, J. W. 128
Wu, T. Y. xiv, 222, 297, 340
Wyld, H. W., Jr. xv, 223

Yip, S. 67
Yvon, J. 340

Zwanzig, R. W. 18, 43, 67, 68
Zweifel, P. F. xv